Basic Techniques for
Transmission Electron Microscopy

Basic Techniques for Transmission Electron Microscopy

M. A. HAYAT

Department of Biology
Kean College of New Jersey
Union, New Jersey

1986

ACADEMIC PRESS, INC.

Harcourt Brace Jovanovich, Publishers

Orlando   San Diego   New York
Austin   London   Montreal   Sydney
Tokyo   Toronto

ACADEMIC PRESS, INC.
Orlando, Florida 32887

United Kingdom Edition published by
ACADEMIC PRESS INC. (LONDON) LTD.
24–28 Oval Road, London NW1 7DX

Library of Congress Cataloging in Publication Data

Hayat, M. A., Date
    Basic techniques for transmission electron
microscopy.

    Includes index.
    1. Electron microscope, Transmission.   2. Electron
microscopy–Technique.   I. Title.
QH212.T7H39   1985   578'.45   85-3957
ISBN 0–12–333925–1 (alk. paper)
ISBN 0–12–333926–X (paperback)

PRINTED IN THE UNITED STATES OF AMERICA

86 87 88 89   9 8 7 6 5 4 3 2 1

Contents

2   Rinsing, Dehydration, and Embedding

3 Sectioning

4   Positive Staining

5 Negative Staining

Appendix

Preface

Transmission electron microscopy has been making profound contributions to the solution of problems in biology and medicine in that the transmission electron microscope gives us the hope of direct examination of biological structures at the atomic level. Individual molecules and their components can be identified—thus the emergence of "molecular microscopy." Such information is of vital importance in understanding the correlation between structure and function at the cellular level. Since methodology is a major constraint in obtaining more detailed and accurate information on cell ultrastructure, the problems of biological specimen preparation need to be attacked with the same converging intensity as that afforded by the improvement of resolving power.

In the preface of my book, "Fixation for Electron Microscopy" (Academic Press, Inc., 1981), I expressed my wish to continue compiling information that might help to improve the quality of specimen preservation as well as the interpretation of the fine structure. This wish is partly fulfilled by publishing the present volume, which covers a vast variety of techniques and alternative approaches used for studying morphological and cytochemical aspects of specimens. Much of this volume is a compilation of methods scattered throughout a large number of scientific journals, and should therefore save readers valuable time in searching through journals for the desired method and/or in contacting various laboratories.

To minimize artifacts and to obtain the required information, each type of specimen requires special, optimal processing. The precise parameters of processing a wide variety of specimens are presented in Chapter 7. It is inappropriate to present only the best variation of a basic method, since for a particular study one variation may give a better result than another; for example, the variation for the preservation of yeast cell wall differs from that ideal for preserving yeast mitotic spindle. Furthermore, the use of a superior variation may not be feasible in a certain laboratory because of the unavailability of a required piece of equipment, lack of funds, or lack of expertise. Variations on methods are therefore given so that readers can choose the most suitable approach to achieve their specific objectives.

Generally, no attempt was made to explain the theory or principles that govern preparatory procedures, because the primary objective of this volume is to present procedures in a step-by-step fashion. Other books (e.g., "Principles and Techniques of Electron Microscopy: Biological Applications," 2nd ed., International Pub., Berkeley Heights, New Jersey) by this author present the principles governing preparatory procedures. Original source references are provided throughout the text so that the reader can obtain additional information.

It is almost impossible to personally test each of the methods included. I am hopeful that the methods presented in the text will be successful. However, preparatory methods are subject to modification depending on the objective of the study and available facilities. In order to achieve the best results, it is imperative that an attempt be made to optimize even a basic method.

Preparatory procedures for both eukaryotic and prokaryotic groups are presented. Optimal preparatory requirements for these two groups differ from each other. Since plant specimens frequently present special problems (cell wall and large vacuoles), optimal protocols for these specimens differ from those used for animal specimens. Special preparatory methods for plant specimens as well as for viruses are also presented. The processing of uncommon specimens and the solution of unusual, individual problems are included.

Electron microscopy is playing an increasingly crucial role in the confirmation of human diseases; important diagnostic information is provided by electron microscopy. Accordingly, I have attempted to include techniques useful for clinical medicine. Presently, diagnostic information can be obtained by electron microscopy within three hours after a biopsy or collection of a surgical specimen. I have not attempted to present every preparatory procedure used in diagnostic electron microscopy, but have focused instead on well-established, routinely used techniques (for example, negative staining for viral diagnosis).

This book departs from tradition in that books on methodology present only the contemporary consensus of knowledge. Here relatively new methods are also presented, provided that they show potential usefulness. Some of these methods are in the developmental stage, and will probably be refined and standardized; all are restricted in their scope to the preparation of specimens for transmission electron microscopy. Because of the availability of an enormous number and variety of techniques for transmission electron microscopy, limited available space did not allow the inclusion of techniques for scanning electron microscopy. Esoteric techniques that require very expensive instruments are also excluded because of space limitations.

This book is intended for students, technicians, teachers, and research scientists in biology and medicine, and is essentially a laboratory handbook that can be used in formal courses and by individuals. I hope that it will yield practical advice to beginners who are learning the preparatory techniques as well as to experienced and busy scientists who cannot afford to spend time searching for procedures in literature. It is suggested that the entire procedure be read and necessary solutions prepared prior to undertaking the processing. An exhaustive list of references with complete titles is provided, as is a detailed subject index.

Because this is a book on methodology, many techniques are included that were extracted and synthesized from scientific journals, personal communications, and laboratory testing. A vast majority of the methods presented were checked for accuracy and updated by their originator. Numerous scientists were very kind to allow me the use of their illustrations, for which I am most grateful. I am also thankful to the publishers of various scientific journals who were more than prompt in granting me permission for reproducing the illustrations.

This volume could not have been completed without the help, encouragement, and inspiration that were graciously extended to me by a large number of very able scientists.

M. A. Hayat

1

Chemical Fixation

INTRODUCTION

Ultrastructural studies of cells and tissues that are carried out with the transmission electron microscope (TEM) usually require chemical fixation. The primary objective of fixation is stabilization and preservation of cell components without visible coagulation of cellular proteins. This process needs to be accomplished as rapidly as possible so that postmortem (autolytic) changes are kept at their minimum. Fixation protects the specimens against subsequent treatments, including rinsing, dehydration, staining, vacuum, and exposure to the electron beam. Fixed specimens show less shrinkage in the last steps of dehydration as well as during infiltration and embedding. Fixation facilitates transmembranous diffusion of substrates and the trapping of agents in enzyme cytochemical studies. Prefixation with aldehydes prevents loss of proteins from specimens prepared for negative staining. In addition, prior fixation with an aldehyde minimizes artifacts in specimens treated with a cryoprotectant. However, all fixatives currently in use do not prevent losses of intracellular ions, and structural modifications of cell components caused by such leakage should not be ignored.

Aldehydes and OsO_4 are the most effective fixatives for routine transmission electron microscopy. Cell stabilization occurs because fixatives cross-link various macromolecules, rendering them immobilized and insoluble. Cross-linking of protein macromolecules is a prerequisite to satisfactory preservation of cell morphology. Fixatives may form cross-links not only between their reactive groups and the reactive groups in the tissue, but also between different reactive

1

groups in the tissue. Fixation may unmask or free certain reactive groups in the tissue for cross-linking, groups that otherwise may not be involved in intermolecular bonding.

Glutaraldehyde introduces both inter- and intramolecular cross-links into protein macromolecules; these cross-links are irreversible. This dialdehyde reacts readily with free amino groups in proteins and phospholipids. Proteins are efficiently cross-linked via the ϵ-amino groups of lysine. It also reacts with the α-amino groups of other amino acids when present in a free form. N-Terminal amino groups of peptides also react with the dialdehyde. Relatively less reactive groups are tyrosinyl and guanidyl and imidazolyl residues. Glutaraldehyde is unable to cross-link low concentrations of protein.

Primary amino groups present in phospholipids also react with glutaraldehyde, resulting in cross-linking between lipids and membrane proteins. Most other types of lipids are not fixed with glutaraldehyde, nor are glycoproteins, although the latter may be immobilized. Glutaraldehyde does not stabilize myelin, which tends to undergo reorganization during dehydration. DNA associated with proteins in the eukaryotic nucleus tends to be cross-linked with glutaraldehyde and other aldehydes. Glutaraldehyde is ordinarily used in concentrations ranging from 2 to 3% for 1–2 hr at 4°C or room temperature. The mechanisms involved in glutaraldehyde fixation have been reviewed in detail elsewhere (Hayat, 1981a,b).

Another aldehyde used less commonly as a fixative for electron microscopy is formaldehyde, obtained by the dissociation of paraformaldehyde powder. Activities of labile enzymes and immunogenic properties are less impaired by formaldehyde than by glutaraldehyde. Being a monomer, formaldehyde penetrates the tissue more rapidly than does glutaraldehyde. Relatively large tissue blocks can be easily fixed with formaldehyde. This property makes it useful for fixing large samples of biopsy or surgical tissue for both light and electron microscopy. Formaldehyde is effective in stabilizing nucleic acids of phages and other viruses. However, the quality of ultrastructure preservation is inferior to that obtained by glutaraldehyde or acrolein fixation. Moreover, protein cross-linking introduced by formaldehyde is slow. The major drawback of formaldehyde fixation, however, is that its reactions with proteins and other cellular substances are completely or partly reversible. Formaldehyde is not recommended for preserving the ultrastructure except in special cases or in combination with glutaraldehyde. The ratio of the two aldehydes in the mixture depends on the objective of the study; a 1 : 1 ratio is recommended as a standard mixture. Formaldehyde is usually employed in concentrations varying from 1.5% to 3% for 2–4 hr at 4°C or room temperature.

Reactions of formaldehyde with amino acids, proteins, and lipids are well understood (Hayat, 1981b). Formaldehyde initially forms an addition product with free amino groups in amino acids and proteins. This product then undergoes condensation by the formation of methylene bridges. This monoaldehyde also reacts readily with guanidyl groups of arginine and somewhat slowly with $\geqslant$CH groups of phenol and imidazole rings. Other groups that form addition products with formaldehyde include hydroxyl, carboxyl, sulfhydryl, and peptide bonds.

Formaldehyde is a poor fixative for lipids and degrades some of them. Most lipids in tissue fixed with formaldehyde are extracted during dehydration. Only lipids containing primary amino groups are partially fixed by formaldehyde. The major reactions of formaldehyde with nucleic acids is largely reversible. This monoaldehyde does not preserve soluble polysaccharides but prevents the extraction of glycogen. It is very effective in fixing mucoproteins.

Acrolein, another monoaldehyde used for fixation, is bifunctional by virtue of its double bond. Acrolein is an extremely reactive, flammable, and hazardous reagent, and must be handled with care in a fume hood. It reacts rapidly with free amino groups and is superior to formaldehyde for protein cross-linking. Acrolein also reacts with carboxyl and imiadazol groups and has a propensity to react with substances that bear the sulfhydryl group or thiols. These sites can be visualized with the Schiff reaction. Acrolein penetrates and reacts faster than other aldehydes and OsO_4, and it is useful where fixative penetration is a problem, that is, in large and/or dense specimens. Solutions of low concentrations are usually used for a short duration (15–30 min) at room temperature or at 4°C. Because of its volatility, acrolein vapor is ideal for the fixation of mineralized tissues such as bone.

Mixtures of aldehydes yield better preservation of the ultrastructure of many types of specimens than that obtained by using a single aldehyde. By careful selection of the proportion of each of the components in the mixture, the most desirable characteristics of various components can be utilized. A case in point is a mixture of glutaraldehyde and formaldehyde. The latter, being a smaller molecule, penetrates and reacts rapidly, while the former, being a more efficient cross-linking agent, provides more stable bonds. A similar rationale applies to the use of a mixture of glutaraldehyde and acrolein. Other bifunctional but nonaldehydic fixatives include imidoesters (Hassell and Hand, 1974), lysine (McLean and Nakane, 1974), and carbodiimide (Yamamoto and Yasuda, 1977). These reagents are useful for immunoelectron microscopy (Hayat, 1981b).

Because aldehydes are organic in nature, they do not impart electron density to the specimen. This deficiency is overcome by using OsO_4 as a postfixative. This heavy metal salt not only reacts with unsaturated lipids and certain proteins but also imparts electron density to the tissue. Thus OsO_4 acts both as a fixative and as an electron stain. Osmium tetroxide is an additive fixative, for it becomes a part of the cell substances it fixes. Unsaturated bonds in the lipid molecule are the primary reaction sites. One atom of osmium oxidizes one lipid double bond in unsaturated lipids. Osmium tetroxide forms osmate monoesters and diesters involving double bonds of acyl chains. The monoesters are not very stable, but the diesters are stable and very resistant to extraction by organic solvents during dehydration and embedding.

Osmium tetroxide is capable of cross-linking proteins to a modest degree. Weak solutions of OsO_4 can form gels with proteins such as albumin, globulin, and fibrinogen. This metal reacts readily with lipoprotein membranes. A considerable proportion of the membrane proteins, which are originally in an α-helix configuration, are unfolded during OsO_4 fixation (Lenard and Singer, 1968). Osmium tetroxide does not react with most of the pentose or hexose sugars or their polymers, and most of the carbohydrates in OsO_4-fixed tissues are extracted during rinsing and dehydration. The possibility that OsO_4 oxidizes certain carbohydrates without the formation of a black precipitate cannot be ruled out.

In the presence of pyridine, OsO_4 reacts with the pyrimidine moieties in polynucleotides, whereas adenosine and guanosine are not oxidized under similar conditions. Thymine is attacked by OsO_4 about 10 times more rapidly than is uracil. Phenol-containing regions in plant cells are osmiophilic. A variety of reactions between OsO_4 and various macromolecules have been presented by Zingsheim and Plattner (1976), Hayat (1981b), and Behrman (1984).

The best overall preservation of the ultrastructure is obtained by primary fixation with glutaraldehyde or a mixture of glutaraldehyde and formaldehyde followed by OsO_4. This sequential fixation is called double fixation. Since aldehydes are potent reductants, excess aldehyde should be removed before osmication. A mixture of glutaraldehyde and OsO_4 is effective in preserving those specimens (e.g., blood cells) that are not well fixed with double fixation. Glutaraldehyde and OsO_4 solutions are mixed immediately before use; a 1 : 1 ratio is recommended. Osmium tetroxide is ordinarily used in 1–2% concentrations in a buffer at 4°C.

Phosphate buffer (0.1 M) is the best vehicle used with 2–3% glutaraldehyde for the fixation of plant tissues. It allows the least amount of extraction of cellular materials, especially amino acids and proteins. Phosphate buffer is more efficient in facilitating the penetration by the fixative during immersion fixation than is cacodylate or PIPES buffer. If phosphate buffer introduces artifactual, fine precipitate in a certain tissue, then it can be replaced with PIPES or cacodylate buffer. For animal tissues 0.03–0.1 M PIPES buffer is recommended; this buffer allows the least amount of extraction of lipids. In immersion fixation the osmolarity of the buffer is more important than that of the fixative. In fixation by vascular perfusion, on the other hand, the total osmolarity of the fixative solution is important. The optimal osmolalities of the buffer and the fixative solution for each type of specimen are determined by trial and error. However, most commonly used osmolalities are presented in the form of tables by Hayat (1981a,b).

HAZARDS, PRECAUTIONS, AND SAFE HANDLING OF REAGENTS

Almost all the reagents used in processing specimens for electron microscopy are potentially hazardous to various degrees. Many chemicals employed for fixation, rinsing, dehydration, embedding, and staining are potentially capable of causing harm to workers. They can be absorbed either by skin contact or by inhalation. A fume hood, which is checked periodically to ensure that fumes are efficiently removed, is essential in the laboratory. The fume hood should have a capacity of maintaining a face velocity (the velocity of air entering a hood) of about 100 fpm.

One of the most commonly used buffers, sodium cacodylate, contains arsenic and this is a health hazard. If inhaled or absorbed through skin, it can cause dermatitis and liver and kidney inflammation. Hands should be protected by disposable gloves made of impermeable material, and eyes should be covered with gas-tight goggles. A fume hood should be used for weighing out the reagent and preparing the buffer solution. To avoid the production of arsenic gas, this reagent should not come in contact with acids. Since cacodylate buffer reacts with H_2S, the two should not be used in the same solution. Sodium barbitone (Veronal) buffer is a poison if taken by mouth, and s-collidine buffer is toxic and foul-smelling and therefore should be used in a fume hood.

Osmium tetroxide volatizes readily at room temperature. It is dangerous because of its toxicity and vapor pressure. Its fumes are injurious to the nose, eyes, and throat. Hands or any other part of the body must not be exposed to this reagent. Osmium tetroxide must be handled at all times in a fume hood, and gas-tight goggles should be worn when handling it (see also p. 10). All aldehydes are hazardous to various degrees and should be handled in a fume hood; breathing of aldehyde vapors is dangerous. Aldehyde solutions may cause dermatitis if permitted to wet the skin. Spillages of aldehydes can be inactivated by using Fehling's solution, which changes its color from blue to red.

A carcinogen can be formed spontaneously in air when formaldehyde and HCl vapors mix. Formaldehyde has been reported to cause nasal squamous cell carcinoma in animals exposed to 11–15 ppm. About 5 ppm is the limit for any instant exposure, while 2 ppm is the limit for prolonged or repeated exposure and is the level at which formaldehyde can just be smelled. The potential carcinogenic hazard of formaldehyde in the human respiratory tract is currently being debated (Perera and Petito, 1982). A variety of pathologic consequences arising from exposure to formaldehyde have been reviewed by Loomis (1979).

Acrolein is a hazardous chemical because of its flammability and extreme reactivity. It is highly toxic through vapor and oral routes of exposure, is irritating to respiratory and ocular mucosa, and induces uncontrolled weeping. Acrolein is moderately toxic through skin absorption and is a strong skin irritant even at low concentrations. Although acrolein is highly toxic by the vapor route, the sensory response to very low vapor concentration gives adequate warning. The physiologic perception of the presence of acrolein begins at 1 ppm, at which concentration an irritating effect on the eyes and nasal mucosa is felt. Thus there is little risk of acute intoxication because the lachrymatory effect compels one to leave the polluted area. A threshold level between 0.05 and 0.1 ppm of air has been suggested. Above this level, acrolein sensitizes the skin and respiratory tract and causes bronchitis and pneumonia in experimental animals. In human beings, when there is contact with acrolein, cutaneous or mucosal local injury may be observed. Gloves and fume hood must be used when handling this chemical. In the preparation of solutions, acrolein should be slowly added to water with stirring rather than the reverse. Any acrolein that contacts the skin should be washed off immediately with soap and water. Waste acrolein should be disposed of by pouring it into 10% sodium bisulfite, which acts as a neutralizer.

The components of all the epoxy resins are potentially dangerous. They can be absorbed by skin contact or inhalation. The hazards associated with the use of resins are carcinogenesis, primary irritancy, systemic toxicity, environmental pollution, and fire hazards (Causton, 1981). Vinyl cyclohexene dioxide (ERL 2406, Spurr mixture) is carcinogenic and is known to produce tumors in animals. The resin should be used only when necessary, for example, for embedding hard animal tissues and plant tissues. Monomeric resin components (liquid or vapor form) can cause skin, mucosal, and eye irritation. For example, corneal injury may be induced by nonylsuccinic anhydride and butyl glycidyl ether, and methyl methacrylate can cause contact dermatitis.

Resins emit vapors especially during polymerization; if the oven is not vented to the outside, vapors will seep into the laboratory. Systemic toxicity can result in heart dysfunction, chest spasms, blood disorders, and neurologic disorders. Contamination can occur by skin absorption, inhalation, or accidental ingestion. These effects may be sudden or gradual. Almost all resins are volatile and form aerosols during polymerization. These vapors can cause headaches and reduce appetite. Most of the resins are flammable. For example, benzoyl peroxide is potentially explosive when dry and is capable of violent reaction with amines. When propylene oxide is allowed to mix with phosphotungstic acid (PTA) for longer than 5 min, the result is an explosive exothermic reaction. Propylene oxide is potentially carcinogenic. Urea–formaldehyde embedding medium may also be a health hazard.

Unpolymerized resins should be handled in fume hoods. They should be mixed with a magnetic stirrer or disposable glass rods, and transferred without spillage, using disposable pipettes. Before disposal, waste resin should be kept in a fume hood for several days at 60°C until polymerized. Resins should not be disposed of in a monomeric form. Items contaminated with resins should be collected in a polyethylene bag in the fume hood and then sent for burial or burning. Every effort should be made to avoid contaminating handles of doors and refrigerators. All containers used for monomeric resins should be disposable.

Gloves should be worn while processing the specimens, and hands must be washed with soap and cold water after gloves are removed; gloves are not impermeable to resins. A major spillage should be covered with sand or vermiculite. Polymerization of blocks should be carried out in an oven vented to the outside or in a small oven kept in the fume hood. No resin block for electron microscopy is completely polymerized. Therefore dust and small chips produced during sawing or filing of "polymerized blocks" are hazardous. Resin dust should not be allowed to remain on a bench, on the floor, or in a wastepaper basket. It should be collected with a damp paper towel or cloth, or a vacuum cleaner. If possible, sawing or filing should be carried out in a fume hood. All bottles containing resins should be unbreakable and clearly labeled. The bottles should be tightly capped during storage, and their exterior should be kept free of resins. Resins must never be washed off the skin with a solvent; soap and cold water should be used to wash contaminated skin.

Uranium and its compounds are radioactive and highly toxic. Uranyl acetate is a dual hazard, being both chemically and radiologically toxic. The maximum allowable concentration of soluble uranium compounds in the air is 0.05 mg/m³ air. But more dangerous is the inhalation of powdered uranyl compounds. A daily uptake of 50 mg is considered lethal. Inhalation may cause disorders to the upper respiratory tract, lungs, and liver. Damage can also occur to the kidneys. Exposure to insoluble compounds of uranium may lead to lung cancer, pulmonary fibrosis, and blood disorders. As already noted, uranium compounds constitute a source of ionizing radiation. One gram of uranium emits 12,500 decays/sec of alpha particles, 25,000/sec as beta emission, and some gamma radiation. Beta emission of uranyl acetate is thought to be close to the aforementioned theoretical value. Uranium compounds, particularly in powder forms, must not be touched by bare skin, inhaled, or ingested. Glasswares containing even minute amounts of these salts should not be left exposed. After use of these reagents, the hands and the work area should be washed. Unfortunately, uranium compounds are not labeled radioactive by the suppliers. Darley and Ezoe (1976) have discussed potential hazards of uranium and its compounds in electron microscopy.

Lead and mercury salts used for staining are highly toxic. Tannic acid is poisonous, and diaminobenzidine may be carcinogenic. Bismuth, silver nitrate, dimethyl sulfoxide, collidine, potassium ferrocyanide, and potassium permanganate are harmful. Lanthanum nitrate is an irritant and readily oxidizes.

Many organic solvents used in the laboratory are fire hazards because of their high flammability and volatility, while others are physiologic hazards owing to their toxic fumes. Some of the solvents are hazardous on both counts. Among these reagents are acetone, propylene oxide, ethanol, carbon tetrachloride, benzene, xylene, chloroform, and amyl acetate. These reagents should not be used near an open flame. All solutions in the laboratory should be pipetted by pipettes that do not require mouth suction. Propylene oxide may explode if used in the same room with an open flame. This reagent should be used only in a fume hood. Acetone and xylene should not be used in a sonicator. Benzene should always be kept in a fume hood, for repeated inhalation of its fumes can result in severe physiologic damage such as reduced circulation of white blood cells, injury to the liver and spleen, and even leukemia. Such damage may not appear until some months or even years have passed. Ether is equally hazardous. Amyl acetate is an anesthetic and so it should be used only in a fume hood. Inhalation of its fumes may cause liver damage.

Lectins and certain enzyme inhibitors are dangerous. Photographic developers may cause dermatitis in susceptible persons. Many dyes (e.g., toluidine blue and methylene blue) are toxic and should not be permitted to come in contact with the skin. Phosphorous pentoxide, a desiccant, reacts violently with water. It should not be discarded in a water sink. Contact of skin or eyes with liquid nitrogen or its gas can cause severe burning. A metal piece cooled with liquid nitrogen is also extremely cold. Liquid nitrogen should always be used in well-ventilated areas so that the buildup of N_2 gas will not deplete the O_2 in the air.

Because of the qualitative similarities of basic biologic processes among various species of mammals, data on the effects of chemicals on animals can be used to predict human response. Therefore chemicals proven hazardous for animals should also be considered hazardous for human beings. Positive human data are not needed to consider a chemical as a likely carcinogenic or hazardous. Health and safety hazards in the electron microscope laboratory have also been discussed by Humphreys (1977), Thurston (1978), Weakley (1981), Ringo *et al.* (1982), and Lewis (1983). EMscope Laboratories Ltd. (Kingsnorth Industrial Estate, Ashford, Kent) has printed a very useful "Safety Chart, Chemicals in Electron Microscopy," for free distribution.

FIXATIVES

Purification of Glutaraldehyde

The Charcoal Method

Approximately 200 ml of commercial glutaraldehyde solution (50%) is added to 30 g of activated charcoal in a flask. The mixture is thoroughly shaken for 1 hr at 4°C and then vacuum-filtered through Whatman no. 42 filter paper mounted in a Buchner funnel. The filtrate is remixed with 20% fresh, activated charcoal and refiltrated. The process is repeated at least twice. The final yield of purified glutaraldehyde is 20–30 ml.

The Distillation Method

In one method, the distillation is carried out at atmospheric pressure. The distillate is collected at 100°C in 50-ml aliquots, which are monitored by measuring pH. Any sample showing a pH lower than 3.4 is discarded. Pure glutaraldehyde with a concentration of 8–12% is obtained. Alternatively, a single-step distillation under moderate vacuum yields a glutaraldehyde of equivalent purity. Approximately 250 ml of commercial glutaraldehyde is charged into a 500-ml vigreux distilling flask heated by an electrical heating mantle and connected to a Liebig condenser. The distillation is performed under vacuum at 15 mmHg. The temperature is raised to 65°C and the distillate, a viscous clear liquid, is collected. Upon interruption of the vacuum, the distillate is immediately diluted with an equal volume of freshly boiled demineralized distilled water by slow addition of the latter (75°C) to the magnetically stirred distillate under a stream of nitrogen.

Another single-step distillation method was introduced by Dijk *et al.* (1982). The purification of glutaraldehyde is expressed as the purification index,

$$Pi = extinction\ at\ 235\ nm/extinction\ at\ 280\ nm$$

At 235 nm the impurities and polymers are measured, while at 280 nm the monomers are measured. An adiabatic fractionation column is used to reduce the heat exchange. The interior of the fractionation column is filled with small glass cylinders leading to a large surface increase. This procedure facilitates the purification of the glutaraldehyde. The vapors condense in a vertical cooling unit and the distillate is led to two flasks. The distillation is carried out under vacuum at a pressure of 10 mmHg. A flask is heated in an oil bath, and 650 ml of 25% glutaraldehyde solution with Pi > 10 is distilled each time. The first distillate, collected at a temperature of 20–40°C after about 2 hr, contains impurities and is discarded. Subsequently, the purified glutaraldehyde with Pi < 0.2 is collected at a temperature of 80°C after 90 min.

Storage of Glutaraldehyde

Purified glutaraldehyde remains relatively stable for several months if stored at 4°C or below, provided the pH is lowered to 5.0. Purified glutaraldehyde can be stored for 6 months at −14°C and for 1 month at 4°C without significant polymerization. The most effective way to minimize the deterioration of purified glutaraldehyde is by storing it as an unbuffered, 10–25% solution at subfreezing temperatures (−20°C). Somewhat different storage conditions are required for purified and unpurified glutaraldehyde. There is no great advantage to storing glutaraldehyde in the dark or under inert gas, because commercial lots already contain sufficient acid to catalyze polymerization. However, the purified glutaraldehyde may be stored under oxygen-free conditions.

Preparation of Formaldehyde Solution from Paraformaldehyde Powder

The formaldehyde solution is prepared by mixing 2.5 g paraformaldehyde powder with 50 ml of 0.2 M buffer at 65°C with stirring. A few drops of 1 N NaOH are added with continuous stirring to clear the cloudiness from the solution. Sufficient distilled water is added to make up to 100 ml. This fixative solution contains 2.5% paraformaldehyde in 0.1 M buffer. The desired pH can be obtained with dilute HCl or NaOH. Certain commercial lots of paraformaldehyde powder do not dissolve in some buffers. In such cases the powder is dissolved in distilled water instead of in a buffer.

Preparation of and Precautions in Handling Osmium Tetroxide Solution

Osmium tetroxide is dangerous to handle because of its toxicity and vapor pressure. Its fumes are injurious to the nose, eyes, and throat. Hands or any other part of the body must not be exposed to this reagent. Accidental spills of OsO_4 on skin should be washed off immediately with copious volumes of water. Any spillage on worktables can be inactivated by a reducing agent such as stannous chloride. It can be used as a 10% aqueous solution or sprinkled on as a dry powder. Alternatively, the spillage can be reduced by using ascorbate powder; it reacts with OsO_4 within 2–3 sec. If a very small amount of OsO_4 solution needs to be disposed down a sink, large amounts of running water should be used. The safer way is to add used OsO_4 solution to an adequate amount of vegetable oil and allowed to remain in a sealed bottle for at least 48 hr. It reacts with the oil and any unreacted portion is precipitated. To break down the oil, the mixture is treated with enough household detergent. Finally, these contents can be washed down the drain with a copious amount of water. However, burial in a guarded area designated for toxic waste is preferable.

This reagent is a strong oxidizing agent and is readily reduced by organic matter and exposure to light. It should be noted that even the smallest amount of organic matter may reduce it to the hydrated dioxide, which is worthless as a fixative. However, reduction can be avoided by the complete exclusion of dust and organic matter and by use of a brown glass bottle. If OsO_4 solution shows violet to light brown coloration because of the formation of osmium dioxide, it can be regenerated by adding a drop of hydrogen peroxide.

Osmium tetroxide must be handled in a fume hood, and because the reagent is rather expensive, unstable, and volatile, solutions should be prepared with utmost care. The first step in the preparation of its aqueous solution is to remove the label (after reading it!) from the glass ampoule containing the OsO_4 crystals. (OsO_4 is also supplied as an aqueous solution in glass ampoules.) The glass ampoule, a glass-stoppered bottle, and a heavy glass rod are carefully cleaned with concentrated nitric acid (to remove all the organic matter), and they are then washed thoroughly with distilled water to eliminate all traces of the acid.

The bottle and the rod should be dried in an oven; they should never be wiped with a paper or cloth towel because these materials invariably leave behind some lint, which would reduce the solution to hydrated dioxide. A measured amount of distilled water, buffer, or another vehicle is added to the bottle, and the glass ampoule, after having its neck scored with a file (ampoules with prescored necks are also available), is gently placed into the same bottle. After the glass ampoule has been broken with a heavy glass rod, the bottle is quickly stoppered and shaken vigorously. Several hours are required to dissolve OsO_4 crystals completely in the vehicle. The solution can be prepared more rapidly by gently heating it on a steam bath over a magnetic stirrer or by using a sonicator for a few minutes.

Since OsO_4 is extremely volatile and its solutions rapidly decrease in concentration, solutions should be prepared in small quantities and stored in a tapered flask fitted with a glass stopper and Teflon sleeve or Teflon tape. The use of ground-glass stoppers is not recommended, for they do not prevent decrease in the concentration of OsO_4 even when maintained at 4°C. The only effective way of keeping OsO_4 solutions is by using Teflon liners on glass stoppers. The flask must be tightly stoppered, wrapped in aluminum foil, and stored in a refrigerator. This flask should be kept inside a closed container such as a tin or glass jar; otherwise osmium vapors escaping from the flask will blacken the inside of the refrigerator. The solution is thought to be stable for several months under the aforementioned conditions of storage. Alternatively, it may be stored in a bottle having a ground-glass stopper, but it is less stable in this condition.

Aqueous solutions of OsO_4 (2%) can be stored in small amounts in vials in a freezer and used after thawing and diluting with a buffer solution when needed. These solutions can be kept in the freezer for at least several months without any adverse effect. Such a practice is helpful when the time available on a certain day is limited.

Regeneration of Used Osmium Tetroxide

The procedure (performed in a fume hood) introduced by Kiernan (1978) involves oxidation of osmium residues in the used solution to OsO_4, which is then extracted into carbon tetrachloride (CCl_4). The OsO_4 solution in CCl_4 is reduced with ethanol, precipitated, and recovered by filtration on the filter paper.

Procedure for Preparing OsO₂

1. First 500 ml of used OsO_4 solution is poured into a 1-l flask. Next, 5 ml of concentrated H_2SO_4 is slowly added while stirring; then 200 ml of aqueous 6% solution of $KMnO_4$ is added in the same manner. After the ingredients are thoroughly mixed, the flask is covered with an inverted small beaker and allowed to remain at room temperature for 30 min. A brown color indicates the precipitation of MnO_2. Additional $KMnO_4$ solution may have to be added until the color becomes purple.

2. The mixture is extracted by shaking vigorously for 3 min in a separatory funnel after three changes (each 150 ml) of CCl_4. The heavier fraction containing CCl_4 and osmium is carefully separated from the aqueous phase (containing MnO_2) into a 500-ml conical flask.

3. On the addition of 10 ml of 100% ethanol to the flask, the color becomes darker, and black OsO_2 is precipitated. The flask is covered and kept at room temperature for 48 hr.

4. The colorless supernatant is decanted, and the precipitate is retained.

5. About 20 ml of 100% acetone is added to the flask containing the precipitate of OsO_2. The precipitate adhering to the sides of the flask is scraped loose with a glass rod. The flask is shaken, and the suspension is poured into a filter funnel containing a dry Whatman no. 1 filter paper. The flask is rinsed out several times with additional acetone, which is poured through the filter until almost all of the OsO_2 has been recovered.

6. The OsO_2 precipitate on the filter paper is rinsed with 3 aliquots of 50 ml of acetone.

7. On drying, the filter paper is placed on a porcelain evaporating dish, transferred to a vacuum desiccator, and allowed to remain overnight. It is preferable to have a desiccant (anhydrous $CaSO_4$) in the desiccator, although drying is due to evaporation of the acetone.

8. Dry OsO_2 powder is collected in a small glass bottle which is tightly stoppered, and it can be stored. The black powder is almost indefinitely stable, though it will be transferred to OsO_4 on contact with oxidizing agents, including prolonged exposure to atmospheric oxygen. About 5 g of dry OsO_2 is obtained from each liter of used OsO_4 (2%) solution.

Preparation of 50 ml of 2% OsO₄ Solution

1. One gram of dry, black OsO_2 powder is transferred into a clean glass bottle that has a glass stopper and a 50-ml mark. Then 45 ml of distilled water is added to the bottle, which is shaken to suspend the powder.

2. One milliliter of 3% H_2O_2 is added, and the ingredients are mixed rapidly. The bottle is covered with a piece of aluminum foil and placed in a refrigerator for 30 min.

3. One milliliter of 30% H_2O_2 is added in drops while the solution is being

continuously swirled. Almost all the OsO_2 should be dissolved; however, the bottle is covered with aluminum foil and returned to the refrigerator for 10–30 min to ensure complete dissolution.

4. Enough distilled water is added to reach the 50-ml mark, and the glass stopper is then inserted.

The procedure just described applies only to OsO_4 solutions containing buffers, electrolytes or nonelectrolytes, and tissue lipids. Osmium tetroxide solutions containing inorganic iodides and potassium and chromium compounds have not been subjected to this procedure.

Preparation of Fixatives

Acrolein

0.2 M Buffer	50 ml
10% Acrolein	20 ml
Distilled water to make	100 ml

The fixative contains 2% acrolein in 0.1 M buffer.

Acrolein–Glutaraldehyde

0.2 M Buffer	50 ml
25% Acrolein	4 ml
25% Glutaraldehyde in water	10 ml
Distilled water to make	100 ml

The fixative contains 1% acrolein and 2.5% glutaraldehyde in 0.1 M buffer.

Acrolein–Glutaraldehyde–Formaldehyde

0.2 M Buffer	5 ml
10% Acrolein	3 ml
6% Formaldehyde	5 ml
10% Glutaraldehyde	6 ml
Distilled water	1 ml

Acrolein–Glutaraldehyde–Dimethyl Sulfoxide

50% DMSO in 0.1 M buffer	42 ml
25% Glutaraldehyde	4 ml
25% Acrolein	4 ml

Acrolein–Glutaraldehyde–Formaldehyde–Dimethyl Sulfoxide

2% Formaldehyde in 0.1 M buffer	44 ml
10% Acrolein	0.5 ml
25% Glutaraldehyde	5.0 ml

DMSO   1.25ml
$CaCl_2$ 0.008 g
The final concentration of $CaCl_2$ is 0.001 M.

Carbodiimide–Glutaraldehyde

Stock buffer
Tris base   1.4 g
Dibasic $Na_2HPO_4 \cdot 7H_2O$ 0.67 g
Monobasic $NaH_2PO_4 \cdot H_2O$ 0.345 g
Dulbecco's PBS   50 ml
Distilled water   50 ml

This buffer is stored at 4°C and brought to room temperature before use.
Fixative mixture
Stock buffer   10 ml
1-Ethyl-3(3-
   dimethylaminopropyl)carbodiimide-HCl   100 mg
50% Glutaraldehyde   10–100 μl

The final concentration of glutaraldehyde is 0.05–0.5%, and pH is adjusted to 7.0 with 1 N NaOH (requiring 1–8 drops). At precisely 4 min after initial mixing, the fixative solution is added to the cell culture dish that has been prewashed with PBS. Fixation time is 7 min, followed by washing with PBS (Willingham and Yamada, 1979).

Dalton's Chrome–Osmium (Dalton, 1955)

Ten milliliters of 2% OsO_4 solution and 5 ml of 3.4% aqueous NaCl are added to 5 ml of chromate buffer (pH 7.2). This buffer is prepared by adding enough 2.5 M KOH to 80 ml of 5% potassium dichromate to obtain a pH of 7.2 and making the final volume 100 ml by adding distilled water. This fixative can be stored for several months at 4°C.

Dimethyl Suberimidate (DMS)

Distilled water   7.8 ml
NaOH (0.1 N)   1.2 ml
Tris base   121–182 mg
DMS   160–200 mg

One milliliter of 0.2 M $CaCl_2$ is added drop by drop. The pH is adjusted to 9.5 with 0.1 N NaOH. The fixative solution is prepared immediately before use. The duration of fixation should not exceed 2–3 hr at room temperature, for DMS is unstable in aqueous solutions.

Formaldehyde–Chromic Acid

0.1 M Cacodylate buffer (pH 7.2)	36 ml
2% Formaldehyde	4 ml
1% Chromic Acid	4 ml

The fixative contains 0.2% formaldehyde and 0.1% chromic acid.

Formaldehyde–Picric Acid

0.1 M Phosphate buffer (pH 7.3)	120 ml
8% Formaldehyde	40 ml
1% Picric acid	3.2 ml

The fixative contains 2% formaldehyde and 0.02% picric acid and has an osmolality of 895 mosmols.

Glutaraldehyde–Phosphate Buffer

$NaH_2PO_4 \cdot H_2O$	3.31 g
$Na_2HPO_4 \cdot 7H_2O$	33.71 g
25% Glutaraldehyde in water	40 ml
Distilled water to make	1000 ml

The fixative contains 2.5% glutaraldehyde and has an osmolality of 320 mosmols; the pH is 7.4.

Glutaraldehyde–Cacodylate Buffer

0.2 M Cacodylate buffer	50 ml
25% Glutaraldehyde in water	8 ml
Distilled water to make	100 ml

The concentration of glutaraldehyde is 2%, and the molarity of the buffer is 0.1 M.

Glutaraldehyde–Alcian Blue

Glutaraldehyde fixative containing 0.5–2.0% alcian blue (or astra blue 6GLL) enhances the preservation and contrast of cell coat and intercellular substances. The staining of these substances can be further enhanced with lanthanum or uranyl acetate and lead citrate.

Glutaraldehyde–Caffeine

When phenolic-containing plant cells are fixed with glutaraldehyde followed by OsO_4, phenolics leach from the vacuoles into the cytoplasm where they subsequently react with OsO_4. The result is a dense, osmiophilic cytoplasm, the details of which are obscured. This leaching can be prevented by fixation with glutaraldehyde containing 0.1–1.0% caffeine and then a rinse with a buffer containing caffeine.

Glutaraldehyde Containing High Concentration of Potassium

Glutaraldehyde containing a high concentration of potassium in the presence of calcium has been reported to cause a near-simultaneous depolarization and fixation of the glomus cells of rat carotid body (Grönblad, 1983). As a result, exocytosis is stimulated and exocytotic profiles are fixed. The animal is fixed by transcardiac perfusion for 15 min at a perfusion rate of 5 ml/min and at a pressure of 110 cmH_2O. The fixative solution consists of 2.5% glutaraldehyde in the following vehicle:

NaCl	64 mM
KCl	60 mM
$MgCl_2$	1.5 mM
$CaCl_2$	1.0 mM
Glucose	10 mM
HEPES buffer	20 mM
$KHPO_4$	0.5 mM
$NaHCO_3$	5.0 mM

The solution is brought to pH 7.4 with NaOH. Postfixation is accomplished with 1% OsO_4 in 0.1 M cacodylate buffer (pH 7.2) for 1 hr at 4°C.

Glutaraldehyde–Digitonin

Free cholesterol is extracted from the tissues after conventional fixation with glutaraldehyde followed by OsO_4. This cholesterol can be retained by fixation with the following mixture:

0.2 M Cacodylate buffer (pH 7.2)	2.5 ml
50% Glutaraldehyde	2.5 ml
4% Formaldehyde	25 ml
0.4% Digitonin in 0.2 M cacodylate buffer	25 ml
$CaCl_2$	0.025 g

Each of the ingredients is dissolved in sequence.

Glutaraldehyde–Formaldehyde

0.2 M Buffer	50 ml
25% Glutaraldehyde in water	10 ml
10% Formaldehyde in water	20 ml
Distilled water to make	100 ml

The fixative contains 2.5% glutaraldehyde and 2% formaldehyde in 0.1 M buffer. The osmolarity can be increased by adding sucrose, glucose, $CaCl_2$, or NaCl.

Glutaraldehyde–Formaldehyde–Nitrogen Mustard N-Oxide

This method has been used for preserving secretory granules containing thyroglobulin in the thyroid (Fujita et al., 1978) and catecholamine granules in the nerve terminals of vas deferens and adrenal medullary cells (Chiba and Murata, 1982). The fixative should be effective in preserving low molecular weight substances lacking primary amino groups that do not react with glutaraldehyde. Nitrogen mustard N-oxide is highly toxic, so the fixative must be prepared in a fume hood. The fixative should be used within 30 min after its preparation. Tissue blocks are fixed with a mixture of 1% nitrogen mustard N-oxide, 1% formaldehyde, and 1% glutaraldehyde in Millonig's phosphate buffer (pH 7.4) for 2–4 hr at room temperature. After a brief rinse in buffer, the specimens are postfixed with 1% OsO_4 in buffer for 1 hr.

Glutaraldehyde–Hydrogen Peroxide

> 0.1 M Cacodylate buffer   25 ml
> 25% Glutaraldehyde   5 ml
> 15% H_2O_2 5–25 drops

Drops of H_2O_2 are added to glutaraldehyde while the solution is continuously stirred. The concentration of glutaraldehyde ranges between 3 and 6%. Hydrogen peroxide should not be used with formaldehyde.

Glutaraldehyde–Lead Acetate

> When conventional fixation is used, soluble inorganic phosphate is lost.
> 0.2 M Cacodylate buffer (pH 7.0)   16 ml
> 8% Glutaraldehyde   16 ml
> 4% Lead acetate   32 ml

Specimens are fixed for 3 hr at room temperature and then transferred to 4% lead acetate in 35% acetate acid for 12 hr at 4°C. Uranyl acetate and OsO_4 should not be used.

Glutaraldehyde–Formaldehyde–Terpenoids

Tissue blocks are fixed for 1–2 hr at room temperature with a mixture of 2% formaldehyde and 0.5% glutaraldehyde in 0.1 M cacodylate buffer (pH 7.4) containing 0.2% $CaCl_2$ (Wigglesworth, 1981). After a rinse in buffer, the specimens are postfixed for 1 hr with 1% OsO_4. This procedure is followed by treatment for 30–60 min with a 1% by volume farnesol ($C_{15}H_{26}O$) (Aldrich Chemical Co.) in 70% ethanol, and then reosmication with 1% OsO_4 for 1 hr. Alternatively, a saturated solution (4%) of monoterpene hydrocarbon myrcene ($C_{10}H_{16}$) can be used for partition; however, farnesol is the preferred reagent. The specimens are dehydrated in ethanol as the sole solvent (propylene oxide is not used) and embedded in a resin.

Masked lipids can be made accessible to farnesol by treating the aldehyde and OsO_4 fixed tissues for 24 hr at 4 or 22°C with saturated (0.09–0.1%) thymol in sucrose (0.34 M). This treatment is followed by exposing the specimens to farnesol and again to OsO_4 as indicated above.

Glutaraldehyde–Malachite Green

Certain cellular lipids are better preserved with the following mixture than with conventional fixation.

0.07 M Cacodylate buffer (pH 6.8)	22 ml
25% Glutaraldehyde	3 ml
1% Malachite green	2.5 ml

The fixative contains 3% glutaraldehyde and 0.1% malachite green. Specimens are fixed for 18 hr at 4°C. After a brief rinse in the buffer containing 0.2 M sucrose, specimens are postfixed with 2% OsO_4.

Glutaraldehyde–Phosphotungstic Acid

Anionic sites in basement membranes and collagen fibrils can be demonstrated with cationic polyethyleneimine (PEI), which acts as a tracer particle for anionic sites. Tissue blocks are immersed in 0.5% solution of PEI (adjusted to pH 7.3 with HCl), whose osmolality has been raised to 400 mosmols with NaCl. After being washed in cacodylate buffer, specimens are fixed in a mixture of glutaraldehyde (0.1%) and phosphotungstic acid (PTA) (2%) for 1 hr at room temperature. Postfixation is carried out in 1% OsO_4. Phosphate buffer is undesirable.

Glutaraldehyde–Potassium Dichromate

This method is useful for the demonstration of biogenic amines (dopamine, noradrenalin, and 5-hydroxytryptamine). The best method is simultaneous aldehyde fixation and chromation. Essential conditions are as follows: lower glutaraldehyde concentration (1%); slightly acidic pH (6.0) during fixation and prolonged incubation in 0.2 M sodium chromate–potassium dichromate buffer before postfixation with OsO_4; short duration of fixation; and the use of sodium chromate–potassium dichromate as buffer.

Glutaraldehyde	1%
Formaldehyde	0.4%
Sodium chromate/potassium dichromate (pH 7.2)	0.1 M

Specimens are fixed in this mixture either by immersion with agitation (1–10 min) or by vascular perfusion (5–15 min) at 4°C. After being stored with constant agitation in 0.2 M sodium chromate–potassium dichromate buffer (pH 6.0) for 18 hr at 4°C, the specimens are postfixed with 2% OsO_4 in the same buffer (pH 7.2) for 1 hr at 4°C. Sections are stained with lead citrate, but not with uranyl acetate.

Glutaraldehyde–Potassium Permanganate

This procedure preserves both microtubules and membranes. The relationship between microtubules and membranes in the mitotic apparatus becomes clear.

Cells are fixed with 2.5% glutaraldehyde in 0.1 M cacodylate buffer for 15 min at room temperature. After being rinsed in buffer, the cells are postfixed with a saturated aqueous solution of $KMnO_4$ for 5 min at room temperature. The durations of fixation and postfixation are critical in obtaining satisfactory preservation of microtubules as well as membranes. The optimal length of each of the two fixation periods for a specific cell type can be determined by trial and error.

Glutaraldehyde–Potassium Permanganate– Phosphotungstic Acid–Hematoxylin

This procedure is useful for preserving and staining heterochromatin in the nucleus. The blood is collected in a test tube containing a small volume of heparin solution and allowed to sediment by gravity at 37°C for 1 hr (Issidorides and Katsorchis, 1981). The leukocyte-rich plasma is aspirated and centrifuged at 250 g for 10 min. The supernatant is decanted, and the leukocyte pellet is fixed with ice-cold 2.5% glutaraldehyde in 0.2 M cacodylate buffer (pH 7.2) for 30 min at 4°C. After being rinsed twice (1 hr each) in an ice-cold buffer, the pellet is postfixed with 0.25% $KMnO_4$ for 5 min. After a 10-min rinse in water, the pellet is treated with 5% oxalic acid and then rinsed in water for 10 min. The pellet is cut into small pieces and immersed in a mixture of phosphotungstic acid and hematoxylin (pH 1.6) for 24–48 hr at room temperature. This mixture consists of 1 g hematoxylin, 20 g PTA, 1000 ml distilled water, and 0.177 g $KMnO_4$. The pellet is rinsed in tap water and dehydrated.

Glutaraldehyde–Tannic Acid–Saponin

Tannic acid is an excellent stain for both membranes and cytoplasmic fibers, and it protects actin filaments from fragmentation by OsO_4 used with or without glutaraldehyde prefixation. When used alone, tannic acid does not penetrate intact cells. The mixture given here allows the tannic acid to penetrate intact cells (e.g., HeLa cells) without disruption of membranes or extraction of the cytoplasmic matrix (Maupin and Pollard, 1983). Most cytoplasmic structures including actin and intermediate filaments, membrane coats, and ground substance are well preserved and stained, allowing the study of very thin sections (dark gray color).

The culture medium is replaced by a mixture of 100 mM (1%) glutaraldehyde, 0.5 mg/ml saponin, and 2 mg/ml tannic acid in buffer A (100 mM sodium phosphate, 50 mM KCl, and 5 mM MgCl, pH 7.0) at 37°C for 30 min. The dish is rinsed briefly with buffer A (pH 6.0) and postfixed with 40 mM (1%) OsO_4 in buffer A (pH 6.0) for 20 min at room temperature.

Glutaraldehyde–Trinitro Compounds

These mixtures yield better preservation of spermatozoa and tissue blocks of large size than does the conventional fixation.

0.2 M Buffer (pH 7.2)	45 ml
25% Glutaraldehyde	5 ml

4% Formaldehyde   50 ml
2% 2,4,6-Trinitrocresol   50 ml

The fixative contains 2.5% glutaraldehyde, 2% formaldehyde, and 0.5% 2,4,6-trinitrocresol. These compounds are potentially explosive.

Glutaraldehyde–Uranyl Acetate

This mixture is useful for bacteria infected with phage. Bacteria are sedimented, and the well-drained pellet is resuspended into a mixture of 0.1% uranyl acetate and 5% glutaraldehyde in Michaelis buffer at a final pH of 5.4. The initial pH of the buffer is 5.9, which, upon addition of a concentrated aqueous solution of uranyl acetate, drops to 5.4. Fixation is completed overnight at room temperature.

About 1 ml of the fixed cells is spun down in a Micro-centrifuge tube used in a small swinging bucket-type top centrifuge. The pellet is suspended in 2% agar in Michaelis buffer (pH 5.7) by pouring 3–4 mm of agar in the tubing and then mixing with a microsyringe while holding the tubing in a bath at 45°C. The agar–bacteria mixture pulled back into the syringe makes a cylinder 6 mm long. The syringe is transferred into the refrigerator for some minutes. Then the solidified cylinder is pushed out of the tubing onto a glass slide and cut into small blocks that contain a sufficiently high concentration of bacteria. These blocks are treated in a saturated aqueous solution of uranyl acetate for 2 hr at room temperature.

Lanthanum Permanganate

La(NO$_3$)$_3$·6H$_2$O   1 g
KMnO$_4$ 1 g
Veronal acetate stock buffer   20 ml
Ringer's solution   6 ml
0.1 N HCl   enough to obtain
                                pH 7.6
Distilled water to make   100 ml

Osmium Tetroxide Quick-Fix Method

This method is effective in fixing rapid forms of cell motility such as metachronal ciliary beating. Cilia on the surface of *Paramecium* beat 20 times/sec. These cilia continue to beat for 30 sec after the *Paramecium* is placed in glutaraldehyde. Thus, primary fixation with aldehydes is too slow to preserve the metachronal wave of cilia. When a brief primary fixation with buffered OsO$_4$ is used, cilia are "quick-fixed," preserving their metachronal wave. The following method was used to fix apical vesicles in chloride cells of fish gills (Bradley, 1981).

The decapitated head of the fish is pinned on a board and exposed gills are washed for 20 sec with 1% OsO$_4$ in 0.05 M cacodylate buffer (pH 7.8). The gills are excised and placed in 2% glutaraldehyde in the same buffer for 1 hr and then in 1% OsO_4 in buffer for 30 min. Alternatively, a 1% solution of OsO_4 in 0.05 M cacodylate buffer (pH 6.0) is perfused through the mouth of the decapitated head using a Pasteur pipette. The OsO_4 solution passes through the oral cavity and over the gills, which turn black within seconds. After 20 sec, the entire head is placed in 4% glutaraldehyde in 0.066 M cacodylate buffer (pH 7.8) and the gills are dissected out. They are cut into small pieces and then placed first in identical glutaraldehyde fixative for 1 hr and then in 1% OsO_4 in buffer for 30 min.

Potassium Permanganate

Acid Permanganate

Small pieces of adrenal medulla (monoamine-storing cells) are fixed in 3% $KMnO_4$ in 0.1–0.2 M acetate buffer (pH 5.0) for 30 min at 4°C. After a brief rinse in the buffer, specimens are stained en bloc with uranyl acetate. Ultrathin sections are viewed without poststaining.

Neutral Permanganate

Specimens are fixed in 3% $KMnO_4$ buffered with Krebs–Ringer glucose (pH 7.0) for 2 hr at 4°C. After fixation, specimens are rinsed in the buffer several times and then allowed to remain in the buffer overnight before dehydration.

Sodium Permanganate

NaCl	2.8 g
KCl	0.4 g
$CaCl_2$	0.2 g
$MgCl_2 \cdot 6H_2O$	0.2 g
NaH_2PO_4	0.16 g
0.83% $NaMnO_4$ in Veronal acetate (pH 7.5) to make	100 ml

Ruthenium Red

Solution A

4% Glutaraldehyde in water	5 ml
0.2 M Cacodylate buffer (pH 7.3)	5 ml
Ruthenium red stock solution (1500 ppm in water)	5 ml

Solution B

5% Osmium tetroxide in water	5 ml
0.2 M Cacodylate buffer	5 ml
Ruthenium red stock solution (1500 ppm in water)	5 ml

Tissue specimens are fixed and stained with solution A for 1 hr at room temperature. Specimens are rinsed briefly with the buffer and then postfixed and stained with solution B for 3 hr at room temperature. Ruthenium red and OsO_4 should be mixed immediately before use.

Ruthenium Tetroxide

Tissue specimens are fixed with buffered 4% glutaraldehyde followed by postfixation with buffered 0.1–0.5% ruthenium tetroxide (pH 7.1) for 1 hr at 4°C. Penetration of ruthenium tetroxide into certain tissues is poor.

PREPARATION OF BUFFERS

2-Amino-2-methyl-1,3-propanediol (Ammediol) Buffer

Stock Solutions
   A: 0.2 M solution of 2-amino-2-methyl-1,3-propanediol (21.03 g in 1000 ml of distilled water)
   B: 0.2 M HCl
      50 ml of A plus x ml of B and diluted to a total of 200 ml with distilled water:

x	pH	x	pH
2.0	10.0	22.0	8.8
3.7	9.8	29.5	8.6
5.7	9.6	34.0	8.4
8.5	9.4	37.7	8.2
12.5	9.2	41.0	8.0
16.7	9.0	43.5	7.8

Acetate Buffer

Stock Solutions
   A: 0.2 M solution of acetic acid (11.55 ml in 100 ml of distilled water)
   B: 0.2 M solution of sodium acetate (16.4 g of $C_2H_3O_2Na$ or 27.2 g of $C_2H_3O_2Na.3H_2O$ in 1000 ml of distilled water)
      x ml of A plus y ml of B and diluted to a total of 100 ml with distilled water:

x	y	pH
46.3	3.7	3.6
41.0	9.0	4.0

x	y	pH
30.5	19.5	4.4
20.0	30.0	4.8
14.8	35.2	5.0
10.5	39.5	5.2
4.8	45.2	5.6

Barbital Buffer

Stock Solutions
   A: 0.2 M solution of sodium barbital (Veronal) (41.2 g in 100 ml of distilled water)
   B: 0.2 M HCl
      50 ml of A plus x ml of B and diluted to a total of 200 ml with distilled water:

x	pH	x	pH
1.5	9.2	22.5	7.8
2.5	9.0	27.5	7.6
4.0	8.8	32.5	7.4
6.0	8.6	39.0	7.2
9.0	8.4	43.0	7.0
12.7	8.2	45.0	6.8
17.5	8.0		

Cacodylate Buffer

Stock Solutions
   A: 0.2 M solution of sodium cacodylate [42.8 g of $Na(CH_3)_2AsO_2 \cdot 3H_2O$ in 1000 ml of distilled water].
   B: 0.2 M HCl
      50 ml of A plus x ml of B and diluted to a total of 200 ml with distilled water:

x	pH	x	pH
2.7	7.4	29.6	6.0
4.2	7.2	34.8	5.8
6.3	7.0	39.2	5.6
9.3	6.8	43.0	5.4
13.3	6.6	45.0	5.2
18.3	6.4	47.0	5.0
23.8	6.2		

Carbonate–Bicarbonate Buffer

Stock Solutions

    A: 0.2 M solution of anhydrous sodium carbonate (21.2 g in 1000 ml of distilled water)

    B: 0.2 M solution of sodium bicarbonate (16.8 g in 1000 ml of distilled water)

        x ml of A plus y ml of B and diluted to a total of 200 ml with distilled water:

x	y	pH	x	y	pH
4.0	46.0	9.2	27.5	22.5	10.0
7.5	42.5	9.3	30.0	20.0	10.1
9.5	40.5	9.4	33.0	17.0	10.2
13.0	37.0	9.5	35.5	14.5	10.3
16.0	34.0	9.6	38.5	11.5	10.4
19.5	30.5	9.7	40.5	9.5	10.5
22.0	28.0	9.8	42.5	7.5	10.6
25.0	25.0	9.9	45.0	5.0	10.7

Citrate Buffer

Stock Solutions

    A: 0.1 M solution of citric acid (21.01 g in 1000 ml of distilled water)

    B: 0.1 M solution of sodium citrate (29.41 g of $C_6H_5O_7Na_3 \cdot 2H_2O$ in 1000 ml of distilled water)

        x ml of A plus y ml of B and diluted to a total of 100 ml with distilled water:

x	y	pH	x	y	pH
46.5	3.5	3.0	23.0	27.0	4.8
43.7	6.3	3.2	20.5	29.5	5.0
40.0	10.0	3.4	18.0	32.0	5.2
37.0	13.0	3.6	16.0	34.0	5.4
35.0	15.0	3.8	13.7	36.3	5.6
33.0	17.0	4.0	11.8	38.2	5.8
31.5	18.5	4.2	9.5	41.5	6.0
28.0	22.0	4.4	7.2	42.8	6.2
25.5	24.5	4.6			

Collidine Buffer (0.2 M)

Stock Solution

    s-Collidine (pure)   2.67 ml

    Distilled water to make   50.0 ml

Buffer
    Stock solution   50.0 ml
    1 N HCl   9.0 ml (approx. for pH 7.4)
    Distilled water to make   100 ml

Glycine–HCl Buffer

Stock Solutions
    A: 0.2 M solution of glycine (15.01 g in 1000 ml of distilled water)
    B: 0.2 M HCl
        50 ml of A plus x ml of B and diluted to a total of 200 ml with distilled water:

x	pH	x	pH
5.0	3.6	16.8	2.8
6.4	3.4	24.2	2.6
8.2	3.2	32.4	2.4
11.4	3.0	44.0	2.2

Maleate Buffer

Stock Solutions
    A: 0.2 M solution of acid sodium maleate (8 g of NaOH plus 23.2 g of maleic acid or 19.6 g of maleic anhydride in 1000 ml of distilled water)
    B: 0.2 M NaOH
        50 ml of A plus x ml of B and diluted to a total of 200 ml with distilled water:

x	pH	x	pH
7.2	5.2	33.0	6.2
10.5	5.4	38.0	6.4
15.3	5.6	41.6	6.6
20.3	5.8	44.4	6.8
26.9	6.0		

Phosphate Buffer (Sörensen)

Stock Solutions
    A: 0.2 M solution of monobasic sodium phosphate (27.8 g in 1000 ml of distilled water)
    B: 0.2 M solution of dibasic sodium phosphate (53.65 g of $Na_2HPO_4 \cdot 7H_2O$ or 71.7 g of $Na_2HPO_4 \cdot 12H_2O$ in 1000 ml of distilled water)
        x ml of A plus y ml of B and diluted to a total of 200 ml with distilled water:

x	y	pH	x	y	pH
93.5	6.5	5.7	45.0	55.0	6.9
92.0	8.0	5.8	39.0	61.0	7.0
90.0	10.0	5.9	33.0	67.0	7.1
87.7	12.3	6.0	28.0	72.0	7.2
85.0	15.0	6.1	23.0	77.0	7.3
81.5	18.5	6.2	19.0	81.0	7.4
77.5	22.5	6.3	16.0	84.0	7.5
73.5	26.5	6.4	13.0	87.0	7.6
68.5	31.5	6.5	10.5	90.5	7.7
62.5	37.5	6.6	8.5	91.5	7.8
56.5	43.5	6.7	7.0	93.0	7.9
51.0	49.0	6.8	5.3	94.7	8.0

Phosphate Buffer (Karlsson and Schultz, 1965)

$NaH_2PO_4 \cdot H_2O$	3.31 g
$Na_2HPO_4 \cdot 7H_2O$	33.77 g
Distilled water to make	1000 ml

The pH is 7.4 and the osmolality is 320 mosmol, which is equal to that of the cerebrospinal fluid of rats.

Phosphate Buffer (Maunsbach, 1966), 0.135 M

$NaH_2PO_4 \cdot H_2O$	2.98 g
$Na_2HPO_4 \cdot 7H_2O$	30.40 g
Distilled water to make	1000 ml

Phosphate Buffer (Millonig, 1961)

Stock Solutions
A: 2.26% $NaH_2PO_4 \cdot H_2O$ in water
B: 2.52% NaOH in water
Buffer (0.13 M)
   Solution A   41.5 ml
   Solution B   8.5 ml

The pH is 7.3. the desired pH can be obtained with solution B without changing the molarity. The buffer is stable for several weeks at 4°C.

Phosphate Buffer (Millonig, 1964)

$NaH_2PO_4 \cdot H_2O$	1.8 g
$NaHPO_4 \cdot 7H_2O$	23.25 g

NaCl   5.0 g
Distilled water to make   1000 ml

Piperazine (PIPES) Buffer, 0.3 *M*

Distilled water   50 ml
Piperazine-*N,N*-bis-2-ethanol-sulfonic acid   9 g

Enough 0.1 *M* NaOH (0.4%) is added to adjust the pH; at pH 5.5–6.0 the powder is completely dissolved. After the required pH is reached, more distilled water is added to make up 100 ml. The stock solution is stable for several weeks at 4°C.

PM Buffer

Guanosine triphosphate   1 m*M*
MgSO$_4$ 1 m*M*
Ethylene glycol-bis (β-aminoethyl ether)-   2 m*M*
  N,N-tetraacetic acid
Piperazine-*N,N*-bis(2-ethanesulfonic acid)   100 m*M*

Ryter–Kellenberger Buffer

Stock Solutions
A: Sodium acetate (CH$_3$COONa·3H$_2$O)   9.714 g
    Sodium barbiturate (C$_8$H$_{11}$N$_2$NaO$_3$)   14.714 g
    Distilled water   500 ml
B: 0.1 *N* HCl
C: 8.5% NaCl in boiled distilled water
D: 1 *M* CaCl$_2$ in boiled distilled water:
                CaCl$_2$ 11 g/100 ml water
                        or
            CaCl$_2$·2H$_2$O   14.7 g/100 ml water
                        or
            CaCl$_2$·6H$_2$O   22.8 g/100 ml water

All solutions should be stored in bottles having ground-glass stoppers.

pH	Solution B (ml)	Boiled distilled water (ml)
3.9	13	5
4.1	12	6
4.3	11	7
4.7	10	8
4.9	9	9
5.3	8	10
6.1	7	11

pH	Solution B (ml)	Boiled distilled water (ml)
6.8	6.5	11.5
7.0	6.0	12.0
7.3	5.5	12.5
7.4	5.0	13.0
7.7	4.0	14.0
7.9	3.0	15.0
8.2	2.0	16.0
8.6	1.0	17.0
8.7	0.75	17.25
8.9	0.5	17.5
9.2	0.25	17.75
9.6	—	18.0

To each of the preceding mixtures, the following are added:
   5 ml of solution A
   2 ml of solution C
   0.25 ml of solution D

Succinate Buffer

Stock Solutions
   A: 0.2 M solution of succinic acid (23.6 g in 1000 ml of distilled water)
   B: 0.2 M NaOH
      25 ml of A plus x ml of B and diluted to a total of 100 ml with distilled
      water:

x	pH	x	pH
7.5	3.8	26.7	5.0
10.0	4.0	30.3	5.2
13.3	4.2	34.2	5.4
16.7	4.4	37.5	5.6
20.0	4.6	40.7	5.8
23.5	4.8	43.5	6.0

Tris(hydroxymethyl)aminomethane Buffer
Stock Solutions
   A: 0.2 M solution of tris(hydroxymethyl)aminomethane (24.2 g in 1000 ml
      of distilled water)
   B: 0.2 M HCl
      50 ml of A plus x ml of B and diluted to a total of 200 ml with distilled
      water:

x	pH	x	pH
5.0	9.0	26.8	8.0
8.1	8.8	32.5	7.8
12.2	8.6	38.4	7.6
16.5	8.4	41.4	7.4
21.9	8.2	44.2	7.2

Tris(hydroxymethyl)aminomethane Maleate Buffer

Stock Solutions
A: 0.2 M solution of Tris acid maleate [24.2 g of tris(hydroxymethyl) aminomethane plus 23.2 g of maleic acid or 19.6 g of maleic anhydride in 1000 ml of distilled water]
B: 0.2 M NaOH
   50 ml of A plus x ml of B and diluted to a total of 200 ml with distilled water

x	pH	x	pH
7.0	5.2	48.0	7.0
10.8	5.4	51.0	7.2
15.5	5.6	54.0	7.4
20.5	5.8	58.0	7.6
26.0	6.0	63.5	7.8
31.5	6.2	69.0	8.0
37.0	6.4	75.0	8.2
42.5	6.6	81.0	8.4
45.0	6.8	86.5	8.6

Veronal Acetate Buffer (Zetterqvist, 1956)

Stock Solution
Sodium Veronal (barbitone sodium)	2.94 g
Sodium acetate (hydrated)	1.94 g
Distilled water to make	100 ml

Ringer's Solution
Sodium chloride	8.05 g
Potassium chloride	0.42 g
Calcium chloride	0.18 g
Distilled water to make	100 ml

Buffer
Stock solution	10.0 ml
Ringer's solution	3.4 ml

Distilled water   25.0 ml
0.1 M HCl   11.0 ml (approx.)

PREPARATION OF TISSUE BLOCKS

Generally, the tissue blocks used for fixation should be as small as possible. If the animal needs to be anesthetized, ether or sodium pentobarbital is recommended. Alternatively, the animal can be sacrificed by spinal dislocation or, if appropriate, beheading. The method of killing may alter the ultrastructure of certain tissues. Tissue specimens are collected immediately after the animal has been sacrificed. The reason for this urgency is that postmortem changes follow the cessation of blood circulation. Tissue pieces are dissected out of the animal's body by using a pair of fine scissors or a new razor blade and are immediately placed in an appropriate fixative solution. After about 10–20 min, the tissue piece is minced into 0.5- to 1.0-mm cubes, and fixation is continued for an additional 2–3 hr at either 4°C or room temperature. If for some reason the tissue cannot be cut into small cubes, it can be cut into thin strips having a thickness of 0.5 mm or less. When a hand-held razor blade is used, the cutting should be accomplished by one quick slashing motion; several back-and-forth movements of the razor blade will result in extensive physical damage, especially to soft tissues.

During dissection and mincing, the tissue is prone to physical damage. This damage can be minimized by bathing the organ of interest with the fixative solution for 5–10 min prior to dissecting. This treatment stabilizes (hardens) the area close to the surface of the organ. Several types of tissue sectioners are available that provide extremely small tissue blocks (5–250 μm) with minimum physical damage. Various instruments and approaches are used to cut tissue slices of small size (Hayat, 1981a,b). Vascular perfusion is the ideal approach for minimizing physical damage and obtaining uniform fixation even in the deeper regions of the organ. However, this approach is not always practical or necessary.

Plant tissues are relatively easy to collect for fixation. The presence of a rigid cell wall minimizes physical damage to the cell. Usually a new one-edge razor blade (after being rinsed with acetone) is used to obtain small slices, which are fixed by immersion. Certain plant tissues tend to float in the fixative vial because of the presence of air in the tissue, thus hampering a rapid and uniform fixation. This problem can be solved by subjecting the tissue (while it is floating in the vial containing the fixative) to negative pressure, using an aspirator. The pressure should be relatively low (1 atm for roots) so that the air is released in the form of a continuous stream of very small air bubbles. Higher pressures may disrupt the cytoplasmic components including separation of the plasma membrane from the cell wall. The aspiration is stopped as soon as the release of bubbles ceases. The success of this treatment is indicated by the sinking of tissue slices.

Some studies may necessitate the perfusion of certain plant tissues such as leaves and stems. Perfusion can be accomplished by injecting the fixative solution into a major vascular channel (vein) and then excising the specimen and further fixing by immersion. Leaves and flowers can be fixed *in situ* by sealing a metal ring (3 mm high and 12 mm in diameter) with lanolin (or with petroleum jelly) onto the surface of the organ; no lanolin should be present within the ring. The ring is filled with the fixative solution by using a syringe and then covered with a glass coverslip. After about 1 hr, the tissue slices are cut out and are further fixed for 1–2 hr.

Surfaces of most plant organs are usually covered by waxes and/or hair. These structures impede the penetration of fixative solutions. This problem can be alleviated somewhat by briefly dissolving the impervious structures in dilute household detergent. Sometimes gently rubbing the surface with a brush while the organ is in the fixative solution helps dislodge the air bubbles and some surface substances. Even vigorous shaking of the specimens in the fixative solution in a vial may help. Total immersion of the tissue slices in the fixative within a short time is a prerequisite for satisfactory fixation. The collection and preparation of unicellular organisms, cells in culture or suspension, and other types of specimens are explained in Chapter 7.

FIXATION

Vascular Perfusion

Recommended Osmolality of Perfusate for Selected Tissues

Tissue type	Osmolality (mosmol)
Rat brain	330
Rat heart	300
Rat kidney	420
Rat (or cat and fowl) liver	450
Rat (or chicken) embryo liver	420
Rat lung	330
Rat skeletal muscle	475
Rat developing renal cortex	828
Rat renal medulla	
Inner stripe of outer zone	700
Outer level of inner zone	1000
Middle level of inner zone	1300
Papillary tip	1800
Rabbit brain	820
Frog liver	300

Recommended Perfusion Pressure for Selected Tissues

Tissue type	Perfusion pressure (mmHg)
Rat (or mouse) heart	103–120
Rat kidney	100–200
Rat (or cat and fowl) liver	100
Rat (or chicken) embryo liver	50
Rat skeletal muscle	100
Rabbit arteries	100–200
Rabbit (or rat) brain	150
Rabbit spleen	110–120
Monkey brain	110
Kitten liver	160–180
Baboon lung	40

Anesthesia

To avoid initiating cell death prior to arrival of the fixative, sublethal anesthetization is preferable to decapitation or cervical dislocation. Several equally effective anesthetics are available for animals of different body weights. Small animals (e.g., rat and guinea pig) can be anesthetized with a volatile anesthetic such as ether or halothane. However, barbital anesthesia is preferred over ether, because incomplete perfusion of the medulla has been shown to occur in animals anesthetized with ether (Bohman, 1974).

For large animals an intravenous or intraperitoneal injection with pentobarbital (Nembutal), Inactin, urethane, or chloral hydrate is used. Pentobarbital anesthesia usually results in better fixation than that obtained after ether anesthesia. Anesthesia and analgesia can also be accomplished by first injecting 30 mg of Pentazocine and then 30 mg of pentobarbital per kilogram of body weight. The type and dose of anesthetic required are species-dependent. The correct dose of pentobarbital for commonly used laboratory animals is given in Table 1.1. Further information on various anesthetics and recommended doses for different animals can be found in books by Lumb (1963), Soma (1971), Barnes and Etherington (1973), and Altmann and Dittmer (1973).

Intravenous injection is preferred for administering drugs in solution. Drugs that are tissue irritants can be safely administered intravenously. Before administering the drug the skin should be shaved. To avoid running the needle through the vein, the skin and vein should be pierced in separate thrusts. Usually two persons are needed to administer the drug intravenously in cats, dogs, and monkeys. On the other hand, intraperitoneal injection is relatively easy and is recommended for administering nonirritating drugs. This technique can be performed by one person even with cats and dogs. The peritoneum of the abdominal

TABLE 1.1.

**Recommended Dose of Pentobarbital
Administered Intravenously for Selected
Animals**[a]

Species	Dose (mg/kg of body weight)
Cat	25
Dog	15
Guinea pig	30
Monkey	25
Mouse	35
Rabbit	30
Rat	25

[a]Pentobarbital has a high water solubility and induces rapid anesthesia. Optimal dose depends on the animal's weight, age, and sex.

cavity presents a large absorptive area. The onset of anesthesia is slower with intraperitoneal than with intravenous injection.

If detrimental effects of chemical anesthetics are suspected, low temperatures as anesthetic may be used. Above-freezing temperatures may also be desirable for very small or newborn animals. Advantages of such temperatures include the production of analgesia, reduction of metabolic rate and motor activity, and elimination of potential harmful effects of chemical agents. This approach may also be useful for puncture perfusion.

Aorta

The animal (e.g., rabbit) is killed by intravenous injection of 10% urethane anesthesia. The dorsal aorta from the arch to about the first intercostal artery is removed and placed in 0.9% saline. Each arterial section is tied to a hydrostatic column preset to a height calculated to produce a pressure head equivalent to either 80 or 125 mmHg. Containers for heparinized saline and for fixative are connected (Fig. 1.1). The tied arterial section is immersed in the saline bath and the T-connection opened to allow heparinized 0.9% saline to flow through. Any leaks or opened side vessels are clamped.

After the leaks have been closed, the perfusion with the saline is terminated, and 3% glutaraldehyde in 0.1 M phosphate buffer (pH 7.4) is allowed to flow through the vessel. The fixative can be dyed with methylene blue to monitor the proper flow of the fixative. The saline bath in which the section of the aorta has been immersed is replaced by 3% glutaraldehyde. The fixative can be recovered as it flows out of the system and poured back into the container for recirculation so as to maintain the desired height of the column (h in Fig. 1.1) of the fixative.

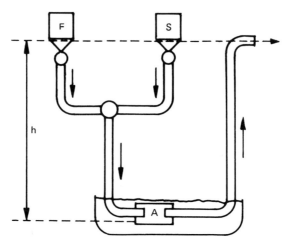

Fig. 1.1.   Diagram of the apparatus used for perfusion of the aorta. Arrows indicate the direction of perfusate flow. F, the fixative; S, saline; A, aorta; h, height of the solution containers and the drainage tube. (From Swinehart *et al.*, 1976.)

The fixation by perfusion is maintained for 45 min. The artery, still maintained under pressure, is immersed in fresh glutaraldehyde bath and left overnight at room temperature. Small pieces of the tissue are postfixed with OsO_4. For further details, see Swinehart *et al.* (1976).

Arteries

The animal is anesthetized and 1 cm of the abdominal aorta near the renal arteries and 1 cm at the bifurcation is dissected, leaving 7 cm of the vessel untouched. The vessel is perfused with 10 ml of Krebs–Ringer's solution through a silastic catheter that has been proximally introduced. During this washing, a second catheter is distally introduced. The outflow of the second catheter is placed 60 cm above the aorta (82 mmHg pressure) (Haudenschild *et al.*, 1972). The washing is followed by perfusion with 1.5% glutaraldehyde in 0.1 *M* phosphate buffer (pH 7.4) having an osmolality of 500 mosmols for 20 min at room temperature. The containers with washing and fixation solutions are placed 100 cm above the aorta (136 mmHg pressure). Small blocks from the perfised vessel are refixed by immersion in the same fixative for 1 hr and then postfixed with 2% OsO_4.

Central Nervous System

The following method is recommended for simultaneously localizing the fluorescence with the light microscope and fixing the tissue for electron microscopy (Furness *et al.*, 1977, 1978). A mixture of glutaraldehyde and formaldehyde produces a fluorophore with catecholamines in the peripheral and central nervous systems. The fluorescence reaction is produced at room temperature and is stable in aqueous solutions.

Perfusate

Formaldehyde   4%
Glutaraldehyde   1%
Phosphate or cacodylate
    buffer (0.1 *M*, pH 7.0)

The animal (e.g., rat) is anesthetized with pentobarbital (40 mg/kg body weight) and injected with heparin (4000 U/kg body weight). The chest is opened and the heart is exposed. The tip of an 18-gauge cannula connected to a perfusion apparatus (Fig. 1.2) is introduced into the aorta via the left ventricle and held in position with a clamp across the ventricles. The blood is flushed out with a 1% solution of sodium nitrite in 0.01 *M* phosphate buffer (pH 7.0). This step is accomplished in 10–30 sec of perfusion at 120 mmHg and is followed immediately by perfusion with the fixative solution for 10 min at the same pressure.

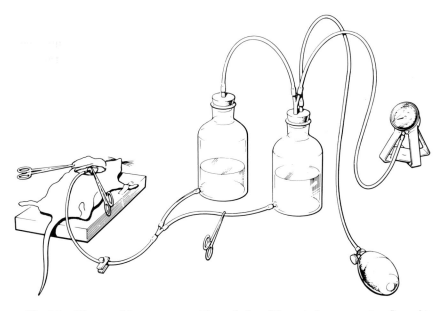

Fig. 1.2. Diagram of the apparatus used for perfusion of the central nervous system. It consists of two aspirator bottles, each holding 1 l, which are connected to a sphygomanometer bulb and gauge to maintain and monitor the perfusion pressure. One bottle contains buffered sodium nitrite solution, and the other carries the fixative solution. Clamps are used to direct the perfusate selectively from one or the other of the bottles. All tubing used to connect the apparatus is kept as short as practicable and is of wide bore (10 mm inside diameter) so that any pressure difference between the gauge and the cannula is kept to a minimum. (From Furness *et al.*, 1978.)

About 200 ml of the fixative solution is perfused. The fixed brain is placed in the fresh fixative at room temperature. Small pieces of the fixed brain are further fixed for 30 min and then postfixed with 1% OsO_4.

Embryo

The pregnant animal (e.g., mouse) is anesthetized by intraperitoneal injection of sodium pentobarbital (30 mg/kg body weight). The animal is laparotomized and the uterus is exposed. An opening showing the yolk sac is dissected in the uterine wall under a stereomicroscope. Care should be taken not to disturb the circulation of the conceptus. The beating heart of the embryo is located, and a micropipette with a tip diameter of 25–50 μm is inserted through enveloping membranes and the precardiac wall into the atrium (Fig. 1.3). The micropipette is immediately retracted, thus leaving an opening in the wall of the atrium through which the blood can flow out. The tip of the pipette is placed in the

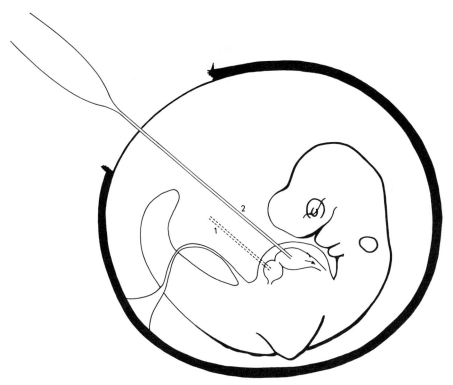

Fig. 1.3. Diagram of microperfusion fixation through the embryonic heart. The pipette is first introduced through enveloping membranes and the precardiac wall into the atrium (1). It is then immediately retracted, thus leaving an outflow opening. In the second step, the tip of the pipette is forwarded into the lumen of the ventricle (2), and the perfusion is started. (From Abrunhosa, 1972.)

lumen of the ventricle, and the microperfusion by the fixative is started. The start of the perfusion is accompanied by the escape of blood through the opening in the atrium. The perfusate consists of 2% glutaraldehyde in 0.1 M cacodylate buffer (pH 7.4) containing 2% polyvinylpyrrolidone (PVP).

The embryo is bleached and becomes yellowish immediately. The speed of fixation can be observed by adding alcian blue or astra blue 6GLL to the fixative; all the capillaries show fixative penetration within the first few seconds. The perfusion is continued for $\frac{1}{2}$–2 min. Small pieces of the embryo are fixed further for 2 hr in the same fixative. For further details, see Abrunhosa (1972).

Fish

The fish (e.g., catfish) is anesthetized by exposing it to a freshly prepared solution of tricaine methane sulfonate (M.S. 222) in water to a final concentration of 1:4000 (McFarland and Klontz, 1969). In 5–10 min the fish ceases all but opercular movement, loses equilibrium, and is insensitive to touch. Gauze sponges (5 × 5 cm) are soaked in the anesthetic solution and are positioned under the operculum and over the gill arches. An applicator bottle of 500-ml capacity is filled with the anesthetic solution.

The fish is removed from the solution and placed in the supine position if it has a depressed body form. A triangle is inscribed: the sides are made by percutaneous incisions along the inferomedial body wall near both opercula, and the transverse cut is made by a similar incision anterior to the bases of the pectoral fins. A scalpel and blunt forceps can be used to reflect the skin from the surgical area and to reflect the skeletal muscle to expose the parietal pericardium. This layer is removed to uncover the bulbus arteriosus of the aorta. The triangle is enlarged, and the heart and aorta are exposed by placing heavy surgical scissors transverse and superficial to the aorta and cutting away the cranial apex of the pectoral girdle. The aorta is exposed adequately without causing extensive trauma to gills and vessels supplying the body wall and liver. The applicator bottle is positioned and squeezed so that a continuous flow is introduced over the gills.

To start the perfusion, a moist cotton thread is passed dorsally to the aorta and is tied in a loose knot between the bulbus arteriosus and the first branchial arch artery. The lateral margin of the bulbus arteriosus is grasped with blunt forceps, and a small opening is made with scissors in the wall of the vessel. Immediately, a cannula of polyethylene is inserted into the aorta to a point distal to the thread. Perfusion is begun at once with a mixture of glutaraldehyde (2%) and formaldehyde (2%), and the thread is tied securely around the vessel and cannula to prevent loss of perfusate from the vessel aperture. A transverse incision is made through the ventricular wall to permit escape of blood from the vascular system upon displacement of the fixative. Sufficient pressure (60–100 mmHg) is maintained to give a flow rate of 10 ml/min. A total of 20–100 ml of the fixative is perfused per fish, depending on its size.

The gravity feed device comprises a ring stand, a 50-ml syringe (perfusate reservoir), and an 18-gauge needle over which polyethylene tubing (internal diameter 0.1 cm; external diameter 0.2 cm) is fitted. A height of 1–2 cm allows sufficient pressure.

After the perfusion, the fish is covered with moist towels, and 30–60 min is allowed for *in situ* fixation. Yellowish coloration in the tissue indicates the completion of fixation. For further details, see Hinton (1975).

Heart

Method 1 (Fetus)

The pregnant animal (e.g., ewe) is operated on under low spinal analgesia (Dae *et al.,* 1982). The fetus is delivered through a hysterotomy, and a saline-filled glove is placed over the head to avoid expansion of the lungs. After local anesthesia, a polyvinyl catheter (no. 8F feeding tube) is inserted into the fetal carotid artery through a neck incision, and while the arterial pressure is monitored, the tip is advanced to the aortic root. A left thoractomy is performed to expose the heart, and the position of the catheter tip is confirmed by palpation. The heart is arrested in diastole by perfusing a cardioplegic solution consisting of 2.5% potassium citrate in Hank's buffer (pH 7.4). Perfusion is maintained at the same level as the measured mean aortic pressure with the use of a simple apparatus consisting of two glass reservoirs connected with latex tubing to an air pressure gauge and a one-way valved rubber bulb. The right atrial appendage is cut to facilitate flow. After the capillaries have been cleared of blood, the perfusate is switched to a mixture of 2.5% glutaraldehyde, 1% formaldehyde, and 0.0013% picric acid in Hank's buffer (pH 7.4).

Perfusion is continued for 15 min, and then the heart is excised and tissue pieces of 1 mm^3 are immersed in a mixture of 5% glutaraldehyde, 4% formaldehyde, and 0.0023% picric acid in 0.2 M cacodylate buffer (pH 7.4). Specimens are stored in this mixture overnight at 4°C (277 K), rinsed in buffer, and postfixed with 2% OsO_4 in buffer for 2 hr.

Method 2 (Adult)

Perfusates

1. Ringer's solution containing 0.1% procaine. The pH is adjusted to 7.2–7.4 with 1 N HCl, and the osmolality to 300 mosmols with NaCl. Perfusion takes 3–5 min.

2. 2% Glutaraldehyde in 0.045 M cacodylate buffer. The pH is adjusted to 7.2–7.4 and osmolality to 300 mosmols. Perfusion takes 5 min.

Procedure. In Fig. 1.4, containers (B) filled with various perfusates are placed in a thermoregulated bath (A). The height of the containers from the worktable should be adjustable to regulate the rate of perfusion. The perfusates are main-

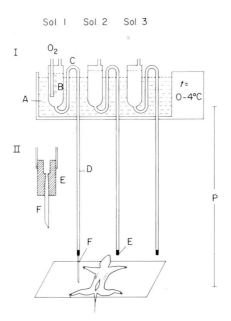

Fig. 1.4. Diagram of the apparatus used for perfusion of the heart. (I) A, thermoregulated bath; B, perfusate containers; C, siphon; D, plastic tubing; E, middle piece; F, cannula. (II) Details of the middle piece. (From Forssmann *et al.*, 1967.)

tained at 4°C through a siphon (C) made of soft plastic tubing. The cannula (F) made of Teflon is attached to the tubing (D) by a middle piece that has a conical-shaped distal opening in which the cannula slides. This system allows rapid change from one solution to another, and the same cannula can be used for all perfusates. The distal end of the cannula is obliquely pointed.

Since the tubing (D) is not heat insulated, the temperature of the perfusates will rise to 6°C. With adequate flow and pressure, the temperature of the perfusate coming out of the animal does not exceed 10°C. The containers are placed 140 cm above the worktable, taking into account the average arterial pressure of 103 mmHg in rats. During perfusion, a flow rate of 40 ml/min is necessary.

The animal is anesthetized by intraperitoneal injection of 125 mg urethane/100 g of body weight. The animal is tied on its back to the worktable, and the ventral abdominal wall is opened widely with an incision along the white line. The intestine is gently pushed aside and the abdominal aorta is exposed. Two silk threads are slipped under the aorta, the first under the renal arteries and the second above the bifurcation of the iliac arteries. The two threads separated by a distance of 5 mm delineate the segment of the aorta where a slit is made for the insertion of the cannula.

The cannula, filled with Ringer's solution containing procaine, is inserted into the lumen of the aorta. The cannula is then tied in place to the aortic wall with the first thread. The second thread is used to tie the aorta so as to prevent hemorrhage. After the cannula is in place, perfusion is started. The proximal end of the cannula is slipped into the conical opening of the middle piece (D in Fig. 1.4), and the solution flows into the circulatory system of the animal. Simultaneously, the inferior vena cava is cut to allow the evacuation of blood and perfusate. For further details, see Forssmann *et al.* (1967).

Kidney

Method 1 (Animal)

Perfusate
25% Glutaraldehyde   40 ml
0.135 *M* Phosphate buffer to make   1000 ml
The solution should have a pH of 7.2 and an osmolality of 429 mosmols.

Procedure. The animal is anesthetized intraperitoneally with 35 mg sodium pentobarbital/kg of body weight. The animal is tied on its back to the operating table. The abdominal cavity is opened wide by midline incision, and the intestine is moved to the left side. The aorta below the origin of the renal arteries is carefully exposed for retrograde perfusion. A simple intravenous infusion set (a bottle, a drip chamber, and intravenous tubing with a squeeze-type flow regulator) can be employed.

A hypodermic needle (gauge 10–16), bent at a right angle, is inserted into the abdominal aorta proximal to its distal bifurcation. The needle is connected to a tubing that, in turn, is connected to a bottle containing the fixative. A hydrostatic pressure equal to 150 cm water is adequate. Immediately after the perfusion has started, the aorta is clamped just below the diaphragm. After a 10-min perfusion, small pieces of the tissue are fixed by immersion in the same fixative for several hours. For further details, see Maunsbach (1966).

Method 2 (Human)

The following method is useful for vascular perfusion of surgically removed human kidneys (Møller *et al.*, 1982). Simultaneous clamping of the renal artery and vein is desirable because it prevents increased vascular resistance. The kidney is cleared of blood by perfusion with a Tyrode solution (see Appendix) of pH 7.3 and osmolality of 285 mosmols. This solution contains 0.006% papaverine, which reduces spasms of the renal arteries. The fixative consists of 2% glutaraldehyde in 0.1 *M* cacodylate buffer (pH 7.3) containing 2% dextran; total osmolality is 400 mosmols. Light green or Lissamine green is added to the fixative to indicate the perfused area of the kidney.

The kidney is perfused with the rinsing solution through the arterial branch with a blunt-ended needle (outer diameter is 3 mm) in order to avoid injuring the intima. Before the needle is inserted, all air bubbles in the tubes and the needle are removed. The needle is secured in the position by two artery forceps. This perfusion is continued for 3–5 min using 100–200 ml solution until the emerging solution is essentially free of blood. The switch from rinsing to fixative solution is done without interruption of flow (Fig. 1.5). About 500 ml of fixative is perfused over a period of 10 min at a continuous pressure of 90–150 mmHg (the pressure should be equal to diastolic blood pressure). The kidney is cut into 1-mm pieces (areas showing signs of satisfactory fixation) and fixed by immersion in the fixative without the dextran for an additional period of 24 hr. After a rinse, the specimens are postfixed with 1% OsO_4 in Veronal buffer for 1 hr and stained *en bloc* with 2% uranyl acetate in maleate buffer (pH 5.2).

Liver

Method 1 (Adult)

Perfusate

Glutaraldehyde (25%)	40 ml
0.15 *M* Phosphate buffer (pH 7.2) to make	500 ml

The solution should have an osmolality of 450 mosmols, which can be adjusted with sucrose.

Procedure. The anesthetized animal (e.g., rat) is tied to the operating board with its back down. The abdominal cavity is opened by a midline incision with lateral extensions, and the intestine is gently moved to the left side. The portal vein is exposed and two ligatures are passed behind it. The distal ligature is tied

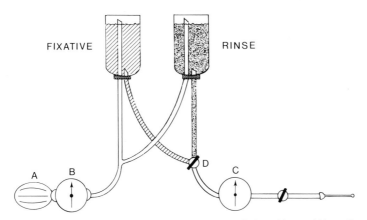

Fig. 1.5. Schematic drawing of the apparatus used for perfusion of human kidney. Pressure in perfusion flasks is created by a manual rubber pump (A) and recorded on a manometer (B). Outflow pressure is recorded on another manometer (C) close to perfusion needle. A three-way stopcock (D) facilitates the switch from rinse to fixative. From Møller *et al.*, 1982. Reprinted by permission of Hemisphere Publishing Corporation.)

to block the flow of venous blood from the portal vein and the flow of hepatic arterial blood to the liver. At the site where the portal vein is branched to different liver lobes, a 1-cm, 20-gauge syringe needle is inserted and secured by the second ligature. Before the needle is inserted the perfusion pressure is adjusted to 20 mmHg and subsequently falls to 10 mmHg during perfusion. The needle has been previously connected via an ordinary clinical intravenous infusion set to a flask containing the perfusate.

The inferior vena cava below the diaphragm is cut open to relieve the pressure in the right heart. The flask containing the fixative is hung 25–30 cm above the animal. The rate of flow should be 5–10 ml/min. Care should be taken to prevent air bubbles from entering the portal vein or the perfusion system.

Within 3 min after the start of the flow of fixative through the portal vein, the thoracic cavity of the animal is opened, and the right thoracic wall is removed. A syringe needle is inserted into the main stem of the hepatic vein and secured with a ligature. The needle has been connected previously to a second flask of fixative, which is hung 15 cm above the animal. The retrograde perfusion should begin 5 min after the start of flow of the fixative through the portal vein. The flow rate of the retrograde should be 5 ml/min.

The perfusion is usually completed in 10–15 min, at which time only clear fixative solution comes out of the portal vein. Within 30–40 sec after the start of perfusion, the color of the liver changes from dark reddish brown to light brown. The completion of fixation can be detected grossly on the surface of the liver. The consistency changes from soft to rather stiff, resembling that of a boiled egg. Uniform overall fixation can be checked by immersing the liver slices in distilled water; poorly fixed areas will show white discoloration. The uniformity of fixation can also be checked by examining 0.2- to 0.5-μm-thick sections stained with 1% toluidine blue in 1% borax with a light microscope. Immediately after the completion of perfusion, the liver is removed and cut into small segments which are immersed in glutaraldehyde for 2 hr and then postfixed with OsO_4.

Method 2 (Embryos and Very Small Animals)

Perfusate

Glutaraldehyde (2.5%) is buffered with calcium acetate (pH 7.2) having a final osmolality of 420–450 mosmols.

In very small animals the blood vessels are likewise very small, and thus consecutive perfusion through the portal vein and the hepatic vein is difficult to perform. In animals such as adult fowl, the liver and its blood vessels are hidden behind the thick muscles of the chest wall. Because of the lengthy time required for dissection, it is necessary to apply artificial respiration until perfusion can begin. In embryos having patent ductus venosus, perfusion through the umbilical, portal, or hepatic veins may not yield good results because of the long bypass through the ductus venosus. Vascular perfusion of the liver in the aforementioned specimens can be performed directly through the hepatic substance.

Procedure. The apparatus consists of two containers (carrying saline and fixative solution separately) coupled to a gas tank containing compressed oxygen. The perfusion pressure can be regulated by means of a valve between the gas tank and the fluid containers. The perfusate is led through a piece of plastic tubing to a hypodermic needle. Connection with the liver circulation is achieved by inserting the needle superficially into the hepatic parenchyma. A branch of the superior mesentric vein is cut open with scissors, and the perfusion is immediately started with saline. The perfusion pressure is raised from 0 to 50 mmHg. The rinse with saline should not last longer than 40 sec. It is followed by perfusion with glutaraldehyde, and the pressure is raised to 150–175 mmHg. The perfusion with glutaraldehyde is continued for 10–15 min. Small tissue blocks are cut from the liver and fixed further in glutaraldehyde for 1–2 hr. The specimens are postfixed with 1% OsO_4. The tissue parts adjacent to the perfusion needle are discarded. For further details, see Sandström (1970).

Method 3 (Biopsy)

Liver biopsy and surgical samples can be preserved by transparenchymal fixation. Vonnahme (1980) introduced this method for the fixation of human biopsies weighing 2–8 g. Immediately after the wedge biopsy is collected, it is immersed in Ringer's solution containing heparin so as to prevent the accumulation of neutrophils in the sinusoids. A needle (0.9 mm diameter) connected to a perfusion apparatus by means of a plastic tube is inserted from the capsule side into the sample. To avoid precipitation of serum proteins, the heparinized Ringer's solution is perfused for 4 min at a rate of 1.5–10 ml/min, depending on the sample size. The sample is then perfused with 2% glutaraldehyde in 0.1 *M* cacodylate buffer (pH 7.4, 340 mosmols). During perfusion, the cut surface of the sample should face upward. In this way, the perfusate (restricted on either side by the capsule) builds up and flows out over the cut surface. After removal of the needle, a thin slice containing the puncture part is discarded. The remaining tissue is further fixed for 1 hr.

Lung

The following method uses a simple perfusion apparatus for pulmonary perfusion of lungs (Coalson, 1983). Expensive equipment such as peristaltic pumps, pressure transducers, and recorders is not needed. The pressure is provided by a constant inflow of air from an air inflow valve present in any routine laboratory. Isosmotic–isoncotic electrolyte solutions and fixatives are employed to prevent edema in the lung. The experimental setup uses a Harvard respiratory pump, which should be set for a tidal volume delivery appropriate to the species under study.

The perfusates are transferred to quart-size, wide-mouthed canning jars placed in a water bath at 37°C. The jars are fitted with size 12 rubber stoppers. One of the stoppers is bored with two holes. One piece of the 5-mm glass tubing should extend about 3 cm below the stopper. The other piece of tubing should extend to within 2 cm of the bottom of the jar. Both pieces of glass tubing should extend 5–8 cm above the top of the rubber stopper. The pieces of glass are sealed with silicone grease to prevent air leaks. The jar that receives the air flow from an air inlet valve is constructed in a similar manner, except that there is a third bored hole to accommodate one additional piece of glass tubing. The short glass tubes are designed to deliver the air and build enough pressure to drive the perfusates into the longer pieces of glass tubing.

Tygon tubing is then used to interconnect the system. One piece of Tygon tubing should be run from a compressed air source through the air inlet valve to one of the short pieces of glass tubing in the bottle with the three bored holes. Two additional pieces of Tygon tubing are connected to the two short pieces of glass tubing within the two canning jars. These are joined by a Y connector to a gas control valve with a needle point stem (0.25 in.) and to a mercury sphyg-momanometer gauge (Tycos, Taylor Instrum. Arden, South Carolina). This gauge allows for direct monitoring of the pressure which is controlled by regulating the air egress from the system with the gas control valve.

The two pieces of Tygon tubing are attached to the longer pieces of glass tubing that extend below the level of the perfusates. These two pieces of Tygon tubing are connected to a three-way stopcock that allows for an immediate sequential switch from one perfusate to another during the perfusion. A single piece of tubing is then constructed that is adapted with appropriate intramedic polyethylene tubing for a suitable cannula to fit the main pulmonary artery. Owing to the diameter changes in the tubing system as it enters the heart and its possible change in resistance, a needle should be inserted into the line at the level of the inflow into the main pulmonary tract. This allows monitoring of infusion pressure with a linear water manometer. Infusion pressure can be regulated by altering the perfusate flow by adjusting the gas control valve. The lines should be cleared of air bubbles and filled totally with the perfusate prior to perfusion.

The animal (cat) is anesthetized and restrained in a supine position. The trachea is cannulated and lungs are ventilated on air with a Harvard respiratory pump (Fig. 1.6) with a tidal volume of 30 ml and 2 cmH_2O PEEP. A midline incision extending from the upper neck to the lower abdomen is made. The thorax is opened and the sternum is reflected back. The pericardium is opened and the apex of the heart is secured by suture to the diaphragm. A ligature is loosely placed around the main pulmonary artery. After a small incision is made in the right ventricle, the perfusate cannula is quickly inserted into the main pulmonary artery and secured by the ligature.

An isoncotic–isosmotic Krebs–Henseleit solution (see Appendix) containing 1% sucrose, 3.5% dextran (mol. wt. 70,000), and heparin (5000 U-USP/l) is perfused under 20 mmHg pressure for 1 min to flush the blood vessels. Krebs–Ringer solution can be used with equal success. The pressure is created by

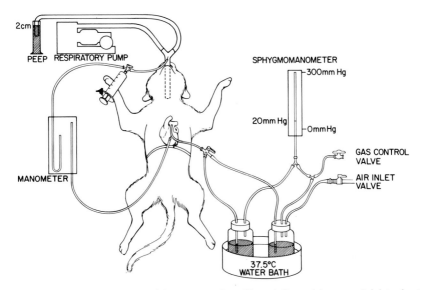

Fig. 1.6. Schematic diagram of the apparatus that utilizes air from a laboratory air inlet valve to provide the pressure for the system. (From Coalson, 1983.)

forcing air from an air inlet valve into a closed reservoir system. The pressure is monitored by a sphygmomanometer gauge and adjusted with an air control valve. The left auricle is cut to allow the perfusate to drain from the lung. The abdominal aorta and inferior vena cava are transected to decrease the blood flow into the right heart. This transection is followed by perfusion with 1% glutaraldehyde in 0.1 M cacodylate buffer (pH 7.4) containing 2% sucrose and 6% dextran (mol. wt. 70,000), with total osmolality of 348–363 mosmols for 15 min at 37°C. Simultaneously, the ventilator is stopped and the airway pressure is adjusted to 10 cmH$_2$O (or any desired pressure). The perfusates are aspirated from the posterior thoracic cavity. Both the airway and perfusion lines are clamped, and the lungs are removed intact. Care should be taken that the lungs are not compressed or deformed. The lung is floated in the fixative for 24 hr, cut into specimens for light and/or electron microscopy and further processed accordingly.

Muscle (Skeletal Muscle of Rat Hind Limb)

Perfusates
A. Locke's solution containing 0.1% NaNO$_2$

NaCl	9.2 g
KCl	0.42 g
CaCl · 2H$_2$O	0.24

NaHCO$_3$	0.15 g
Sucrose	1.0 g
NaNO$_2$	10.0 g
Distilled water to make	1000 ml

The final solution contains 1 mM CaCl$_2$.

B. 12% Glutaraldehyde

| 25% Glutaraldehyde | 60 ml |
| 0.1 M Cacodylate buffer (pH 7.2) to make | 500 ml |

The final solution contains 2 mM CaCl$_2$ and has an osmolality of 474 mosmols.

Procedure. The animal is anesthetized intraperitoneally with 40 mg of Nembutal/kg of body weight. The abdominal aorta is cannulated just above the bifurcation to the hind limbs, and the vena cava is cut to allow free flow of solutions. The temperature of the perfusate is maintained at 4°C in a water bath. The perfusate is pumped with the aid of a peristalic pump (Hughes Hi-Lo) at a constant pressure of 100 mmHg into the hind limb. Perfusion with solution A is carried out for 2–3 min to dilate the vascular bed and remove most of the blood. This is followed by perfusion with solution B for 10 min. The tissue is cut into small pieces and fixed in solution B for an additional 1 hr, rinsed in the buffer, and postfixed in 1% OsO$_4$ for 1–2 hr. For further details, see Bowes *et al.* (1970).

Ovary

Perfusate

Glutaraldehyde	1%
Formaldehyde	1%
2,4,6-Trinitrocresol	0.01%
Cacodylate buffer (pH 7.2)	0.1 M

Procedure. The animal (e.g., guinea pig) is anesthetized by an intraperitoneal injection of sodium pentobarbital (30–35 mg/kg of body weight). Each of the two ovaries is perfused independently of the other. A Holter peristalic pump (Extracorporeal Medical Specialties Inc., King of Prussia, Pennsylvania; model 911, fitted with size D pump chamber) can be used to deliver the perfusate. The perfusion can be started at a pressure of 35–50 mmHg. The abdomen is opened with a U-shaped incision extending from the pubis to the ribs, a ligature (4–0 surgical silk) is placed around both the ovarian artery and vein and tied loosely in an overhand knot. A length of the uterine artery is carefully separated from its companion vein, and a second ligature is tied loosely around the artery. Ligatures are placed in similar locations around the blood vessels of the other ovary.

The uterine artery is grasped somewhat caudal to the ligature with fine forceps, lifted, and pulled slightly taut. The cannula (26-gauge needle), with Krebs–Ringer–bicarbonate buffer (containing 2 mg glucose/ml, pH 7.4, 300 mosmols) flowing from it at a rate of 2 ml/min, is inserted into the uterine artery, guided past the ligature and secured into place by tightening the ligature. A nick is immediately made in the uterine vein to provide an outflow for the perfusate, and the ligature around the ovarian artery and vein is tightened, isolating the ovary from the circulatory system.

After a 2–3 min wash with the buffer, fixation is begun and continued for 10–30 min. Shortly after the fixative has reached it, the ovary begins to harden, indicating a favorable perfusion. After the perfusion is underway on one side, perfusion of the other ovary is started by using the same sequence. The ovary is cut into small pieces that are placed directly in 0.1 M cacodylate buffer containing 5% sucrose for a wash and are postfixed with 1% OsO_4 in the same buffer for 1 hr in the cold. For further details, see Paavola (1977).

Region at Transition between Peripheral and Central Nervous Systems

The animal (e.g., cat) is perfused with 5% glutaraldehyde in Millonig's phosphate buffer (pH 7.2, 300 mosmols) containing 2.7% low molecular weight dextran and sucrose (200 mosmols) (Berthold, 1968; Carlstedt, 1977). The total and effective osmolalities of this perfusate are 1000 and 500–700 mosmols, respectively. The latter is about twice that of serum. Small pieces of the perfused tissue are refixed with the same fixing solution for 4 hr. After being washed overnight in the same buffer containing sucrose (200 mosmols), specimens are postfixed with 2% OsO_4 in phosphate buffer (pH 7.2) for 4 hr. The specimens are dehydrated in acetone (starting with 10% solution) via a prolonged and extensive stepwise procedure and then embedded in Vestopal W, which is a reliable embedding medium for nerve fibers.

Spinal Cord

Method 1 (Goldfish)

Perfusate: 4% Glutaraldehyde

25% Glutaraldehyde	32 ml
Distilled water	168 ml
$NaH_2PO_4 \cdot 2H_2O$	3.75 g

The pH is adjusted to 7.4 by adding 0.9 g of NaOH.

Procedure. A 6–9-cm-long goldfish is anesthetized by placing it in a 1:1000 solution of MS-222 (Sandoz, Hanover, New Jersey) in tap water. Anesthesia is maintained by perfusing a 1:2000 solution of MS-222 through the mouth and out of the gills. The fish is then placed ventral side up in a special stand, and an incision is made in the midline. A syringe is used to perfuse 5 ml of the perfusate through the heart, and the tail is cut off to allow its escape. The perfusion may last up to 4 min, depending on the resistance offered by the vascular system of the fish. The spinal cord is removed and cut into 1-mm segments while immersed in the perfusate. The specimens are rinsed briefly in the buffer and postfixed in 1% OsO_4 for 1 hr at 4°C.

Method 2 (Rat) (Hill et al., 1982)

The animal (rat) is anesthetized with a 36 mg/kg intraperitoneal injection of Nembutal (sodium pentobarbital). The abdominal aorta is ligated immediately above its bifurcation, and 3 ml of 1% sodium nitrate–heparin solution (1000 U/ml) is injected into the inferior vena cava for vasodilation and anticoagulation. Five minutes later, a midline thoractomy is performed, and a 16-gauge metal cannula is inserted into the left cardiac ventricle. The cannula is advanced into the ascending aorta and secured tightly with a ligature.

A Harvard infusion pump model (no. 975) is used to deliver 600 ml of a mixture of 4% glutaraldehyde and 6% formaldehyde in 0.1 M cacodylate buffer (pH 7.4) containing 0.2% $CaCl_2$ at room temperature. The perfusion rate is 24 ml/min for the first 3 min and 33 ml/min for completion of prefixation. The spinal cord is removed, and the desired part is minced into 1-mm^3 pieces and further fixed in the perfusate for 2 hr at 4°C. After being rinsed in buffer overnight at 4°C, the specimens are postfixed with 1% OsO_4 in cacodylate buffer for 1 hr at room temperature.

Spleen

Perfusates
   A. Modified Ringer's solution

KCl	0.3 g
NaCl	8.78 g
Distilled water to make	1000 ml

Procaine hydrochloride is added to a final concentration of 0.1%. The pH is adjusted to 7.4. Procaine is added to prevent arteriolar spasm and thus allow free flow of the perfusates.
   B. Fixation fluid
   This consists of 1.7% buffered glutaraldehyde with a vehicle osmolarity of 0.08 M and total osmolality of 310 mosmols. Procaine hydrochloride (5%) in water is added to a final concentration of 0.1%.
   Procedure. The animal (e.g., rabbit) is anesthetized by intravenous pentobarbitone sodium (30 mg/kg body weight). A nylon catheter with a three-way stopcock is inserted in the right femoral artery with the tip between the left renal and coelic arteries. The aorta is dissected free so as to control the position of the tip of the catheter and to prepare the clamping of the aorta above the origin of the coelic artery, just below the diaphragm.
   Modified Ringer's solution containing 0.1% procaine is perfused at a perfusion pressure of 110 mmHg, which is equal to the systolic blood pressure. The aorta is clamped above the coelic artery. Then either a small cannula is inserted into the inferior vena cava or one or two mesentric veins are severed to allow free blood flow. After 20 min of perfusion (300–400 ml of Ringer's solution), the

splenic vessels appear pale, and the perfusate is changed to the fixation fluid at the same perfusion pressure. The perfusion–fixation is maintained for 15–20 min. After fixation, the spleen is removed, and small tissue blocks are immersed in 2.5% glutaraldehyde for 1–2 hr, and then postfixed in 1–2% OsO_4 for 1–2 hr at 4°C. For further details, see Elgjo (1976).

Testes

Rinsing Solution

NaCl	9 g
Polyvinyl pyrrolidone	25 g
Heparin	0.25 g
Procaine-HCl	5 g
Distilled water	1 L

The pH is adjusted to 7.35 with 1 N NaOH and the solution is filtered twice through Millipore filters (pore size 3 μm or less).

Fixative Solution A

0.2 M monosodium phosphate (NaH_2PO_4)	45 ml
0.2 M disodium phosphate (Na_2HPO_4)	405 ml
Formaldehyde (25%)	60 ml
Glutaraldehyde (25%)	60 ml
Polyvinyl pyrrolidone	25 g
Distilled water to make	1 L

The pH is adjusted to 7.35 and filtered like the rinsing solution.

Fixative Solution B

The formula is the same as that for solution A, except that 120 ml each of formaldehyde and glutaraldehyde is used and 0.5 g of picric acid is added.

Procedure. The perfusion apparatus and instruments required for the method described here are shown in Fig. 1.7. The details of the abdominal aorta and its main branches, as well as the location and function of ligatures, are shown in Fig. 1.8.

The animal is anesthetized and its abdominal cavity is exposed. The viscera are displaced to the animal's right, and the aorta is cleared from connective tissue and fat. Two ligatures of cotton surgical threads are placed loosely around the lower part of the aorta about 6 mm apart. A ligature is placed around the aorta above the left renal artery. The first ligature of the aorta is made by tightening the lowermost knot; the middle ligature is then lifted to stop the blood flow.

The aorta is incised obliquely with fine scissors below the middle ligature, and the cannula is quickly inserted in a retrograde manner. The blood flow is established by releasing tension on the loose middle ligature. The cannula is made from polyethylene tubing (Intermedic P.E. 160); one end is cut obliquely and filled with a solution of 0.9% heparin, while the other is closed with a clamp and left with a trapped air space. Simultaneously, the cannula is pushed further into

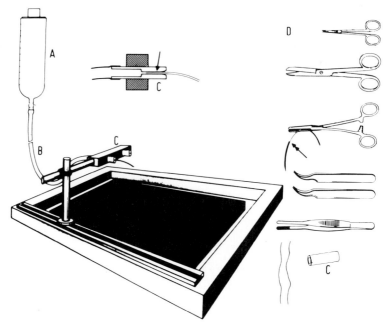

Fig. 1.7. The perfusion apparatus and instruments needed for fixation of the testis. A graduated reservoir (A) to contain the perfusates is connected to a tubing 200 cm long and 0.4 cm in diameter (B). The adaptor system (C) is used to join the cannula to the tubing. A cutaway view of the adaptor is seen above C. Instruments and material used in setting up the perfusion apparatus are shown at D. At the double arrow, note the partially filled (dark) portion of the cannula containing the heparin solution. (From Forssmann *et al.*, 1977.)

the aorta beyond the point of the ligature and secured by tightening the middle ligature.

The vena cava is first exposed between the right renal vein and the liver and then opened with a large incision. The aortic cannula is cut 3 cm from its insertion; the blood flow is then reversed by connecting the cannula to the adaptor of the rinsing solution. The ligature of the aorta is closed above the left renal artery. The rinsing solution is perfused for 1–1.5 min at a pressure of 100 mmHg. This perfusion is followed by the fixative solution A for 2–2.5 min and finally by the fixative solution B at a pressure of 140 mmHg for 2–5 min. For further details, see Forssmann *et al.* (1977).

Uterus

Azaperonum (Sedaperone) (2 mg/kg) and atropine (2–3 mg/100 kg) IM are given to the pregnant animal (e.g., swine) (Björkman *et al.*, 1981). The animal is anesthetized with pentobarbitone sodium (8 mg/kg) through a permanent cannula in an ear vein; this anesthesia is maintained during the operation to a total of 12–14 mg/kg body weight. The animal is tied to an operating table. The dorso-ventral axis is tilted to the horizontal plane by supporting the belly with wooden blocks. Laparotomy is performed on the left side under local anesthesia with lidocaine–noradrenalin along the incision line shown in Fig. 1.9. The left uterine artery and vein(s) are exposed, and collateral arteries are clamped (Fig. 1.9). At this stage the animal is heparinized at 40 IU/kg IV. The widest possible cannula (2–4 mm inner diameter depending on the stage of pregnancy) is introduced into the uterine artery as close to the aorta as feasible. The perfusion is started 0.5 sec later, and the corresponding uterine veins are opened. Immediately before inser-tion of the perfusion cannula, a deeper anesthesia is obtained, eventually fol-

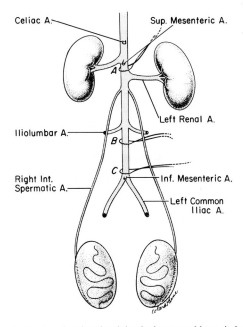

Fig. 1.8. Schematic drawing showing the abdominal aorta and its main branches. Note the right and left internal spermatic arteries originate immediately below the renal arteries. The three ligatures (A, B, and C) are indicated. The cannula is inserted retrogradely into the aorta between the two lowermost ligatures, and the tip is pushed beyond the point of the middle ligature (B). During the perfusion procedure the three ligatures are tied (see text). Ligature A prevents fixative from reaching the GI tract and thorax; ligature B secures the cannula in the aorta and prevents fixative from reaching the iliac vessels and legs; ligature C secures the cannula in the aorta and prevents displacement. The only portion of the aorta that receives fixative is the 2–3-cm region between ligatures A and B; the left renal artery, the two internal spermatics (testicular artery), and the lumbar vessels originating from this part of the aorta. (From Forssmann *et al.*, 1977.)

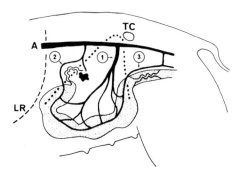

Fig. 1.9. Schematic drawing of lateral view of arterial supply of uterine horn with fetuses (late in pregnancy). Abdominal aorta (A), curvature of last rib (LR), tuber coxae (TC). Dotted line indicates laparotomy. The perfusion cannula is inserted in the uterine artery at 1. The uterine branch of the ovarian artery can be clamped at 2, and the uterine branch of the vaginal artery at 3. (From Björkman *et al.*, 1981.)

lowed by an overdose of anesthetic to avoid convulsion during perfusion and to put the animal to death.

The perfusion is carried out with 3% glutaraldehyde in 0.07 M phosphate buffer (pH 7.4) containing 6% PVP with a total osmolality of 575 mosmols. The perfusion is accomplished at 20°C in 15 min using about 15 L of the fixative. As a rapid perfusion flow is desirable, the container holding the fixative is held at a height of 1–2 m above the animal. Tissue cubes of 1 mm^3 are excised from the yellow and hard tissue and fixed in the same fixative for an additional period of 1 hr. They are then rinsed in the buffer and postfixed with 1% OsO$_4$ in 0.1 M cacodylate buffer for 1 hr.

Vapor Fixation and Staining

Method 1

BEEM capsules (no. 00) are modified by cutting the lid off at the point of attachment to the capsule, leaving the polyethylene hinge attached to the lid to facilitate subsequent handling of the capsule chamber (Shepard and Mitchell, 1980) (Fig. 1.10A). Following removal of the lid, the tip of the preshaped pyramid is cut to produce an opening (no larger than 2.8 mm in diameter) that will prevent standard 3-mm grids from falling through. About 10 mm of the cylindrical part of the capsule is cut away (Fig. 1.10A). Assembly of the cap and remaining portion of the pyramidal section produces the BEEM chamber (Fig. 1.10B).

Osmium tetroxide fixative solution is pipetted through the cut opening of the BEEM chamber in a fume hood. Care should be taken not to contaminate the opening with the solution. The chamber is transferred with forceps to a hot plate set at 37°C. Grids are placed over the opening of the BEEM chamber (Fig. 1.10C) to accomplish fixation and/or staining with OsO_4. The duration of exposure depends on the objective of the study. Staining is usually completed in 30–120 sec. Grids are placed in covered 100-mm glass Petri dishes for a short period to allow dissipation of residual osmium vapor before their poststaining and storage. Crystals of OsO_4 or iodine can also be used in the BEEM chamber.

Method 2

For vapor fixation, tissues are placed in a desiccator containing an open 0.25-g vial of OsO_4 crystals. Osmication seems to be completed in 8 hr. The tissue turns greyish with these vapors and turns black after introduction of a fluid of some sort. For certain studies, vapor fixation is carried out in a chamber that is maintained at −90°C and contains an open ampoule of OsO_4 crystals. After an overnight fixation, the specimens are gradually brought to 4°C in a refrigerator. It should be noted that OsO_4 vapor improves the preservation of the surface association of bacteria and yeast with the skin. Acrolein vapor is more effective in preserving the surface mucus and associated microorganisms.

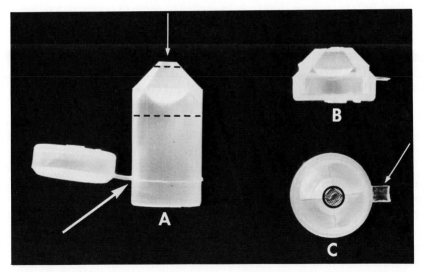

Fig. 1.10. Device for vapor fixation and staining. (A) The lid of a BEEM capsule is removed (arrow) and the tip of the pyramid (vertical arrow) cut along the dotted line (upper), prior to cutting (lower dotted line) on the cylindrical portion. (B) Cross section of an assembled BEEM chamber. (C) The BEEM chamber is picked up by the hinge (arrow) for placement on the hot plate; a grid covers the cut opening. (From Shepard and Mitchell, 1980. ©1980 The Williams & Wilkins Co., Baltimore.)

MICROWAVE FIXATION

The main advantage of microwave fixation is its exceedingly fast speed (1–2 min). Tissue blocks up to 2 cm × 2 cm × 2 cm can be fixed in a few minutes. Such rapid fixation may eliminate poor fixation in the core of a tissue block of a large size and preserve diagnostic features such as mitosis. Mitosis is known to reach completion immediately after the tissue has been excised from the body; chemical fixation fails to stop this completion. However, the damaging effects of irradiation on biologic macromolecules under the conditions of fixation are not known. Red blood cells are largely destroyed. Although the quality of ultrastructure preservation is less than satisfactory, this method is useful for routine histopathology.

Microwave irradiation has been used for fixation of tissues for light (Login, 1978) and electron microscopy (Chew *et al.*, 1983; Hopwood *et al.*, 1984). Optimum temperature for microwave or heat fixation is 45–70°C. Specimens can be microwave-fixed for a few minutes without chemical fixation. Alternatively, specimens can be placed in formaldehyde or glutaraldehyde solution in a vial, which is then placed in the center of the microwave oven for about 10 sec with the power selector on "high." The fixative temperature reaches 50°C after irradiation. An oven with a power output of 700 W at a frequency of 2450 MHz (philips 7910 microwave oven) can be used. The depth of penetration of microwaves is 55–60 mm in any direction.

ARTIFACTUAL ELECTRON-DENSE GRANULES OCCURRING DURING DOUBLE FIXATION

Occasionally, ultrathin sections of certain plant or animal tissues fixed either sequentially or simultaneously with glutaraldehyde and OsO_4 exhibit artifactual electron-dense granules of varying size. There does not seem to be any definite pattern to their distribution within a cell or tissue. These granules may occur in poststained or unpoststained sections, as well as with or without *en bloc* staining with uranyl acetate. Apparently these granules are not a staining artifact, nor are they osmium blacks; but they are caused by some defect in the fixation procedure. One prerequisite for the formation of this artifact seems to be the presence of reactive groups of glutaraldehyde and and/or its decomposition products and reduced osmium in the specimen. Another prerequisite is the presence of phosphate buffer under certain conditions. Phosphate ions act as intermediates between the bound glutaraldehyde and OsO_4 (Hendricks and Eestermans, 1982); the type of phosphate buffer is not important.

The granules most likely are composites of glutaraldehyde, OsO_4, and phosphate. The final phosphate concentration in the tissue specimen during postfixation is a critical factor in the occurrence of the granules. The possibility that other inorganic substances also act as intermediates cannot be disregarded. Since such granules are not seen in cell suspensions and the presence and distribution of the granules are inconsistent, it is likely that local conditions arising in the tissue specimen during fixation contribute to the formation of granules.

The formation of this artifact can be prevented either by not using a phosphate buffer or by employing a phosphate buffer having a concentration not exceeding 0.1 M. When this artifact is already present, the most effective way of removing it from sections is by using oxidizing agents such as periodic acid and hydrogen peroxide. These oxidizing agents are routinely used for removing bound osmium to obtain specific staining of polysaccharides (Hayat, 1975). Thin sections on grids or Marinozzi rings are treated with either freshly prepared 1% periodic acid or 3% hydrogen peroxide for 5–10 min at room temperature. After a thorough washing with distilled water, sections are poststained with 2.5% uranyl acetate in 50% methanol, and then with lead citrate. Copper grids should not be used, because they are not resistant to these oxidizing agents. Periodic acid is preferred over hydrogen peroxide. Longer times of rinsing between primary fixation and postfixation may minimize the formation of granules in certain cases. The inclusion of sucrose (0.05 M) or glucose (3%) in the buffer used for preparing the OsO_4 solution may also reduce the occurrence of this artifact.

2

Rinsing, Dehydration, and Embedding

INTRODUCTION

For double fixation with aldehydes followed by OsO_4, rinsing between fixations is desirable to avoid any reaction between the fixatives. After fixation, most of the excess fixative should be rinsed off before dehydration so as to minimize the possible reaction between the fixative and the dehydration agent. If possible, rinsing should be accomplished in the same buffer as that used with the fixative, to minimize differences in effective osmolarity. Two or three rinses in buffer for a total period of 20 min are adequate; longer durations are required for removing the fixative before incubation for enzyme cytochemistry.

Dehydration is necessary because commonly used embedding resins are immiscible with water. Therefore all the free water from the fixed and rinsed specimens must be replaced with a suitable organic solvent prior to embedding. The removal of water is unnecessary when the specimen is to be embedded in water-miscible resins. The water is removed by passing the specimens through a series of solutions of ascending concentrations of either ethanol or acetone. The use of propylene oxide at the last stage of dehydration is necessary when ethanol is used as the main dehydration agent. When acetone is employed, the use of propylene oxide as a transitional solvent is not necessary. Propylene oxide extracts PTA after *en bloc* staining during or before dehydration. Undesirable effects, including extraction of cellular substances, of dehydration have been discussed by Hayat (1981a,b).

Since specimens are extremely small, they must be embedded in a suitable material for safe handling. Furthermore, specimens are very fragile, porous, and brittle, so they must be permeated with an appropriate material for ultrathin sectioning. Attempts have been made to section unembedded but glutaraldehyde-fixed tissues (Pease, 1982). The resinless section technique, in which tissues are embedded in a Carbowax that is removed after sectioning, has also been used (see p. 85). However, these techniques are of limited usefulness. The most suitable materials for embedding are resins for reasons that have been presented by Hayat (1981a). Both water-immiscible and water-miscible resins are in use, although the former are most commonly employed. Water-immiscible resins are polymerized by heat, while water-miscible resins are cured either by heat or by ultraviolet (UV) irradiation at low temperatures.

Resins that are miscible with water can be employed for dehydration, thereby avoiding the use of ethanol or acetone and preserving certain cellular components that would otherwise be extracted. These resins also save the specimens from denaturation by heat during polymerization. Their role in preserving immunoreactivity is well known. However, these resins also extract cellular substances. Recently developed water-miscible embedding media, Lowicryl K4M and HM20 (Kellenberger et al., 1980), used at low temperatures seem promising (see p. 67). Commonly used formulations of both water-immiscible and water-miscible resins are given later. General properties of embedding media are presented in Table 2.1.

STANDARD PROCEDURE FOR FIXATION, RINSING, DEHYDRATION, AND EMBEDDING

Materials needed are as follows: buffered fixatives in glass stoppered bottles (appropriately labeled); complete buffer (recently checked for pH); two plastic Petri dishes (absolutely clean); tweezers; dissecting needles; new single-edge razor blade (or scalpel) that has been cleaned with acetone; crushed ice; small vials; pipettes; saturated solution of uranyl acetate in Veronal acetate buffer; container for discarding fluids; complete embedding mixture; embedding capsules or containers for flat embedding; holder for capsules; labels; plastic syringe without the needle; acetone series of ascending concentrations.

After a relatively large piece of tissue is removed from the organism, it is immediately placed directly in the aldehyde fixative in a small vial in a fume hood for about 15 min; enough fixative is used to completely cover the tissue. The tissue is transferred to a plastic Petri dish and then cut into small pieces (2–3

TABLE 2.1.

General Properties of Frequently Employed Embedding Media

		Epoxy			Polyester, Vestopal	Methacrylate
	ERL 4206	Epon 812 (Epok 812) LX-112, Poly/Bed	Araldite	Maraglas		
Penetration	Very rapid	Rapid	Very slow	Slow	Slow	Very rapid
Viscosity in cps at 25°C	7.8	150–210	1300–1650	500	1000	0.7; variable
Extraction	Slight	Slight	Slight	Slight	Slight	Avoidable
Polymerization temperature	70°C	40–60°C	60–80°C	50°C	60°C	40–80°C; 4°C (UV)
Shrinkage	4%	4%	4%	4%	7%	10–20%
Shrinkage damage	Slight	Slight	Slight	Slight	Slight	Moderate but avoidable
Polymerization type	Network	Network	Network	Network	Network	Chain-linked
Solubility of polymers	NRS[a]	NRS	NRS	NRS	Insoluble	Soluble
Final hardness	Variable	Variable	Variable	Variable	Variable	Variable
Sectioning ability	Adequate	Adequate	Adequate	Adequate	Adequate	Very good
Specimen contrast	Slight	Slight	Slight	Slight	Slight	Good
Loss of mass (vol)	} At lower current		30%		20%	50%
Carbon residue			70%		69%	67%

[a]Not readily soluble.

mm^3) with a razor blade and transferred (using a fine brush) to a second dish containing several drops of the fixative. The pieces are further cut clearly until they measure less than 1 mm^3. A tissue chopper is preferred for obtaining tissue slices of 0.5 mm^3 or less. These tissue pieces are picked up with a fine brush and dropped into prelabeled vials containing the fixative. The fixative is at least 12 times greater in volume than the specimens. The vial is kept standing in crushed ice or at room temperature. The primary fixation with aldehydes is completed in 1–3 hr at 4°C or 0.5–2 hr at room temperature.

The aldehyde fixative is removed from the vial with a fine pipette and discarded into a suitable container in the fume hood. The buffer is immediately poured into the vial, and the specimens are rinsed twice for a total period of 10–20 min. The buffer is replaced with 1% OsO$_4$ in 0.1 M cacodylate buffer and postfixed for 1–2 hr at 4°C or at room temperature. After a thorough rinse in Veronal acetate buffer, the specimens are treated with 0.5% uranyl acetate in Veronal acetate buffer (pH 5.0) for 15–30 min. (Glycogen staining is adversely affected by this treatment.) Throughout these procedures, specimens must remain wet.

Specimens in the same vial are gradually dehydrated while being stirred in one of the following series:

I. 1. 10% Acetone   5 min
   2. 20% Acetone   5 min
   3. 40% Acetone   5 min
   4. 60% Acetone   5 min
   5. 80% Acetone   5 min
   6. 95% Acetone   5 min
   7. 100% Acetone   5 min
   8. 100% Acetone   5 min
            or
II. 1. 10% Ethanol   5 min
   2. 20% Ethanol   5 min
   3. 40% Ethanol   5 min
   4. 60% Ethanol   5 min
   5. 80% Ethanol   5 min
   6. 95% Ethanol   5 min
   7. 100% Ethanol   5 min
   8. 100% Propylene oxide   5 min
   9. 100% Propylene oxide   5 min

Specimens are gradually infiltrated (while being continuously stirred) at room temperature with the following mixtures:

1. Acetone plus embedding mixture (3 : 1)   30 min
2. Acetone plus embedding mixture (1 : 1)   30 min
3. Acetone plus embedding mixture (1 : 3)   30 min
4. Embedding mixture   30 min

Most epoxy resins are polymerized at 45°C in an oven for 18–24 hr and then at 60°C for an additional 24 hr. Alternatively, polymerization can be accomplished directly at 60°C for 48 hr. Vinylcyclohexene dioxide (Spurr mixture) is polymerized at 70°C in an oven for 8 hr. Gravimetric measurements of resins are preferred over volumetric measurements; an accuracy of ±0.1 g is acceptable. In spite of a few reports to the contrary, 2,2-dimethoxypropane (DMP) is inferior to acetone for preserving lipids; the former is not recommended for dehydration.

The use of Reichert EM Tissue Processor capsules (Fig. 2.1) is recommended to avoid mechanical damage to and loss of specimens during processing, including rinsing, dehydration, staining en bloc, and infiltration. These capsules are especially useful for processing thick (about 50 μm) slices of fixed tissue encapsulated with 7% agar and cut with a tissue sectioner for enzyme cytochemical studies. These flat-bottomed, porous capsules (12 mm in diameter and 10 mm high) are made of sintered high-density polyethylene and have tightly fitting caps. They are freely permeable to aqueous solutions, organic solvents, incubation media, and resins. Specimens can be processed in these capsules through the final step of infiltration with 100% resin. Low-viscosity resins pass through the pores of these capsules more rapidly than do high-viscosity resins. Since reagents are carried over to the next step involving the walls of the porous capsules, extra changes of 100% dehydration agent and 100% resin are recommended. Although these capsules are permanently blackened by OsO_4, they can be cleaned with

Fig. 2.1.   Reichert EM Tissue Processor capsules are shown with their holder used in the unit.

acetone for reuse. These Flo-Thru capsules are commercially available (Normco, Inc., Silver Spring, Maryland).

GRADUAL, PROGRESSIVE DEHYDRATION AND EMBEDDING

There is some evidence indicating that standard methods of dehydration with a solvent and infiltration with a resin introduce artifacts including considerable specimen shrinkage (~30–40% for soft tissues). At least part of the volume shrinkage through the procedures of embedding is caused by osmotic gradients set up in the tissue (Rostgaard and Tranum-Jensen, 1980). Maximum shrinkage during dehydration occurs in the steps between 70 and 100% of the solvent. The only aspect of routine specimen processing that will be discussed here is the adverse effect of incremental increases in the concentration of solvents and resins; advantages and disadvantages of all other aspects of specimen processing have been presented elsewhere (Hayat, 1978, 1981a,b). This adverse effect is especially serious in regard to fragile specimens for transmission electron microscopy and soft specimens processed for scanning electron microscopy. The specimen surface is particularly vulnerable to osmotic changes.

Superior preservation of cellular components, especially specimen surface morphology, can be obtained by employing progressive, gradient-free protocols of dehydration and infiltration. Since very slow and continuous changes of concentration of the preparative media are accomplished, introduction of steep osmotic gradients into the tissue is avoided. Faithful preservation of diffusible cellular substances is not the primary objective in this approach. It has also been shown that the shape and size of erythrocytes are preserved better when glutaraldehyde concentration is slowly increased during fixation (Eskelinen and Saukko, 1982).

Various types of apparatus and procedures have been introduced for processing soft specimens without subjecting them to steep osmotic gradients (Peters, 1980; Rostgaard and Tranum-Jensen, 1980; Jensen et al., 1981; Marchese-Ragona and Johnson, 1982; Brown, 1983a). Another advantage of some of these systems is the saving of manpower because the procedure can take place automatically. Gradient-free dehydration can be accomplished by using a simple apparatus (Fig. 2.2) introduced by Jensen et al. (1981). In this procedure, fixed tissue blocks are pinned to a rubber ring in a beaker filled with buffer which is gradually replaced by acetone over 24 hr; the tissue blocks are pinned to the rubber ring by a 26-gauge needle. Magnetic stirring ensures complete mixing of buffer and acetone. This system is useful for the dehydration of relatively large tissue blocks for scanning electron microscopy.

Progressive infiltration (in contrast to incremental increases) by the embedding

ACETONE

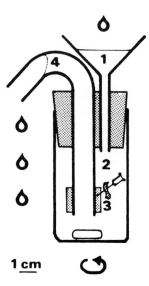

Fig. 2.2.   Apparatus for gradient-free dehydration for scanning electron microscopy; it can be modified for dehydrating small specimens for transmission electron microscopy. Apparatus: 1, funnel-intake for acetone; 2, beaker; 3, specimen pinned to rubber ring; 4, outlet of mixture. (From Jensen *et al.*, 1981.)

medium can be accomplished by placing a 4- to 6-mm layer of 100% acetone (or ethanol) on top of a 15- to 20-mm layer of pure resin in a glass vial (Marchese-Ragona and Johnson, 1982). The vial is shaken very gently to cause the boundary between the acetone and resin to become diffuse. The result is an acetone–resin gradient 1–2 mm deep. The dehydrated specimen is transferred from 100% acetone to the acetone layer in the vial. The specimen will sink until it reaches the acetone–resin gradient. As the acetone–resin exchange proceeds, the specimen sinks into progressively increasing concentrations of resin until it reaches the 100% resin. The tissue specimen (1 mm^3) usually sinks in the pure resin in 30–60 min at room temperature. The acetone–resin gradient and resin are decanted and discarded, and the specimen is immersed in two to three changes of fresh resin and then polymerized.

The semiautomatic fluid exchange apparatus (Fig. 2.3) introduced by Brown (1983a) allows the completion of all specimen processing from fixation to embedding in the same exchange chamber, with minimum osmotic stresses. The following description should help to construct such apparatus. The apparatus consists of a glass fluid exchange chamber with a built-in perforated platform to support the specimen during processing. A captive stirrer located under the platform gently rotates, mixing the contents of the chamber. The stirrer is operated from outside by a rotating magnet. A reservoir is clamped over the exchange chamber. The reservoir is a 50-ml dropping funnel provided with a Rotaflo GP stopcock, underneath which a glass side arm is formed for the attachment of a vacuum pump. A 10-ml syringe is used to supply the necessary vacuum.

The fluid exchange chamber is constructed by closing and flattening one end of a Pyrex glass tube, and perforations are formed with a hot needle. The stirring chamber is fused onto the perforated platform and the glass is drawn out to contain and seal the stirrer within. A length of iron wire sealed inside a thin glass tube forms the stirrer. The outlets of the exchange chamber and reservoir should have the same internal diameter. The magnetic stirrer consists of a horseshoe magnet attached to a 7-cm-diameter wheel which is mounted horizontally onto a hollow shaft and belt-driven by a small variable speed motor. Inserting the outlet of the exchange chamber through the shaft places the captive stirrer within the rotating magnetic field.

The buffer-rinsed (or fixed) specimen, mounted in or on a suitable carrier, is quickly placed into the exchange chamber. The reservoir is clamped over, and the apparatus is attached to the mixer. Using the syringe as a vacuum pump,

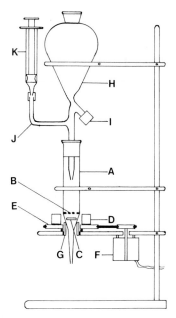

Fig. 2.3. Cross-section diagram of fluid exchange apparatus: A, exchange chamber; B, perforated platform; C, captive stirrer; D, magnet; E, rotating wheel; F, motor; G, hollow spindle; H, reservoir; I, Rotaflo GP stopcock; J, sidearm; K, syringe. (From Brown, 1983a.)

sufficient fixative is drawn into the exchange chamber to completely cover the specimen. The mixer is switched on and the speed of the rotating magnet is adjusted to obtain gentle stirring of the contents without causing turbulence. After fixation, the fixative is gently expelled and replaced with buffer solution. Subsequent steps of postfixation with OsO_4, rinsing, and uranyl acetate staining en bloc are carried out in a similar way.

During the last buffer (or distilled water) rinse, the reservoir is filled with acetone (about 10 times the volume of the contents in the exchange chamber). The stopcock is opened, and the contents of the reservoir are allowed to drop slowly into the exchange chamber over 30 min, mixing with and diluting the buffer. Since the apparatus is airtight, the contents of the exchange chamber remain constant in volume. As the acetone enters the chamber in drops, an equal amount at the same rate will be expelled from the chamber. Thus there will be a progressive increase in the acetone concentration, the end result being pure acetone. Three further changes of pure acetone are drawn into the chamber at 20-min intervals, completing the dehydration.

To infiltrate, 10 volumes of Spurr resin are placed into the reservoir, and the progressive exchange process is repeated. The specimen is removed and transferred into embedding capsules for polymerization. Cleaning is accomplished by flushing the apparatus with pure acetone, soaking overnight in Taab resin solvent (or any other appropriate solvent), and rinsing with distilled water.

PROCEDURE FOR MINIMIZING CHEMICAL HAZARDS DURING SPECIMEN PREPARATION

Exposure to hazardous chemicals during routine preparation of specimens can be drastically reduced by using a device (Fig. 2.4) designed by Redmond and Bob (1984). All the materials needed to construct the device are available in the laboratory. Specimens can be processed in this device from the time of initial fixation up to final embedding. The spout is removed from the cap of a plastic squeeze bottle (2 oz) having a sealable spout. The two minute air holes located on either side (2–3 mm from the end) of the base of the spout are sealed with epoxy; care should be taken not to obstruct the opening in the flared end. The BEEM capsule (size 00) is prepared by cutting off the cap and then puncturing a hole (slightly smaller in diameter than that of the bottle spout) in the apex of the capsule. The puncture is made by pushing a heated metal needle through the plastic, producing a hole of the desired diameter. The pointed end of the spout is inserted through the open end of the capsule until it passes through the hole in the apex. The spout is pushed firmly to make a tight seal between the capsule and the flared end (base) of the spout. This puncture is sealed from the outside with epoxy to make it airtight.

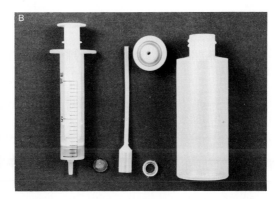

Fig. 2.4. The device that minimizes human exposure during specimen processing to harmful chemicals (A) The assembled device. (B) Parts of the device. See text for details. (From Redmond and Bob, 1984.)

The center of the BEEM capsule cap is removed with a cork borer, leaving the circular band. A small square of metal mesh is placed over the open end of the capsule. The mesh is held in place by using the circular band obtained earlier from the cap. The capsule assembly is placed into the squeeze bottle, allowing it to rest on the bottom. The spout is cut 3–4 mm below the rim of the bottle. The capsule assembly is inserted into the hole in the cap, and the height of the assembly is adjusted so that it hangs about 1 mm above the bottom of the bottle. The level of the band holding the mesh is marked with a grease pencil. The device is now ready for use.

The fixative solution is poured into the squeeze bottle to the level of the grease pencil line. The specimen is placed in the capsule, the mesh is secured in place, the assembly is placed into the bottle, and the top is firmly tightened. The syringe plunger is withdrawn to the 2- to 3-mm mark, and then its tips inserted into the hole of the bottle top. To cover the specimen, the fluid from the bottle is drawn into the capsule by pulling up on the syringe plunger. Care should be taken that the fluid just fills the capsule; if the fluid is drawn into the spout and syringe, it may get contaminated. The fluid is removed from the capsule by pushing the plunger down. The air column within the syringe prevents backflow of the fluid. The bottle is gently swirled to disperse the diluted fluid, and the fluid is drawn again over the specimen.

A separate squeeze bottle may be used for each of the steps: fixation, rinsing, postfixation, dehydration, and infiltration. The specimen is transferred easily by unscrewing the cap (with syringe in position) and then lifting the entire capsule assembly out and screwing it in place in the next bottle. Little fluid is transferred because of the air column in the assembly. While one bottle is in use, the other can be cleaned. The entire procedure should be carried out in a fume hood.

LOW-TEMPERATURE INFILTRATION AND EMBEDDING

Extraction of labile cell components during fixation, rinsing, dehydration, and embedding is minimized at low temperatures. Low-temperature infiltration facilitates preservation and localization of antigens and enzyme activity in plant and animal tissues. Tissue specimens are fixed at 0°C with formaldehyde and/or glutaraldehyde prior to dehydration. Polymerization of the resin is accomplished with UV irradiation at about −50°C (223 K). Since radiation does not penetrate OsO_4-fixed specimens, postfixation with OsO_4 is undesirable when polymerization with UV irradiation is required.

Glycol Methacrylate (2-Hydroxyethyl Methacrylate)

Specimens are fixed with 3% glutaraldehyde in 0.1 M cacodylate buffer (pH 7.2) for 1 hr at 0°C. After two rinsings in the buffer at 0°C, the specimens are simultaneously dehydrated and infiltrated with gradually increasing concentrations of aqueous glycol methacrylate (GMA) mixture according to the following protocol. The monomer mixture consists of 95% GMA, 5% polyethylene glycol, and 0.15% azobisisobutyronitrile. These ingredients are mixed for 30 min and filtered prior to use.

1. GMA mixture (5%)   20–30 min at 0°C
2. GMA mixture (10%)   20–30 min at 0°C
3. GMA mixture (20%)   20–30 min at 0°C
4. GMA mixture (40%)   20–30 min at 0°C

5. GMA mixture (60%)   20–30 min at 0°C
6. GMA mixture (80%)   1.5–2 hr at 0°C
7. GMA mixture (90%)   1.5–2 hr at −20°C
8. GMA mixture (95%)   1.5–2 hr at −20°C

The specimens are left overnight in 100% GMA mixture at −20°C and then transferred to a fresh mixture for 2 hr at −20°C. The specimens are transferred again to a fresh mixture and polymerized in a UV chamber for 36 hr at −20°C.

Lowicryl K4M (Polar) (Kellenberger et al., 1980)

K4M monomer   26 g
K4M crosslinker 4 g
Initiator   150 mg

1. Fixation at 0°C
2. 30% Ethanol   30 min at 0°C
3. 50% Ethanol   60 min at −20°C
4. 70% Ethanol   60 min at −35°C
5. 90% Ethanol   120 min at −35°C
6. 100% K4M and 90% ethanol (1 : 1)   60 min at −35°C
7. 100% K4M and 90% ethanol (2 : 1)   60 min at −35°C
8. 100% K4M   60 min at −35°C
9. 100% K4M   12 hr at −35°C

Lowicryl HM20 (Nonpolar) (Kellenberger et al., 1980)

HM20 monomer   25.5 g
HM20 crosslinker 4.5 g
Initiator   150 mg

1. Fixation at 0°C
2. 30% Ethanol   30 min at 0°C
3. 50% Ethanol   60 min at −20°C
4. 70% Ethanol   60 min at −50°C
5. 90% Ethanol   120 min at −50 to −70°C
6. 100% HM20 and 90% ethanol (1 : 1)   60 min at −50 to −70°C
7. 100% HM20 and 90% ethanol (2 : 1)   60 min at −50 to −70°C
8. 100% HM20   60 min at −50 to −70°C
9. 100% HM20   12 hr at −50 to −70°C

At temperatures below 0°C, the residual water in the specimen should not be allowed to freeze during dehydration. The resin is polymerized overnight at

$-50°C$ by indirect UV irradiation (360 nm) using 2×15 W (Philips TLAD 15W/05 or equivalent) at a distance of 30–40 cm. Either BEEM or gelatin capsules can be used. The resin mixture should be prepared in brown glass containers. Gloves should be used to avoid skin contact (see also p. 6).

INCOMPLETE DEHYDRATION FOR LIPID PRESERVATION

70% Ethanol in water	5 min at 0°C
70% Ethanol in water	5 min at 0°C
95% Ethanol in water	5 min at 0°C
95% Ethanol in water	5 min at 0°C
Resin	1 hr at 0°C
Resin	1 hr at 0°C
Resin	1 hr at 0°C
Resin mixture	Overnight at 4°C
Infiltrate and embed in resin	

WATER-IMMISCIBLE EMBEDDING MEDIA

Epoxy Resins

Epon 812 (LX-112)

Since LX-112 (Ladd Res. Ind.) is a generic replacement for Epon 812, the proportions of DDSA, NMA, and DMP-30 for LX-112 are the same as for Epon 812.* Since the W.P.E. (weight per epoxide equivalent) values of the epoxies vary from lot to lot, the only reliable way to ensure reproducible hardness of the block is by using the W.P.E. for the embedding formulations. The W.P.E. value can be utilized for embedding on a weight basis from the following equation:

$$
\begin{array}{c}
\text{Weight of anhydride} \\
\text{(DDSA and NMA)} \\
\text{required}
\end{array}
= \frac{100}{\text{W.P.E.}} \times
\begin{array}{c}
\text{Anhydride} \\
\text{molecular} \\
\text{weight}
\end{array}
\times
\begin{array}{c}
\text{Ratio of anhydride} \\
\text{to epoxy resin} \\
\text{equivalents}
\end{array}
$$

where 100 = grams of epoxy resin (100 is arbitrary, being used only as an example), and W.P.E. = weight of epoxy resin containing one equivalent weight

---

*Other replacements for Epon 812 include Poly/Bed 812, Polarbed 812, and Epok 812. DDSA is dodecenyl succinic anhydride; NMA or MNA is nadic methyl anhydride; DMP-30 is 2,4,6-tri-dimethylaminomethyl phenol.

of epoxide. Molecular weights of NMA and DDSA are 178 and 266, respectively. The ratio of anhydride equivalent to epoxy resin equivalent is 0.7:1. Table 2.2 indicates the W.P.E. values and the required proportions of DDSA and NMA. The higher the W.P.E. value, the lower the proportions of DDSA and NMA.

For accuracy of measurement, the gravimetric rather than the volumetric measurement is recommended for the preparation of embedding formulations. If, for example, the W.P.E. of LX-112 resin is 166, the following formulation is recommended:

Mixture A

| LX-112 | 80 g |
| DDSA | 90 g |

Mixture B

| LX-112 | 100 g |
| NMA | 75 g |

TABLE 2.2.

Proportions of DDSA and NMA Determined by the Weight per Epoxide Equivalent (W.P.E.) of Epoxy Resins (LX-112, Epon 812)[a]

Resin W.P.E.	Weight of DDSA, mixture B	Weight of NMA, mixture B	Resin W.P.E.	Weight of DDSA, mixture A	Weight of NMA, mixture B
140	106	89	158	94	79
141	106	88	159	94	78
142	105	88	160	93	78
143	104	87	161	93	77
144	103	86	162	92	77
145	103	86	163	91	76
146	102	85	164	91	76
147	101	85	165	90	76
148	101	84	166	90	75
149	100	84	167	89	75
150	99	83	168	89	74
151	99	83	169	88	74
152	98	82	170	88	73
153	97	81	171	87	73
154	97	81	172	87	72
155	96	80	173	86	72
156	95	80	174	86	72
157	95	79	175	85	71

[a]The amount of the resin is 100 g in each mixture, and the weights of DDSA and NMA are in grams.

The components of mixture A are put in an 8-oz bottle, which is capped and shaken vigorously until the two components are completely mixed. Mixture B is prepared in a similar way. Mixtures A and B are mixed in the ratio of 3 : 2 to obtain a block of medium hardness. Then 0.14 g (ml) of DMP-30 is added per 10 g of this mixture. Even a slightly incomplete mixing may produce blocks of poor sectioning properties. To enhance infiltration of the specimen, polymerization is carried out at 45°C for 20 hr and then at 60°C for an additional 24 hr. If necessary, polymerization can be accomplished directly at 60°C.

Mixtures A and B can be stored for as long as 6 months under inert gas (Freon or argon) and continuous refrigeration. They should be warmed to room temperature before the container is opened; otherwise, water will condense in the resin, causing poor sectioning. Also, the presence of water may produce slow hydrolysis of the anhydrides to free acids, resulting in defective blocks. Freshly prepared embedding mixtures are preferred.

The hardness of the block can be controlled by adjusting the ratio between mixture A containing DDSA and mixture B containing NMA. The higher the proportion of NMA, the harder the block (Table 2.3).

Araldite

1.	Araldite 502	10 ml
	DDSA	7.8 ml
	DMP-30	1.5%
2.	Araldite 502	68 ml
	DDSA	19 ml
	DMP-30	3 ml
	Triallyl cyanurate	10 ml
3.	Araldite CY 212	10 ml
	DDSA	10 ml
	Benzyl dimethylamine	0.4 ml

TABLE 2.3.

Proportions of DDSA and NMA and the Resulting Quality of the Block (LX-112, Epon 812)

Mixture A, weight (g)	Mixture B, weight (g)	Total weight (g)	Accelerator DMP-30 (g)[a]	Relative hardness
10	0	10	0.14	Softest
7	3	10	0.14	Soft
5	5	10	0.14	Medium
3	7	10	0.14	Hard
0	10	10	0.14	Hardest

[a]Where 1 g = 1 ml.

TABLE 2.4.

**Proportions of Araldite 502 and DDSA
Determined by the Weight per Epoxide
Equivalent (W.P.E.) of the Resin**[a]

Araldite 502 (W.P.E.)	Araldite 502 (g)	DDSA (g)
232–234	100	80
235–237	100	79
238–240	100	78
241–243	100	77
244–246	100	76
247–249	100	75
250	100	74

[a]Prior to use, 0.136 g (0.14 ml) of DMP-30/10 g of Araldite 502–DDSA mixture is added.

Table 2.4 shows the W.P.E. values of Araldite 502 and the required proportions of DDSA; similar information is given for Araldite 6005 in Table 2.5.

Maraglas

1. Maraglas   48 ml
   Cardolite NC-513   40 ml
   Triallyl cyanurate 10 ml
   DMP-30   2 ml

TABLE 2.5.

**Proportions of Araldite 6005 and DDSA
Determined by the Weight per Epoxide
Equivalent (W.P.E.) of the Resin**[a]

Araldite 6005 (W.P.E.)	Araldite 6005 (g)	DDSA (g)
182	100	91
183–184	100	90
185–186	100	89
187–188	100	88
189–190	100	87
191–192	100	86

[a]Prior to use, 0.411 g (0.46 ml) of benzyl dimethylamine/20 g of Araldite 6005–DDSA mixture is added.

2. Maraglas 655   36 ml
   DER 732   8 ml
   Dibutyl phthalate   5 ml
   Benzyl dimethylamine   1 ml

LR White

LR White is a recently introduced (London Resin Co.) monomer polyhydroxy-lated aromatic acrylic resin capable of thermal or cold curing when an accelerator is used. Tissues postfixed with OsO_4 should not be cold-cured because this process is strongly exothermic, and the dark color of the tissue leads to a focal heat accumulation that can cause local problems in and around the tissue. Dehydration with ethanol is preferred when embedding in LR White, because traces of acetone left in the tissue can interfere with its polymerization.

LR White has a very low toxicity, since it contains only monomers used in medicine and dentistry. It has a viscosity of 8 centipoise (cps) which is lower than that of most other resins and easily infiltrates large tissue blocks. Because of its low viscosity and toxicity, LR White may be a desirable replacement for highly toxic VCD resin. The former may be especially useful for plant and fungal specimens in which resin penetration is a problem and for combined light and electron microscopy. Stains prepared in alcohols should be avoided, for the solvents tend to soften the resin.

Polymerization is carried out in predried (at least for 1 hr at 60°C) gelatin capsules at 60°C for 12–24 hr. After the specimen is placed into it, the capsule is filled completely and covered with the lid. During polymerization the surface of the resin must remain covered to prevent contact with oxygen. If, however, polymerization is carried out in a vacuum oven, no lids are needed. For resin infiltration, the specimen chamber is pumped down to 30 psi, and then the chamber is shut off from the pump. The accelerator is not necessary for thermal polymerization, but is required for cold curing. The resin is sold in premixed batches, and in this form it has a shelf life in excess of 12 months at 4°C. It has some tolerance for water. The resin is available in hard, medium, or soft grades from Polaron, Ernest F. Fullam, and Pelco.

Quetol 651

Quetol 651   33 ml
Nonenylsuccinic anhydride (NSA)   67 ml
DMP-30   1.5–2.0 ml

Dehydrated specimens are infiltrated according to the following procedure:

1. Ethanol + n-butyl glycidyl ether   20 min
   (n-BGE) (1 : 1)

2. *n*-BGE   20 min
3. *n*-BGE + Quetol mixture (1 : 1)   1 hr
4. Quetol mixture   2 changes, 1 hr each
5. Polymerization for 24 hr at 60°C in the vacuum oven. This resin will not polymerize in the presence of air.

Dow Epoxy Resins

DER 334

DER 334	48 ml
DDSA	40 ml
Triallyl cyanurate (TAC)	10 ml
DMP-30	2 ml

Small quantities of the above ingredients should be placed under a vacuum for 30 min at 70°C before mixing. Ingredients should be mixed for 1 hr before the infiltration. Infiltration is facilitated by subjecting the capsules containing the specimen to a negative pressure equivalent to 25 mmHg. Polymerization is completed at 70°C. Harder blocks can be obtained by decreasing the proportion of TAC. TAC has a freezing point of 27°C and may be warmed to the liquid state by placing it in an oven for 10 min at 70°C.

DER 332

DER 332	7 g
DDSA	8 g
DER 732	3.2 g
DMP-30	0.3 g

DER 332 is heated to 60°C to reduce its viscosity prior to mixing it with other components. As the proportion of DER 732 is increased, so is the hardness of the block.

Vinylcyclohexene Dioxide (VCD, ERL 4206, Spurr Mixture)

Standard embedding medium

Vinylcyclohexene dioxide	10 g
DER 736	6 g
NSA	26 g
Dimethylaminoethanol (DMAE)	0.4 g

Variations in these compositions can be obtained by reference to Table 2.6. The exact weight of each of the components should be used for best performance. If

TABLE 2.6.

Suggested Variations of Vinylcyclohexene Dioxide Embedding Formulations

	Mixtures				
	A	B	C	D	E
Ingredients (ml)					
ERL 4206	10	10	10	10	10
DER 736	6	4	8	6	6
NSA	26	26	26	26	26
DMAE	0.4	0.4	0.4	1	0.2
Polymerization time (hr) at 70°C	8	8	8	3	16
Hardness	Firm	Hard	Soft	—	—
Pot life (days)	3–4	3–4	3–4	2	7

bubbles form, they can be drawn off with a gentle vacuum applied to the mixing container. NSA and DMAE should never be mixed together alone due to the possiblity of a rapid exothermic reaction. (Benzyldimethylamine or DMP-30 may also be used if a faster cure is needed.) Minimum exposure to atmospheric moisture is recommended. Continuous mild agitation is desirable during infiltration. If specimens are difficult to infiltrate, vacuum treatment may help. Curing should be accomplished in a BEEM capsule with the cap sealed to prevent moisture contamination. The standard cure is 8 hr or more at 70°C in an oven. This medium is not completely miscible with ethanol.

Vinylcyclohexene Dioxide—*n*-Hexenyl Succinic Dioxide (HXSA)

Vinylcyclohexene dioxide	5 g
HXSA	10.5 g
DER 736	1.5 g
Dimethylaminoethanol	0.2 g

A relative increase of DER 736 will result in softer blocks.

Polyester Resins

Polymaster 1209AC

Polymaster 1209AC (Bondaglass-Voss Ltd., Beckenham, Kent) is superior to glycol methacrylate for embedding large pieces (10 × 10 × 3 mm) of mineralized bone for light microscopy (Mawhinney and Ellis, 1983). It is a polyester resin incorporating a styrene monomer stabilized by hydroquinone. Addition of an organic peroxide catalyst polymerizes the resin by addition reactions between unsaturated C=C bonds in the polyester polymer, cross-linked with the C=C bonds in the styrene. Thick sections (2–6 μm) of the polymerized resin allow most staining reactions commonly used in histologic laboratories. The resin itself does not take up stains.

Tissues are fixed for 2–7 days either with 10% formalin in a buffer or with 100% ethanol. Dehydration is accomplished in four changes with cellosolve at 3, 5, 16, and 24 hr. No water should be left in the tissue prior to infiltration. All buffer salts should be removed from the formalin-fixed tissues during dehydration. These salts can be detected by adding 1 ml of 2% aqueous silver nitrate to 2 ml of the used dehydration cellosolve. A faint white cloudiness in the fluid can be seen with salt concentrations down to those associated with 0.2% water. Infiltration is carried out in cellosolve/resin mixtures of increasing resin concentrations of 75/25, 50/50, 25/75, and 100%, with 24 hr allowed for each step while the mixture is agitated. Further infiltration is done in three changes of 24 hr each at 37°C and 700 mm vacuum in a 19:1 mixture of the resin and dibutylphthalate. Final infiltration is accomplished in 8 hr while agitating in the following mixture:

Resin	95%
Dibutylphthalate	5%
Butanox 50 catalyst	1%
1% Hydroquinone in ethanol	1%

Embedding is done in siliconized glass molds by leaving them in a water-bath heat sink at room temperature. This step at the early polymerization stage prevents bubble formation. In the absence of this step, bubbles will form owing to vaporization of cellusolve by the exothermic polymerization reaction. Hardening of the resin is accomplished in an oven at 56°C for 48 hr. Thick sections can be cut with a tungsten carbide knife, using 70% ethanol as a flotation fluid. The sections are stained free floating through staining solutions, dehydrated, cleared, and mounted in DPX or any other suitable medium.

Poly-N-vinylcarbazole

The photoelectric semiconductor poly-N-vinylcarbazole (PVK) was introduced by Grund et al. (1982) as an embedding resin for studies of thin sections (50 nm) in the photoemission electron microscope (PEEM). This resin permits the imaging of structures without heavy metal staining and provides an extended usable observation period in the PEEM. PVK has a lower viscosity than that of Epon. For embedding, PVK is melted at 80°C, and dibutylphthalate and benzoyl peroxide are added in concentrations of 6 and 1%, respectively. Individual components must be thoroughly mixed, using a heated magnetic stirrer. Dehydrated specimens are infiltrated in a 1:1 mixture of propylene oxide and PVK for 1 hr at 60°C. Polymerization is carried out at 70°C for 2 days.

Rigolac

Rigolac 2004	75 ml
Rigolac 70F	25 ml
Benzoyl peroxide paste	1 g

Vestopal W

Vestopal W	100 ml
Benzoyl peroxide	1 ml
Cobalt naphthenate	0.5 ml

Vestopal, benzoyl peroxide, and cobalt naphthenate should be kept in a cold and dark place to prevent their decomposition. Vestopal is miscible with acetone but not with ethanol. Benzoyl peroxide and cobalt naphthenate must not be mixed with each other alone; they may explode.

Methacrylates

Mixture of *n*-butyl and methyl methacrylate (80:20)	100 ml
Divinyl benzene	5 ml
Benzoyl peroxide or bis(4-*tert*-butylcyclohexyl)-peroxydicarbonate	1 ml

Styrenes

Monomeric styrene	10 ml
Methyl ethyl ketone	0.4 ml
Dibutyl phthalate	0.15 ml

WATER-MISCIBLE EMBEDDING MEDIA

Acrylamide–Gelatin–Jung Resin

A mixture of polymerized acrylamide, gelatin, and Jung resin (Jung, Heidelberg) was introduced by Hartmann (1984) for cryo-sectioning of eggs (snails, fishes, and insects) without rupture of the yolk- and lipid-containing areas. This procedure improves preservation of the ultrastructure and sectioning of large eggs for both light and electron microscopy.

Stock solution

Acrylamide (3.938 M) containing 0.048 M bis(N,N'-Methylenebisacrylamide)	20 ml

Sörensen phosphate buffer (0.2 M, pH 7.1)   10 ml
Gelatin (16%)   35 ml
Jung embedding resin   20 ml
N,N,N',N'-Tetramethylenediamine (TMED)   0.2 ml

This solution is continuously mixed with a magnetic stirrer in a water bath at 30°C before use to prevent phase separation. The solution can be stored in the refrigerator for at least 8 weeks. Double-distilled water should be used to prepare all aqueous solutions.

The specimen is placed in 8.5 ml of the well-mixed stock solution, which is overlayered in sequence with 0.5 ml of 2% ammonium peroxodisulfate and 0.1 ml of fixative (a mixture of 10% formaldehyde and 6.25% glutaraldehyde in 0.1 M phosphate buffer at pH 7.1). All components are rapidly mixed and then poured into a paraffin oil-coated plastic box. The specimen is oriented and kept in position until the mixture solidifies within 30–60 sec. The block with embedded specimen has rubber-like consistency. It is rapidly frozen in isopentane cooled by liquid nitrogen. Using a cryotome at -18 to -22°C, 5–20 μm thick sections are cut, which are mounted on a glue-coated glass slide. The glue contains 0.8% agarose, 1.8% glycerol, and 0.5% glutaraldehyde (or 0.75% formaldehyde). The slides are coated by vertical dipping in the glue at 80°C and stored in a dust-free box until use.

Postfixation is accomplished by exposing the sections mounted on glue-coated slides to vapors of an aldehyde in a moist chamber at 30°C for 5 min, followed by immersion in 4% formaldehyde in PBS or 2% glutaraldehyde in PBS for 30 min. Before staining, the aldehyde is removed by extensive washings in PBS for 2 days. Staining is carried out either with hematoxylin–eosin or with 2–4% OsO_4 in 0.1 M phosphate buffer (pH 7.2) followed by counterstaining with Feulgen.

Durcupan

Durcupan   5 ml
DDSA   11.5 ml
Accelerator 960   1 ml
Dibutyl phthalate   0.3 ml

The procedure for dehydration, infiltration, and embedding is as follows:

1. 50% Durcupan mixture in distilled water   30 min
2. 70% Durcupan in distilled water   30 min
3. 90% Durcupan in distilled water   30 min
4. 100% Durcupan   60 min
5. 100% Durcupan   60 min

Impregnated specimens are placed in gelatin capsules with fresh embedding mixture, and polymerization is completed in 24 hr at 45°C.

Gelatin

10% Gelatin solution in water   30 min
20% Gelatin solution in water   30 min
30% Gelatin solution in water   30 min

The 30% solution containing the tissue specimens is poured into a Petri dish and solidified in a refrigerator at 4°C. When the gelatin becomes hard but not brittle, it is cut into small cubes measuring 4 mm³, each containing one specimen. These cubes are dried by placing them in a vacuum desiccator under a pressure of 10^{-3} mmHg for 5 hr. After drying, the cubes are trimmed and glued to blocks of Plexiglas or hard resins of suitable size for sectioning.

Glutaraldehyde–Carbohydrazide (GACH)

The medium is prepared by adding a 3.3 M solution of carbohydrazide to 50% aqueous glutaraldehyde. It can be employed at neutral pH and cured at 37°C. The speed of polymerization is increased at acid or alkaline pH even at low temperatures.

Glycol Methacrylate (GMA)

Method 1

1. GMA (95%)   66.5 ml
   Distilled water   3.5 ml
   n-Butyl methacrylate   28.5 ml
   Ethylene dimethacrylate   5 ml
   Benzoyl peroxide   1.5 ml

2. GMA (100%)   66.5 ml
   n-Butyl methacrylate   28.5 ml
   Ethylene dimethacrylate   5 ml
   Benzoyl peroxide   1.5 ml
   Polyethylene glycol 400   1 ml

Prepolymerization is accomplished by heating the embedding mixture to 40°C on a magnetic stirrer hot plate while stirring continuously until the benzoyl peroxide is dissolved. After removal from the heat, the mixture is transferred to an Erlenmeyer flask with a Teflon-coated magnet. The flask is stoppered with a two-hole stopper and a thermometer is inserted into the stopper at an angle to permit the movement of the magnet. The thermometer should almost reach to the bottom of the flask. The other hole of the stopper permits ventilation.

The flask is heated on the hot plate to 98°C at the rate of 8°C/min while the mixture is being continuously stirred. The flask is then plunged into a dry ice–ethanol bath and swirled rapidly to quickly cool the mixture to 2°C. The procedure for dehydration and polymerization is as follows:

1.	80% GMA in water	15 min
2.	100% GMA	4 changes, 15 min each
3.	Embedding mixture	2 changes, 1 hr each
4.	Prepolymerized mixture	48 hr

The specimens are embedded in gelatin capsules, which are filled to the top with fresh prepolymer mixture. The capsules should be left uncapped for 30 min to eliminate air bubbles. They must then be capped with as little air as possible trapped within the capsule. Alternatively, before capping, the capsules are placed in a vacuum chamber, and air bubbles are removed for 10 min. Polymerization is effected by long-wavelength UV light (>315 nm) and is completed in 16–20 hr at 4°C. The distance between the lamp and the blocks is 10–20 mm.

Method 2

In the following formulation a toxic tertiary amine accelerator is replaced by a barbituric acid derivative in combination with chloride ions and dibenzoyl peroxide (Gerrits and Smid, 1983). This medium has been used successfully for embedding soft tissues and obtaining serial sections of 1–2 μm thickness.

Infiltration medium

Glycol methacrylate containing co-catalyst XC1 (Technovit 7100)	40 ml
Polyethylene glycol 400	4 ml
Dibenzoyl peroxide (moistened with 20% water)	0.5 g

Embedding medium

Infiltration medium	15 ml
Technovit 7100, hardener II (accelerator)	1 ml

Although GMA is miscible with water, tissue blocks are dehydrated in ethanol and infiltrated with a 1 : 1 mixture of 100% ethanol and the infiltration medium for 2 hr. Complete infiltration is then carried out in the infiltration medium for up to 24 hr, depending on the size of the tissue. The specimens are placed in molding cups that are fitted with aluminum block holders (Du Pont) and contain about 1 ml of the embedding medium. A complete bubble-free enclosure of the tissue by the embedding medium is necessary. An initial polymerization is carried out for 2 hr at room temperature, and further polymerization for 2 hr at 37°C. Semithin sections can be stretched on a hot plate at 60°C.

Hydroxypropyl Methacrylate (HPMA)

Method 1

To prepare the embedding medium, HPMA and distilled water (in a ratio of 4 : 1 by volume) are vigorously shaken at 20°C. HPMA containing 0.1% azonitrile is partially polymerized by heating it to 120°C while stirring constantly. The required polymerization is completed within 5 min; however, the viscosity should be checked occasionally by rapid cooling in ice. Water is then added to make up an HPMA mixture of concentrations of 3–20%. The final embedment of the tissue is carried out in the cold in a partially prepolymerized HPMA having the consistency of a free-flowing syrup.

The dehydration and infiltration procedure is as follows:

1.	HPMA (85%)	1 hr
2.	HPMA (85%)	1 hr
3.	HPMA (97%)	1 hr
4.	HPMA (97%)	1 hr
5.	Prepolymerized HPMA (97%)	1 hr

The dehydration and infiltration are carried out at 2°C, and the tissues must be kept in continuous agitation in the infiltration mixture. Final polymerization is achieved in 12–24 hr after the capsules are exposed to UV radiation (>315 nm) in the cold. Thermal polymerization is completed in 48–72 hr at 56°C. Thin sections must be picked up on Formvar- or carbon-coated grids.

Method 2

HPMA	95 ml
Polyethylene glycol	5 ml
Azonitrile (catalyst)	0.5 g

The prepolymerized mixture is obtained by heating HPMA monomer with the catalyst at 90°C. After the mixture has cooled to room temperature, it is continuously stirred while polyethylene glycol is added. If the prepolymer becomes too viscous, the monomer mixture is added with the catalyst. The prepolymer can be stored safely in a freezer. The specimens are infiltrated as follows:

1.	HPMA and catalyst	1 hr
2.	HPMA and catalyst plus HPMA, catalyst, and polyethylene glycol (1 : 1)	1 hr
3.	HPMA	1 hr
4.	Prepolymer	1–2 hr

Infiltration is carried out at room temperature, preferably on a shaker. Gelatin capsules are filled completely and polymerized at 55°C for 15 hr.

LEMIX

LEMIX is a water-soluble epoxy resin and is claimed to be stable under the electron beam; it is of sufficiently low viscosity (70.8 cps) to infiltrate large tissue blocks (5 mm) (EMscope Lab. Ltd.). It allows the penetration of a number of aqueous histologic stains (e.g., methylene blue–basic fuchsin and toluidine blue). LEMIX and JB4 resin (Polysciences, Inc.) are ideal embedding media to take advantage of a new combined light and electron microscope, the LEM 2000 (ISI Instruments). The embedding medium used for this instrument must allow the production of ultrathin as well as thick sections. Furthermore, the resin must allow the penetration of a wide range of stains for both light and electron microscopy.

Lowicryl K4M

Lowicryl K4M is a polar acrylate–methacrylate resin; it is used in immunoelectron microscopy at low temperatures (Roth *et al.*, 1981; Carlemalm *et al.*, 1982). Before the tissue is immersed in Lowicryl for infiltration, the resin should be degassed using a vacuum pump (Fryer *et al.*, 1983). Oxygen has an inhibitory effect on its polymerization. Infiltration and embedding including polymerization of the dehydrated tissues are carried out at −30 to −50°C:

1. Lowicryl plus dehydration agent (1 : 1)   30 min
2. Lowicryl plus dehydration agent (2 : 1)   30 min
3. Lowicryl (two changes)   1 hr each
4. Lowicryl   Overnight

To initiate cross-linking, 0.6% benzoin methylester (v/v) is added to the resin. Polymerization is carried out overnight at −30°C by UV irradiation from a Philips TLAD 15W/05 fluorescent lamp (wavelength peak at 360 nm). Irradiation then continues for another 2–3 days at room temperature. Gelatin capsules should be filled and capped during polymerization. Alternatively, polymerization is accomplished by placing the gelatin capsules in a cryostat chamber whose reflective metal interior gives all round irradiation (Fryer *et al.*, 1983). The UV lamp (UVSL-58, Ultraviolet Products Inc.) is mounted on the outside of the cryostat window so that the UV light enters through the Perspex access port. Irradiation is given for 18–24 hr. The blocks are further cured by exposure to daylight for 2–3 weeks. The polymerized resin blocks should be kept under partial vacuum and over desiccant because the polymerized resin is hygroscopic. Some evidence indicates that sectioning properties improve when desiccated blocks are used (see also p. 67).

Melamine Mixture

Method 1

Melamine mixture has been used for embedding nervous tissue (Shinagawa and Shinagawa, 1978).

Mixture A

Melamine	28 g
Paraformaldehyde	30 g
Urea	10 g
25% Glutaraldehyde	60 ml
Distilled water	10 ml

The pH is adjusted to 8.0 by adding 1–2 drops of ammonium hydroxide solution. The ingredients are heated at 80°C for 3–5 hr, being stirred occasionally, and then cooled to room temperature. When the mixture appears transparent at room temperature, it is ready for use. If the mixture becomes semitransparent, it is heated again at 80°C for 1–2 hr.

Mixture B

Paraformaldehyde	30 g
Urea	10 g
25% Glutaraldehyde	30 ml
Distilled water	7 ml

The pH is adjusted and the mixture prepared the same way as mixture A. The final mixture is prepared by mixing mixtures A and B in the ratio of 1 : 1. The larger the proportion of mixture A, the harder the block. Fixed specimens are transferred directly into this mixture for infiltration at room temperature for 2–3 hr without dehydration with organic solvents. Polymerization is completed at 55°C for 2 days. The block is not easy to section. Acetone (12%) is used as flotation fluid, the meniscus is kept low, and a rather high sectioning speed is required to avoid wetting the block. Ultrathin sections are negatively stained with PTA (pH 7.0).

Method 2

Melamine–formaldehyde can be modified into a water-soluble embedding medium (Bachhuber and Frösch, 1983). Hexamethylol–melamine–ether (MME) is synthesized by adding 6 ml of formaldehyde (37%) to 1 ml of melamine (Fluka) and then carrying out slight etherification with methanol at pH 5–6. After one wash in chloroform, a clear solution of pure MME is obtained, which is completely miscible with water, free of salts, and stable for many months at room temperature. Thin sections of unstained but fixed cells, tissues, and protein pellets embedded in this medium exhibit a pronounced electron-phase contrast. These images favorably compare with those of corresponding preparations which have been stained with heavy metals (Westphal and Frösch, 1984). Why unstained specimens embedded in melamine produce electron-phase contrast while epoxy-embedded specimens do not remains to be answered.

Fixed tissue blocks are placed into glass vials containing 0.01 mol melamine and 0.06 mol formaldehyde (37%); the pH is adjusted to 8.5 with 0.1 N NaOH. About 40% polyethylene glycol (mol. wt. 10,000) from a 50% aqueous stock solution is added as an elastifying agent. The vials are closed and slowly rotated for 1–1.5 hr in a water bath at 55°C. At this temperature melamine dissolves in about 20 min, and its viscosity is 48–52 mPa/sec.

With progressing condensation, the medium becomes insoluble in water, as indicated by a fine precipitate after 2–2.5 hr. The specimens are transferred to silicone molds that have a raised edge and contain a 50% aqueous solution of MME (pH 5–6 adjusted with phosphoric acid), which has been evacuated previously in the molds to avoid air bubbles. The hardening is accomplished in three steps: 2 days at 40°C (in a glass desiccator of 20 cm diameter, containing silica gel), 2 days at 60°C (without desiccator), and 4 hr at 80°C. Liver shrinkage amounts to about 25% during hardening. This embedding medium seems to be useful for X-ray microanalysis studies. Further testing is needed before it can be recommended for wider use.

Polyacrylamide

Polyacrylamide (PAA) was introduced as an embedding medium for immunohistochemical studies using conventional ultramicrotomy (Yamamoto *et al.*, 1980). It is more suitable for immunocytochemistry on ultrathin cryosections (Slot and Geuze, 1982). The stock embedding medium consists of 50% acrylamide and 1.5% bisacrylamide in distilled water. The preparation of final embedding medium is as follows:

Stock embedding medium	4 ml
Formaldehyde (16%)	0.6 ml
Sodium phosphate buffer (1 M, pH 5.7)	0.4 ml
N,N,N',N'-Tetramethylethylenediamine (TEMED)	10 μl
Riboflavin (100 mg/ml)	10 μl

The acid buffer is added to neutralize TEMED. The above recipe indicates how to prepare an embedding mixture of maximal acrylamide concentration (40%); if desired, this can be diluted to a lower concentration. The addition of formaldehyde to the embedding mixture is optional and possibly useful only when very mild fixation is required.

Fixed tissue specimens are transferred to the final embedding medium and agitated overnight at 4°C in the dark. They are then transferred in drops of the embedding medium to a 9-cm Petri dish. A 3-cm dish is placed rapidly on top of these drops so that each drop is squeezed to a slab. The thickness of this slab is determined by the small rim (0.5 mm) that sticks out from the bottom of the 3-cm dish. The preparation is placed on a light box with fluorescent bulb for polymerization while being cooled by ice. A PAA gel slab forms within a few minutes, but exposure to the light source is continued for at least 1 hr. The presence of formaldehyde in the embedding medium accelerates the polymerization, which is an exothermal reaction and therefore needs proper cooling.

Polyampholyte (Polyamph 10)

The polyampholyte is prepared by distilling monomer anionic methacrylate acid (MA) at 10 mmHg pressure, and the fraction that boils at 60°C is collected and stored at 4°C. After being washed with buffer, fixed tissue specimens are gradually dehydrated and infiltrated with 1 : 1 mixture of MA and cationic dimethylaminoethyl methacrylate (DMA) in the following manner. The specimens are immersed in 10, 20, 40, and 80% of the mixture for 30 min at 4°C at each step. They are then immersed in three changes of 100% monomer mixture for a period of 2 hr, followed by three more changes in the embedding mixture for 12 hr at 4°C. The specimens are transferred to freshly prepared embedding mixture in gelatin capsules. The embedding mixture is prepared as follows:

Methacrylic acid	20 ml
Dimethylaminoethyl methacrylate	10 ml
Tetramethylene dimethacrylate	3.3 ml
Azodiisobutyronitrile	0.25 ml

MA should be added drop by drop to DMA to avoid thermal polymerization. Polymerization is completed by UV irradiation within 48 hr at 4°C.

Polyethylene Glycol (PEG; Carbowax)

Fixed tissue specimens are immersed directly into 4-g vials filled with warm (58°C) 50% PEG (mol. wt. 4000 or 6000) in distilled water. After being placed in a temperature-regulated shaker bath maintained at 58°C, the vials are agitated during infiltration for 1–6 hr. The specimens are removed and, after excess medium is drained on warm filter paper, are transferred to vials containing 70% PEG in distilled water. The specimens are infiltrated on the shaker bath for 1–8 hr. This process is repeated twice more with 100% molten PEG. The specimens are transferred to a small paper cup, together with liquid PEG mixture, and solidified in the refrigerator. Sections about 40 μm thick are cut with a sliding microtome and embedded in Epon 812.

For immunoelectron microscopy, after fixation with a mixture of picric acid and formaldehyde, specimens are incubated at 45°C in PEG (mol. wt. 1500) with 3–5% (by volume) distilled water (two changes of 15 min each) with occasional stirring (Bosman and Go, 1981). The tissue blocks are left to harden at room temperature. After at least 1 hr, 6-μm-thick sections are cut at room temperature on a rotary microtome and mounted on glass slides coated with gelatin–chromealum. A single section is gently pressed onto each slide with a thin paint brush until it adheres slightly. The slide is heated at 45°C on a hot plate, and after the PEG has melted, a small drop of distilled water is applied around the tissue section with a paint brush to stretch the section. After careful removal of excess fluid with filter paper, all the slides are dried overnight in an incubator at 37°C.

Following a wash in phosphate-buffered saline (PBS) for 15 min to remove the remaining PEG, the sections are immunostained. They are then washed for 30 min in PBS and refixed with 2% glutaraldehyde for 20 min. After incubation for 30 min in a 0.005% diaminobenzidine (DAB) in 0.15 M Tris-HCl buffer (pH 7.6), the sections are treated for 2 min with the same solution but containing 0.01% hydrogen peroxide. The sections are postfixed with 2% OsO_4 for 30 min, dehydrated, and embedded by using the inverted capsule method (p. 119). The resin is polymerized at 37°C for 1 week. The blocks are separated from the slides by briefly heating the slides on a gas flame.

PEG is also useful for producing resin-free sections of plant tissues for high-voltage electron microscopy (Hawes et al., 1983). These sections show adequate contrast without poststaining because the resin is absent. Low contrast caused by the electron-scattering properties of the resins at high accelerating voltages is well known (Humphreys, 1976). PEG has been used to show the presence of a network of fine filaments in the cytoplasm in plant tissues (Hawes et al., 1983) and microtrabeculae in animal cells (Guatelli et al., 1982). The following method was used by Hawes et al. (1983) for embedding roots.

PEG solution is prepared by melting solid PEG 4000 at 60°C in distilled water. Drops of 20% PEG are added to the fixed tissues in distilled water over a 40-min period to bring the concentration to 10%. The specimens are transferred to 25% PEG, where they remain for 1 hr, followed by 12 hr in 50% PEG at 60°C, 2 hr in 100% PEG at 60°C, and 2 hr in 100% PEG 6000 at 60°C. These steps are carried out in a shaking water bath. Final embedding is accomplished in PEG 6000 in preheated flat molds, and blocks are solidified at room temperature. Limitations of this method include section fragmentation during the subsequent processing, shrinkage of the cytoplasm, and poor adherence of sections to the grids.

Another version of this method was used by Ross-Canada et al. (1983) for embedding nerve–muscle tissues. Improved preservation of these tissues was obtained by using DAB as a ligand binding OsO_4 to tissue components, and PEG was solidified in swirling liquid nitrogen.

Diethylene Glycol Distearate

Diethylene glycol distearate (DGD) can be used as an embedding medium for obtaining thin or thick embedment-free sections (Capco et al., 1984). DGD has several advantages over PEG: brittleness is reduced; sections can float on the water-filled knife trough; ribbons of sections can be produced; sections produce interference colors, allowing determination of their thickness; and sectioning is easy to perform. The following procedure was used by Capco et al. (1984) for the study of interphase and mitotic cells.

Fixed cells are dehydrated through ethanol and then through 2 : 1 and 1 : 2 mixtures of ethanol and n-butyl alcohol (BA) for 15 min each. The cells are immersed in four changes of BA for 15 min each and then transferred to 2 : 1 and 1 : 2 mixtures of BA and DGD at 60°C for 10 min each. (DGD is a solid at room temperature.) The cells are infiltrated with three changes of 100% DGD for 1 hr each at 60°C and then removed from the oven and allowed to harden. Thin or thick sections are cut by using glass knives at an angle of 10° and transferred to grids coated with collodion-carbon and 0.1% poly-L-lysine. These grids are stored overnight in a desiccator. The DGD is removed by immersing the grid in 100% BA at 23°C. Thin sections require three changes of BA for 1 hr each to completely remove the DGD. The grid is returned to 100% ethanol through a graded series of BA and ethanol mixtures, treated three times with 100% ethanol, and then dried through the CO_2 critical point.

If the objective is to study only a cell's cytoskeleton, the cell's soluble components are extracted in the following mixture for 3–10 min at 4°C:

NaCl	100 mM
PIPES buffer (pH 6.8)	10 mM
$MgCl_2$	3 mM
Sucrose	300 mM
Phenylmethyl sulfonyl fluoride	1.2 mM
Triton X-100	0.5%

This extraction is carried out before fixation. Although postfixation with OsO_4 is unnecessary because image contrast is formed by electron scattering of the embedment-free specimens, a brief postfixation (5 min) produces greater stability of the specimen.

Polyvinyl Alcohol

This procedure of embedding consists of infiltrating the specimens with an aqueous solution of polyvinyl alcohol (PVA), which is then concentrated to a hard gel (Muñoz-Guerra and Subirana, 1982). The gel is cross-linked with glutaraldehyde. Fixed tissue specimens are infiltrated with 1, 5, 10, and 20% aqueous solutions of PVA (mol. wt. 14,000; degree of hydrolysis is 99%) at room temperature for 30 min in each step. Two or three specimens are transferred to the bottom of a glass tube (6 mm diameter and 50 mm height) with one end sealed with a dialysis membrane. About 1 ml of 25% PVA solution is added to the tube, making sure that the specimens rest on the membrane surface.

The tube is partially buried in Aquacide and placed in an oven at 40°C for 2–3 days. The hard gel formed is cut into small cubes, each containing one specimen. These cubes are suspended above a 10% aqueous solution of glutaraldehyde in a small sealed vial for 2 days at 40°C. The cubes are immersed into the glutaraldehyde solution for 2 days, washed with distilled water, dried on silica gel, and mounted onto a resin or Lucite block for sectioning. Sectioning is easy; the highly hydrophilic character of PVA requires a fast speed of sectioning and a careful adjustment of the water meniscus in the trough. Carbon-coated grids should be used. Poststaining is accomplished in a relatively short time. Among the limitations of PVA are that a long time is needed to complete the embedding, thin sections may show holes, and there is poor embedding of certain specimens, such as nervous tissue.

Protein Aldehydes

Fixed specimens are washed with buffer and transferred to an albumin solution (0.2 g/ml of buffer solution) in a small tube (1–2 cm in diameter) with one end sealed with a dialysis membrane. The concentration of the albumin solution is raised to 100% by partially immersing the tube in Aquacide II (Calbiochem). The concentrated albumin, together with the specimens, is cut into narrow strips and cross-linked with glutaraldehyde (3%) over a period of 6 hr. After being washed with water, the cross-linked strips of albumin are dried in a desiccator. For thin sectioning, the strips are glued to the end of a Lucite or Plexiglas rod of appropriate size.

MIXED-RESIN EMBEDDING MEDIA

DER 332–DER 732

Mixture A
DER 332	7 ml
DER 732	3 ml
DDSA	5 ml
DMP-30	0.3 ml

Mixture B
DER 332	7 ml
DER 732	2 ml

DDSA	5 ml
DMP-30	0.28 ml
Mixture C	
DER 332	6 ml
DER 732	3 ml
DDSA	10 ml
DMP-30	0.38 ml

Before mixing, DER 332 is heated to 60°C to reduce viscosity. Mixture A is relatively soft and is recommended for embedding soft tissues. Mixture B yields a harder block and is suitable for embedding collagenous tissues. Mixture C is about the same hardness as mixture A but is tougher because of the greater content of DDSA, and thus more difficult to section. The hardness of these mixtures is controlled by the amount of DER 732 used: the larger the amount of this resin, the softer the final block.

Epon 812 (LX-112)–Araldite 502

Stock mixture
Epon 812 (LX-112)	31 ml
Araldite 502	25 ml
Dibutyl phthalate	4 ml
Final embedding mixture	
Stock mixture	8 ml
DDSA	20 ml
DMP-30	28 drops
or	
Benzyl dimethylamine	56 drops

All the components of these mixtures must be thoroughly mixed, shaking being better than stirring. The accelerator should be added just before using a mixture.

TABLE 2.7.

Block Hardness of Mixtures of Epon 812 and Araldite 502

Mixture[a]	Hard	Rigid	Soft
Epon 812 (LX-112)	21.37 g	15.88 g	15.88 g
Araldite 502	2.8 g	8.96 g	8.96 g
DDSA	28.8 g	26.34 g	26.34 g
Dibutyl phthalate	—	1.0 g	2.0 g

[a]Ten drops (0.18 g) of DMP-30 are added to 10 ml of each of the mixtures. Blocks of all the mixtures can be sectioned with a glass knife.

TABLE 2.8.

Composition and Properties of Embedding Mixtures

Properties	Mixture 1	Mixture 2	Mixture 3
Ingredients (ml)			
Epon 812 (LX-112)	25	62	—
DDSA	55	100	—
Araldite M	15	—	—
Araldite 506	—	81	50
Cardolite NC-513	—	—	25
Dibutyl phthalate	2–4	4–7	1–2
Relative hardness	Medium	Soft–medium	Soft–medium
Image contrast	High	Medium	Low
Tissue preservation	Good	Excellent	Excellent

Tables 2.7 and 2.8 give the composition and properties of embedding mixtures containing different proportions of Epon 812 and Araldite 502.

Epon 812–DER 736

DER	40 ml
Epon 812	10 ml
NMA	45 ml
DMP-30	1.4–2 ml

This mixture is less viscous than the final embedding mixture of Epon. The final hardness of the block can be adjusted by changing the ratio of DER 736 and Epon; a ratio of 4 : 1 is desirable.

Epon 812–Thiokol LP-8

Stock mixture

Epon 812	100 ml
NMA	85 ml
Thiokol LP-8	15–30 ml

Final embedding mixture

Stock mixture	100 ml
DMP-30	3 ml (20–30°C)
	or
	1.5 ml (50°C)

The polymerization is completed in 7 days at 20°C, in 4 days at 30°C, and in 24 hr at 50°C. The greater the amount of Thiokol LP-8, the softer the final block.

The viscosity of this mixture is lower than that of the final embedding mixture of Epon without Thiokol LP-8.

Epon 812 (LX-112)–Maraglas

Epon mixture

Epon 812 (LX-112)	21.88 g
DDSA	12.60 g
NMA	11.90 g
DMP-30	0.7–0.8 g

Maraglas mixture

Maraglas 655	36 ml
DER 736	8 ml
Dibutyl phthalate	5 ml
BDMA	1 ml

A 1:1 mixture of Epon and Maraglas is recommended; a higher proportion of Epon results in a harder block. This mixture is less viscous than the Epon–Araldite mixture.

ERL 4206–Quetol 653

ERL 4206	23 ml
Quetol 653	14 ml
NSA	63 ml
S-1	0.5 ml

Methacrylate–Styrene

n-Butyl methacrylate (stabilizer removed)	70 parts
Styrene (stabilizer removed)	30 parts
Benzoyl peroxide	1 part

Rigolac–Styrene

Rigolac 70F	3 parts
Styrene	7 parts
Benzoyl peroxide	1%

Silicone (Rhodorsil 6349)–Araldite CY 212

Araldite CY 212	20 ml
Rhodorsil 6349	10 ml

Araldite HY 964   30 ml
Araldite DY 964   1.2 ml

The first two components are mixed thoroughly for 20 min before adding the third component. A similar mixing is carried out before adding the fourth component. Polymerization is completed overnight at 95°C.

Vinylcyclohexene Dioxide–*n*-Hexenylsuccinic Anhydride–Araldite RD-2 (Mascorro *et al.*, 1976)

Vinylcyclohexene dioxide (epoxy resin)   5 g
n-Hexenylsuccinic anhydride (hardener)   10 g
Araldite RD-2 (modifier)   0.75 g
Dimethylaminoethanol (catalyst)   0.15 g

An increase in Araldite RD-2 will result in a softer block. The following modification is especially useful for standard or rapid embedding of plant specimens having large vacuoles and/or hard, thick cell walls impregnated with a variety of substances or covered with mucilage (Oliveira *et al.*, 1983).

Vinylcyclohexene dioxide   10 g
n-Hexenylsuccinic anhydride   20 g
Araldite RD-2   0.3 g
Dimethylaminoethanol   0.3 g

The viscosity of this mixture is 20 cps at 25°C, whereas those of Spurr's and Epon 812 mixtures are 60 cps and 200 cps, respectively. Dehydration is carried out with methanol, and the use of propylene oxide as a transitional solvent is unnecessary. Polymerization is accomplished at 60°C for 6–12 hr.

EMBEDDING

Viscosity of Embedding Media

The penetration of a specimen by the embedding media depends in part on their viscosity. The rate of penetration varies inversely with the size of the molecule and the viscosity of the embedding medium, and an increase in temperature lowers the viscosity until polymerization sets in. The viscosity in centipoise (cps) at 25°C of various media, ranging from high to low, follows. It is worthwhile to remember that the viscosity of water at 20.2°C is 1.0.

Araldite 502   3000
Araldite CY 212   1300–1650
DER 334   500–700
Maraglas 655   500
DDSA   290

NMA (MNA)	175–275
Epon 812	150–210
LX-112	150–180
Poly/Bed 812	150–180
NSA	117
Durcupan	100
DER 732	55–100
LEMIX	70.8
DER 736	30–60
Glycol methacrylate	50.5
EMbed	15–20
Quetol	15
LR White	8.0
ERL 4206 (VCD)	7.8
Styrene	0.7

Agar Preembedding (Encapsulation)

Preembedding in agar is desirable for particulate specimens because their pellets disintegrate during processing. It is also useful for orientation of specimens. Since specimens are easily visible in agar, they can be oriented in the capsule for resin embedding. An 8% agar solution is prepared by adding 2 g of dry powdered agar to 23 ml of distilled water in a small flask. The mixture is heated while stirring by placing the flask in a water bath (46°C). Buffers or saline solutions can be used instead of water for preparing the agar solution.

Particulate specimens or microorganisms are fixed with aldehydes and then concentrated by centrifugation in a centrifuge tube. This suspension is mixed with an equal volume of the 8% agar solution at 45°C, resulting in a 4% agar solution. The pellet–agar mixture is hardened into a block by placing the tube in a refrigerator. The block is removed with a needle after adding a few drops of 70% ethanol. The cubes cut from this agar block can be postfixed with 1% OsO_4, dehydrated, and embedded in a resin. If the agar block is excessively soft, it can be hardened with formaldehyde. If necessary, agar can be stained with basic fuchsin to make small cubes more visible.

Bovine Serum Albumin Preembedding

Pellets can be preembedded in bovine albumin without raising the temperature. A 2% solution of bovine albumin is prepared in 0.05 M Tris buffer (pH 7.5) and filtered through a 0.45 μm pore-size Millipore filter (Shands, 1968). The pellet of fixed specimens is suspended in 0.25 ml of 2% bovine albumin and then transferred to a small cellulose centrifuge tube. One drop of 25% glutaraldehyde is added and mixed with the suspension and immediately centrifuged for 5 min to form an opaque gel. The tube is cut into 1-mm-thick slices and placed on filter paper to remove excess moisture. Rims of cellulose are cut to release the disks of gel, which are dehydrated and embedded by routine procedures. Postfixation with OsO_4 is undesirable because a reaction takes place with albumin.

Fibrin Clot Preembedding

Fibrin clot can be used for encapsulation without the need for a higher temperature. The clot is more permeable to dehydrating and embedding solutions than is agar. The use of cacodylate buffer is not desirable. Commercial thrombin solution is diluted with distilled water and stored at a concentration of 1000 units/ml in 1-ml aliquots in vials (Furtado, 1970). This solution is further diluted to a concentration of 50–100 units/ml prior to use. Fibrinogen solution is prepared by dissolving 320 mg of fibrinogen, 160 mg of sodium citrate, and 850 mg of sodium chloride in 100 ml of distilled water.

The pellet is fixed and washed in a centrifuge tube, and 0.2 ml of fibrinogen solution is added. After being stirred, it is centrifuged to concentrate the cells. An equal volume of thrombin solution is slowly added to the supernatant. A sharpened applicator stick is twisted in the fibrinogen solution until a clot becomes apparent. The stick is then twisted in the pellet until the cells are entrapped on the stick. The stick is used to transfer the clot through the remaining preparatory procedures. The clot is cut into small segments when in 70% ethanol.

Direct Pelleting of Cells (Achong and Epstein, 1965)

Soft glass capillary tubes (2-mm bore, 8 cm long) are sterilized in batches by using dry heat. A sterilized tube is held close to horizontal with sterile forceps and dipped into a tilted culture vessel so that the cell suspension flows up inside. When the tube is three-quarters full (containing 0.18 ml) (Fig. 2.5), the empty end is sealed and rounded off in a small flame. The tube is placed, sealed end down, on a cotton-wool support in an horizontal centrifuge and subjected to 100 g for 2 min. The tube is broken short about 2.5 cm from the bottom, and a syringe is used to withdraw the supernatant from above a small, loose pellet of cells. The tube containing the pellet is filled with ice-cold fixative. After fixation, the tube is centrifuged at 9000 g in an horizontal centrifuge, resulting in the formation of a firm pellet. The fixative is removed with a syringe, the base of the tube is snapped off, and the firm pellet (1 mm^3) is expelled with a vaccine teat into the next reagent.

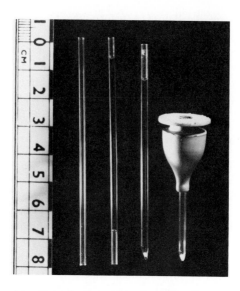

Fig. 2.5. Capillary tubes (2-mm bore) used for cell sampling shown at various stages during the preparation of a pellet. From left to right: empty sterile tube; tube filled with an appropriate amount of cell suspension; filled tube with sealed lower end after light centrifuge to form a cell pellet; centrifuged tube with pelleted cells above broken-off base; and vaccine teat on cut-down open end. (From Achong and Epstein, 1965.)

Pelleting of Cells in Agar

The cells are harvested into 1.0–1.5 ml of Dulbecco's balanced salt solution, either by scraping them off with a rubber policeman or by using glass beads, and transferred into a conical glass tube (60 × 7 mm) (Gowans, 1973). It is convenient at this stage to estimate the total number of cells in the sample, either by using a hemocytometer or by considering the average numbers of cells harvested from the culture.

The cells are fixed in the glass tube, rinsed in buffer, and centrifuged at 250 g for 5 min. This procedure ensures adequate separation of cells from the supernatant and a cellular deposit easily miscible with agar. The supernatant is removed with a Pasteur pipette, and 4% agar, kept molten at 60°C, is added in the correct volume from a 1.5-mm-bore capillary tube (Table 2.9). A vaccine teat on the same tube is used to rapidly and thoroughly mix the cells and agar by drawing up and expelling them until the cells are uniformly distributed throughout the agar. The mixture is drawn up the capillary tube and allowed to solidify. The mixing must be performed rapidly in view of the small volumes of agar used. The solidified agar can be expelled from the capillary tube and cut into small pieces for further processing and embedding. If less than 2 × 10^6 cells are available, the centrifugation method of Achong and Epstein (1965) is recommended, although it is time-consuming.

TABLE 2.9.

**Proportions of Agar and Cells Required to Yield Optimal
Cell Concentrations for Sectioning**

Total number of cells	Volume of agar (cm³)	Length of capillary tube filled (mm)
2×10^6	0.006	4
3×10^6	0.0075	6
4×10^6	0.01	10
6×10^6	0.018	15
8×10^6	0.025	20
10×10^6	0.03	25

Pelleting of Cells in Agarose (Yuan and Gulyas, 1981)

This method utilizes agarose instead of agar as the initial embedding medium prior to postfixation with OsO_4 and embedding in a resin. Agarose is a neutral constituent of agar, has a low gelling temperature, and is a preferred medium for fragile dissociated cells. This method is relatively rapid and gentle because repeated centrifugations are eliminated and as few as 1×10^4 cells can be processed without substantial loss. Heat-sensitive cell components seem to remain undamaged.

Cells are fixed with 2.5% glutaraldehyde in 0.1 M sodium cacodylate buffer (pH 7.4) for 1 hr, pelleted by centrifugation at 100 g for 5 min at room temperature, and washed three times in the same buffer. Cells suspended in buffer are brought to 37°C in a water bath and thoroughly mixed with an equal volume of 3% agarose. The cells are transferred to conical BEEM capsules, using Beckman Micro-centrifuge test tubes as holders. The cells are centrifuged for 5 min at 33–37°C and then cooled on ice for 5–10 min to gel. The gelled agarose–cell blocks are removed gently by cutting the BEEM capsules open with a razor blade. The tips of the agarose blocks containing the cells (Fig. 2.6) are cut off and further sliced into 1-mm³ or smaller cubes for further processing.

General Embedding Methods

Capsule Embedding

Predried and prelabeled polyethylene or gelatin capsules are placed in a holder. Capsule holders can be made by drilling holes in a cardboard lid or in a 5 × 10-cm board; capsule holders are also available commercially. A very small drop

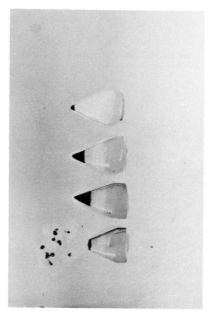

Fig. 2.6. Three different concentrations of dispersed spleen cells embedded in 1.5% agarose blocks. From top to bottom: 1×10^4 cells; 5×10^4 cells; 1×10^5 cells; agarose block with cells is cut into smaller cubes. (From Yuan and Gulyas, 1981.)

of embedding mixture is placed in the capsule, and then a toothpick is used to set the specimen at the bottom of the capsule. A disposable 5- to 10-ml pipette or syringe (without the needle) is used to fill the capsule about half full with the embedding mixture. If necessary, the specimen is gently guided back with a toothpick to the bottom of the capsule. The holder with the capsule is placed in an oven for 48 hr at 60°C for polymerization. Polyethylene capsules can be stripped off by making two longitudinal cuts in the opposite sides of the capsule with a razor blade. Gelatin capsules can be removed by soaking in hot water or by chipping with a razor blade.

Embedding of Buoyant Specimens

A strip of adhesive tape is placed across a glass plate of convenient size. The infiltrated tissue specimens are positioned about 2 cm apart on the tape, with the area to be sectioned facing down. The capsules are completely filled with the embedding mixture, are overturned exactly over the specimens, and are then centered. The glass plate with the capsules is transferred to an oven for 48 hr at 60°C. The tape easily comes off the glass plate, and the latter is easily detached from the capsules.

Flat Embedding

Tissue specimens in batches can be embedded in flat silicone rubber molds that are commercially available. Alternatively, specimens can be embedded in shallow, flat containers such as polyethylene weighing trays, minicube ice trays, small disposable polyethylene beakers cut off 1 cm from the bottom, small flat-bottomed vials, vinyl cups, and caps of vials. A thin (1–2 mm) layer of embedding mixture is made in the flat mold, and then additional embedding mixture is poured into the mold to a maximum level of about 8 mm. Infiltrated tissue specimens are transferred to the mold, with adequate distance maintained between them. A label is placed near each specimen, and the flat mold is placed in an oven for polymerization.

Vice-type chucks are available for accepting flat blocks. Alternatively, a segment of the polymerized resin containing the specimen is cut out with a coping saw and glued with Eastman 910 Adhesive in the desired orientation to the end of a short Lucite rod. Lucite or Plexiglas rods with the appropriate diameter to fit in the standard collet-type specimen chuck are commercially available. The surfaces to be glued together are made smooth by rubbing them with sandpaper. A tiny drop of the glue is sufficient to achieve a stable binding.

Labeling

A labeled piece of paper (an index card) 2.5 cm long is rolled around the wooden end of the dissecting needle (which acts as a carrier) and lowered in the capsule (Fig. 2.7A and B). When the carrier is pressed downward and laterally, the label can be fit to the curvature of the side walls of the capsule (Fig. 2.7C). The label is placed in the capsule before embedding is carried out. For flat embedding, a small piece of paper is placed at the bottom of the embedding tray and near the specimen prior to polymerization. The label should be prepared by using a sharp pencil or India ink.

Rapid Embedding

Fixation, dehydration, infiltration, and embedding of biopsy or surgical specimens can be completed in 1–4 hr. The following method is designed for diagnostic laboratories. The size of the specimen should be less than 1 mm^3.

Method 1

The method developed by Hayat and Giaquinta (1970) for processing normal or diseased animal or plant tissues is as follows:

1. Fixation with 4% glutaraldehyde in 0.1 M 30 min
   cacodylate buffer (pH 7.3) at room
   temperature

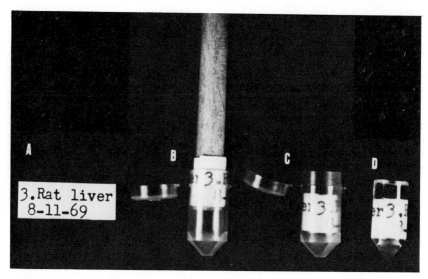

Fig. 2.7. Permanent labeling in the gelatin capsule. (A) Typed label. (B) Label rolled around the wooden end of a dissecting needle being lowered into the capsule. (C) Label fitted to the curvature of the side walls of the capsule. (D) Polymerized specimen block with embedded label. Labels should be written with a pencil, not with ink.

2. Rinse in buffer (twice)	1 min each side
3. Postfixation with 2% OsO_4	30 min
4. Acetone (50%)	4 min
5. Acetone (70%)	4 min
6. Acetone (95%)	4 min
7. Acetone (100%)	4 min
8. Resin mixture plus acetone (1 : 1)	15 min
9. Resin mixture (2 changes)	10 min each
10. Embedment in fresh resin mixture at 100°C	1 hr

All steps except the last one are carried out with continuous agitation, and silicone rubber molds should be used for embedding because BEEM capsules melt when polymerization temperatures approach 100°C. The entire process is completed in less than 3 hr.

Method 2

The method developed by Todd and Burgdorfer (1982) is completed in 1 hr. Very small tissue blocks, including biopsy specimens, are fixed with a mixture of 4% formaldehyde and 2% glutaraldehyde in 0.05 M cacodylate buffer (pH 7.4) containing 0.002 M $CaCl_2$ and 0.02% trinitrocresol for 12 min at 37°C.

After a rinse for 2 min in the buffer, the specimens are postfixed with 2% OsO_4 in 0.01 M cacodylate buffer (pH 7.4) for 10 min at room temperature. Following two rinsings in distilled water, the specimens are dehydrated for a total period of 10 min in two changes of 4 ml of 2,2-dimethoxypropane (DMP). The embedding mixture is prepared just prior to use by adding 3 drops of vinyl cyclohexene to 27 drops of cyanoacrylate in a BEEM capsule. The capsule is capped, and the contents are mixed by vigorous shaking for 5 sec. The specimen is transferred to the capsule, shaken, and heated for 5 min at 55°C; polymerization follows for 10 min at 85°C. In my opinion, the use of phosphate buffer with DMP is undesirable, since these two reagents form artifactual precipitates.

Specimen Orientation

Several methods are presented here so that the reader may choose or modify the one that is most suitable for the specimen under study:

1. Specimens can be preembedded in 2% agar blocks that are subsequently placed in the desired plane in capsules and embedded in the resin.

2. Aldehyde-fixed tissue specimens are oriented in 2% liquid agar maintained at 45°C in a water bath, which is then allowed to solidify in a refrigerator. The solidified agar block containing the specimen is postfixed with OsO_4, dehydrated, and embedded in a standard resin.

3. A strip of exposed film is lowered into the embedding capsule to support and orient the tissue specimen. The film is insoluble in the embedding resins and does not interfere with trimming or sectioning of the block.

4. One end of a minute pin is inserted a short distance into one end of the tissue specimen, and the other end is passed through the center of a 6-mm disk made from an index card. This assembly is lowered into the capsule containing the fresh embedding resin. Since the disk floats in the resin, a relatively precise orientation of the tissue can be achieved by raising or lowering the pin. One must make certain that the pin does not interfere with trimming and, especially, with sectioning.

5. The cap and the pyramidal end of BEEM capsules are severed with a razor blade. A few drops of embedding medium are placed into the cap, and then tissue specimens are manipulated into the desired position. The remaining body of the capsule is inserted carefully into the cap and filled with the embedding medium. The modified capsule is placed in the oven.

6. Tissue specimens are embedded within a sphere of clear plastic (acrylic bead), which is properly aligned, and then cemented with a drop of ethylene dichloride to a plastic rod of suitable size that fits into the chuck of the ultramicrotome.

7. Polymerized slotted castings of the embedding resin can be used for holding

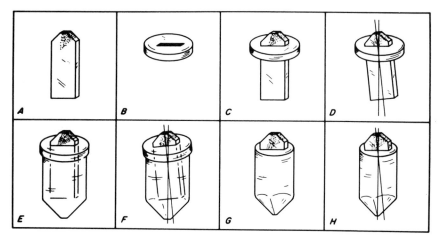

Fig. 2.8. Method for obtaining orientation of specimens embedded in flat molds. See text for details. (From Hwang, 1970.)

thin tissues in position. A slit of the desired size is made at one end of the castings, and the tissue specimen is held in it with a drop of fresh resin. This casting containing the specimen is placed in a new capsule containing a small drop of resin and pressed to expel all air.

8. A piece of aluminum foil is placed obliquely in an embedding capsule, thus dividing it into two unequal parts. The capsule is filled with the embedding resin and polymerized by heat. The polymerized block is broken apart at the plane of the foil. The lower piece (the smaller one) is placed at the bottom of a new capsule containing fresh resin. One tissue specimen is placed against the slant of the polymerized block in the desired direction. Care should be taken to avoid air bubbles at the junction of the new and old embedding resin.

9. Tissue specimens are embedded in a flat mold and then polymerized. A rectangular flat block is removed from the mold (Fig. 2.8A). The cap of the capsule is cut off, and a hole as large as the flat block is made in the cap (Fig. 2.8B). A part of the capsule may have to be trimmed if it is longer than the desired length. The flat block is inserted into the hole by pushing it from the inside of the cap (Fig. 2.8C). Some of the inaccurate orientation can be corrected by changing the block–cap angle (Fig. 2.8D). The capsule is filled with the embedding medium, and the cap with the flat block is fitted over the capsule (Fig. 2.8E). At this stage the orientation can be further corrected by changing the angle of the cap and/or the position of the flat block (Fig. 2.8F). The capsule is transferred to an oven for polymerization, and the final block is obtained (Fig. 2.8G and H).

10. Simultaneous orientation and embedding of thin, flat specimens (e.g., cell

monolayers and fungal mycelia) can be accomplished by using flat-bottom containers modified from standard gelatin capsules. Such containers are obtained by exposing the lower part of the capsules to water for 5 min. At 3 min after they are removed from the water, the lower parts are pressed into flat ends. Specimens embedded in such flat-bottom containers can be cut parallel to the flat surface as well as to the long axis of the specimen. However, embedding in flat molds is preferred when one needs sections perpendicular to the flat surface of a thin specimen.

Embedding of Hard Plant Tissues (Warmbrodt and Fritz, 1981)

The following method is useful for embedding "hard to embed" plant tissues for light and electron microscopy. The technique requires 6 days to complete. Tissue preservation does not seem to be adversely affected. The apparatus (Fig. 2.9) consists of a T-shaped piece of glass tubing (inner diameter 10 mm) with one end closed permanently by melting, another closed by a plastic valve stem and screw cap (this part of the tubing is 16 cm long), and the third (the side leg, 6 cm long) open for the convenient addition or removal of tissue and embedding medium. The plastic valve stem can be screwed in, thus closing off one end of the glass tubing and forming a closed chamber (Fig. 2.9, the pressure chamber) of 8 ml capacity.

After fixation and dehydration, the tissue is infiltrated with propylene oxide under high pressure for 1 hr by placing it in the pressure chamber, which is filled with propylene oxide. The plastic valve stem is screwed in, leaving a small

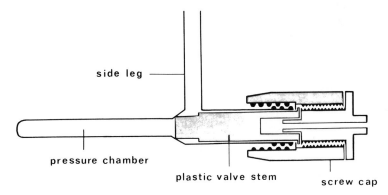

Fig. 2.9.   Sectional diagram of modified Quickfit Rotaflo device used to overcome difficulties encountered during embedding of hard-to-embed plant tissues. It uses a valve and low heat to generate high pressure to aid in the infiltration and embedding of tissues. (From Warmbrodt and Fritz, 1981. © 1981 The Williams & Wilkins Co., Baltimore.)

opening between the pressure chamber and the side leg. To prevent excess evaporation of the propylene oxide, the side leg should be stoppered with a rubber cork. The pressure chamber (which is now completely filled with propylene oxide) is placed in a beaker filled with water at 20°C. After 2 min the chamber is removed from the water bath and closed immediately by screwing in the plastic valve stem, thus creating an overpressure inside the chamber. The rubber cork is removed from the side leg and the excess propylene oxide is discarded. The apparatus with the sealed pressure chamber containing the tissue and propylene oxide is placed in an oven at 26°C for 1 hr. At this temperature a high pressure is formed inside the chamber because the propylene oxide cannot expand.

When the plastic valve stem is unscrewed, the propylene oxide in the pressure chamber is poured out through the side leg and replaced with a 1 : 1 mixture of propylene oxide and resin. The high pressure is generated by cooling the pressure chamber to 20°C with an open valve and then closing the valve and warming the chamber in an oven at 26°C. After 6–8 hr of infiltration, the 1 : 1 mixture is replaced with a 1 : 3 mixture of propylene oxide and resin and placed under the same high pressure as described for the 1 : 1 mixture. After 6–8 hr the tissue is infiltrated with pure resin under high pressure for 24 hr using the same procedure as described for the propylene oxide–resin mixture.

The tissue specimens are transferred to 25-ml liquid scintillation vials (plastic) containing 10 ml of fresh resin. The vials are placed in an oven for 24 hr at 50°C and kept on a slow rotator. Fresh $CaCl_2$ or silica gel is placed in the oven. After 24 hr, 4–6 drops of DMP-30 are added to each vial. The result is a final concentration of 1–1.5% DMP-30 in the resin mixture. The vials are hand-held and slowly rotated for 4 min to ensure uniform mixing of the DMP-30 and resin. The vials are placed on a rotator in an oven at 50°C for 4–6 hr. The tissue and resin are poured into suitable molds and placed in an oven at 60°C for 24 hr and then at 80°C for 2 days.

Orientation and Embedding of Small Specimens (Trett, 1980)

The following method prevents the loss of small specimens (e.g., nematodes, <1 mm) during processing and saves time. A small amount of 1–2% pure agar is poured onto glass slides to obtain a film 2 mm thick. A scalpel or a razor blade is used to cut blocks of agar. The size and shape of these blocks vary depending on the embedding requirements, but in general they should come to a point at one end (Fig. 2.10). A small slit, trough, or hole is made just before the tip of this point. The specimens, which have been in the primary fixative for half of their full time, are carefully placed into the cut in the block and oriented with the aid of a dissecting microscope. A small drop of warm 2% agar solution is placed on

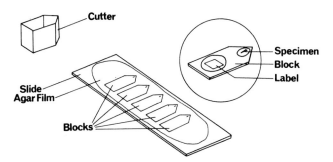

Fig. 2.10. Procedure for handling and orienting small specimens. (From Trett, 1980.)

top to retain the specimen. The entire block is lifted off the slide with a scalpel blade and placed in the primary fixative for the rest of the fixation time.

The agar block is not affected by aldehydes and OsO_4 and can be carried through dehydration, infiltration, and embedding. If the flat agar is shaped to fit exactly within a BEEM capsule with the specimen at the tip, routine handling and sectioning of small specimens in any required plane become simple processes. Small labels can be included in the agar blocks for identification.

Orientation and Embedding of Single-Cell Organisms (Janisch, 1974)

The following procedure facilitates oriented embedding of individual organisms (e.g., paramecium) in various media for both light and electron microscopy. This procedure can also be used in embryology, histology, and cytology. The fixed specimen is placed in a droplet of water on a glass slide (Fig. 2.11A) and covered with a drop of 2% agar at 40°C (Fig. 2.11B). After cooling, excess agar is trimmed off with a microscalpel under a dissecting microscope. The specimen encapsulated in a small agar cube can be oriented in any desired position on the glass slide under the dissecting microscope. Another drop of water can be added to prevent the agar cube from drying.

A narrow strip of black paper beveled toward the same side at both ends is placed close to the agar cube (Fig. 2.11D). The cube and the end of the paper strip are covered with a drop of 2% agar at 40°C (Fig. 2.11E). After cooling, excess agar is again removed, keeping the narrow zone connecting the cube with the strip of paper (Fig. 2.11F). The point of the paper strip can be used to indicate the position of the specimen. The position of a specimen in an opaque medium relative to the paper strip is indicated by the position of the bevel on the free end of the strip. The paper strip also serves as a tab for handling the specimen in further processing. Before dehydration, either a schematic drawing can be made of the position of the specimen with regard to the point of the paper

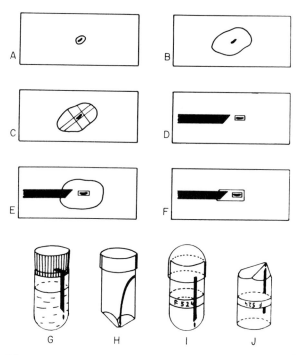

Fig. 2.11. Oriented embedding. (A) A specimen is placed in a droplet on a glass slide. (B) The specimen is covered with a drop of 2% agar at 40°C and then cooled. (C) Excess of gelled agar is cut off. (D) A beveled strip of black paper is placed close to the small agar cube with the specimen in the required position. (E) The end of the black strip, together with the cube, is covered with a drop of 2% agar at 40°C. (F) Excess of gelled agar is removed (see also Fig. 2.12A). (G) The strip-specimen is transferred for dehydration and embedding. (H) Polymerization by heat is carried out in BEEM capsules. (I) Polymerization by UV takes place in gelatin capsules. The strip is held along the wall by a curled strip of paper with a number designating the block (see also Fig. 2.12B). (J) The gelatin capsule is trimmed away (see also Fig. 2.12C). (From Janisch, 1974. © 1974 The Williams & Wilkins Co., Baltimore.)

strip or a photomicrograph can be taken at a higher magnification or in phase contrast (Fig. 2.12A). The paper strip along with the specimen is dehydrated, infiltrated, and then embedded in a resin in a BEEM capsule with a pyramid bottom (Fig. 2.11H). The agar cube fits in the point of the pyramid; trimming is unnecessary.

Embedding of Individual Cells

Single cells (e.g., *Paramecium* or *Discophyra*) can be encapsulated in a thin albumin film (Walker and Roberts, 1982). The advantages of this method are

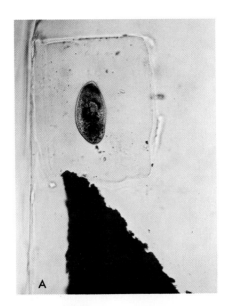

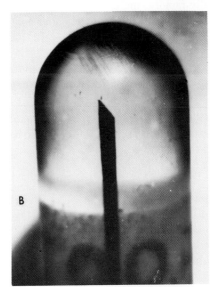

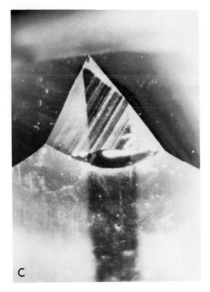

Fig. 2.12.   Oriented embedding of *Paramecium* where the plane of section is perpendicular to the anterioposterior axis of the cell. (A) Photomicrograph of the cell embedded in an agar cube with a strip of black paper. (B) The same cell in a polymerized block at a low magnification. (C) The trimmed block. (From Janisch, 1974. © 1974 The Williams & Wilkins Co., Baltimore.)

that particular cells from a culture can be selected, individual cells can be processed, and cells can be observed at all stages of preparation. Two or three drops of 10% bovine serum albumin (BSA) (prepared in culture medium used for the cells under study) are placed on a clean, uncoated glass slide. Cells are transferred to the slide with a micropipette and gently stirred into the BSA drops. An equal volume of 2% glutaraldehyde (prepared in culture medium) is added and mixed with the cells and BSA, resulting in the formation of a thin layer of the BSA–glutaraldehyde. Slides are supported on glass rods in Petri dishes containing filter papers soaked in 2% glutaraldehyde.

Slides are left in the Petri dishes for 1 hr and then viewed under a dissecting microscope. Small pieces of the resulting gel containing the cells are cut out with a scalpel and transferred into small vials containing 1% OsO_4 in culture medium. After postfixation for 1 hr, the samples are dehydrated in ethanol and propylene oxide and left overnight in a 1 : 1 mixture of propylene oxide and Spurr resin. The samples are transferred to fresh Spurr resin for 4 hr to complete infiltration and then surface-embedded via the following procedure.

The samples are placed in a drop of Spurr resin on a 1-cm strip from a glass slide. BEEM capsules are filled with resin and the pieces of slide are rapidly inverted over these capsules. Then, equally rapidly, the capsule and glass are inverted, leaving the specimen in position on the glass. Immediately after polymerization, the hot blocks with attached glass are placed into crushed solid CO_2 to remove the glass from the block.

Embedding of Ascomycetes (Cell Walls)

The ascus and spore walls of Ascomycetes are difficult to section owing to poor embedding in an Epon medium of average hardness. This difficulty can be overcome by using the following medium (Merkus et al., 1974):

ERL 4206	10 g
DER 736	6 g
Nonenyl succinic anhydride	26 g
DMP-30	0.4 g
Dibutylphthalate	0.8 g

This medium is infiltrated into the specimens via acetone after dehydration with ethanol. Polymerization is accomplished for 12 hr at 70°C. The diamond knife that is used has a clearance angle (β) of 2° and an effective knife angle (α) of 58°; the sections are cut at a speed of 0.5–1.0 mm/sec.

Embedding of Bacteria and Tissue Culture Cells (Narang, 1982)

The following method is useful when small amounts of cells (10^5–10^6 cells/ml) are available. Since cells can be processed without centrifugation at each step, loss of cells is avoided. Bacteria grown on agar plates are flooded with phosphate-buffered saline (PBS), and the cells are suspended in 1 ml of PBS with a wire loop. The suspension is centrifuged in a glass tube (5 × 1 cm) at 1200 g for 15 min in a bench centrifuge. The pellet is resuspended in 1 ml of PBS to give 5 × 10⁹ cells/ml, and 0.5 ml is transferred into the 8-mm truncated cone of a polypropylene embedding capsule, which is placed in a 5 × 1-cm glass tube and centrifuged at 2100 g for 20 min. The supernatant is removed and replaced by an equal amount of 4% glutaraldehyde in 0.2 M PBS (pH 7.2).

All further procedures are carried out without disturbing the pellet in the capsule. Complete removal of each solution before addition of the next is achieved by pipetting off the supernatant fluid and finally draining the capsule on filter paper for 1 min. After 1 hr at room temperature, the fixed pellet is rinsed three times with PBS; then the tube is filled with 1% Dalton's OsO_4. After postfixation with OsO_4 for 90 min at room temperature, the pellet is rinsed three times with PBS. Dehydration is carried out for 15 min in each of 50, 70, and 90% ethanol series and finally in two changes of 100% ethanol. The pellet is exposed to two changes of propylene oxide, 10 min each time, and then left in a mixture (2 : 1) of propylene oxide and a resin overnight. Further infiltration follows for 8 hr in a mixture (1 : 1) of propylene oxide and a resin. After two changes of pure resin for 18 and 8 hr, respectively, fresh resin is added, and the embedding capsule is left at 60°C for 48 hr. The same technique can be applied to pellets derived from broth cultures of bacteria, suspensions of tissue culture cells, or cell fractions.

In Situ Embedding of Cells Grown on Millipore Filters (Codling and Mitchell, 1976)

The following method facilitates embedding tissue culture preparations (grown on Millipore filters) in Araldite. Accurate light microscopic localization of areas suitable for ultrathin sectioning is possible. A drop of concentrated cell suspension is placed on a Millipore filter (GSWP 013.00) in a Petri dish, and cells are allowed to settle and adhere to the filter for 5–10 min. Sufficient fresh tissue culture medium is added to make up the volume in the dish to 5 ml. The culture medium is drained after culture and replaced by 2.5% glutaraldehyde in 0.1 M phosphate buffer (pH 7.4). Fixation for 1 hr at 4°C is followed by thorough washing (2.5 hr) in four changes of buffer at 4°C and postfixation with 1% OsO_4 for 1 hr at 4°C.

The Millipore filters are transferred to a screw-top glass bottle (5–7 ml) and dehydrated in ethanol and then in propylene oxide (two changes of 2 min each). Longer exposure time will lead to fragmentation and eventual dissolution of the filters. The filters are embedded in a thin layer (2–3 mm thick) of accelerated Araldite in an aluminum foil container (Fig. 2.13). After polymerization, the Araldite disk including the filter is easily removed from the container. The filter

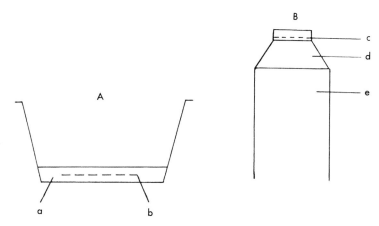

Fig. 2.13. *In situ* embedding of cells. (A) Millipore filter embedded in resin: a, resin; b, Millipore filter. (B) Resin-embedded Millipore filter supported on a larger resin block: c, Millipore filter in resin; d, ester wax; e, resin block. (From Codling and Mitchell, 1976.)

is examined through an inverted microscope, and the surface to which the cells are attached is identified. The cells required for ultrathin sectioning are marked accurately with a felt-tip pen. A jeweler's saw is used to cut this area as a 1- to 2-mm^2 block from the Araldite disk. This Araldite block is mounted on a gelatin capsule filled with polymerized Araldite, using ester wax, so that the cells are oriented as required (Fig. 2.13). When the ester wax is solidified, thick (1 μm) sections are cut until cells are visible on thionin-stained sections. If orientation is satisfactory, ultrathin sections are cut.

In Situ Embedding of Cell Monolayers on Untreated Glass Surfaces (Chien, 1980)

The following is a heat separation technique that, when used together with a silicone-rubber embedding mold (E. M. Specialities Co., 129 W. Orange St., San Gabriel, California), allows cell monolayers to be transferred from an untreated cover glass or glass slide to preshaped tissue blocks. The mold has two types of block configurations: round and rectangular (Fig. 2.14). Each round well forms a standard-size block, 11 mm in depth, with a flat, smooth base. Blocks formed within the rectangular wells have a 4 × 16-mm surface area with the same 11-mm depth. The periphery of this mold forms a depression that accepts a 1 × 3-in. glass slide.

Cell cultures grown on an untreated 22 × 22-mm cover glass are fixed, dehydrated, and infiltrated with a resin. The wells of the mold are slightly overfilled with deaerated epoxy resin. Any area of interest on the cover glass can be embedded (cell side down) onto the proper well. A glass slide is placed atop the cover glass. The resin is polymerized in an oven at 90°C for 2 hr. After cooling to room temperature, the block and the supporting glass are removed from the mold as one unit. This unit is placed on a hot plate (100°C) for 15 sec; the block is then easily separated from the cover glass by using a gradual but firm pressure.

Trimming is best carried out under a dissecting microscope illuminated from beneath. The preshaped round block can be used directly for horizontal sectioning along the plane of cell growth. The rectangular block provides both horizontal and vertical sectioning of the cell monolayer. Free-floating cells are spun onto a glass slide with the Cytospin centrifuge. Cells are deposited in a fixed position on the slide. The slide is placed (cell side down) on the mold. The single round well to the right-hand side of this mold (Fig. 2.14) captures the entire group of cells within a single block.

In Situ Embedding of Virus for Immunoelectron Microscopy (Ohtsuki et al., 1978)

The following method is useful for *in situ* embedding of cells grown in BEEM capsules for immunoelectron microscopy of oncornaviruses. Virus-producing cells are diluted to 4×10^4/ml. After the BEEM capsules are filled with cell suspension, the cells are laid in an inverted position in a Petri dish and incubated for 1–2 days at 37°C. Growth of cells on the lids is checked under an inverted microscope.

The pointed tips of BEEM capsules are cut away and then fastened to a Petri dish with masking tape. Prefixation is carried out with 1% acrolein in 0.1 M phosphate buffer (pH 7.4) for 10 min at 4°C. Cells are washed three times in phosphate-buffered saline (PBS) and incubated for 30 min at room temperature

Fig. 2.14. A silicone rubber mold for monolayer cell embedment. (From Chien, 1980.)

in 30–50 μl of diluted (1 : 10 to 1 : 64) antiserum or with similar amounts of ferritin- or peroxidase-conjugated antiserum. Cells are fixed with 3% glutaraldehyde in 0.1 M phosphate buffer (pH 7.4) for 30 min at 4°C.

The cells for immunoferritin are postfixed with buffered 2% OsO_4 for 30 min at 4°C. For immunoperoxidase reaction, cells are washed briefly with 0.2 M Tris buffer (pH 7.6) and incubated in an incomplete solution of DAB (2 mg in 10 ml of 0.2 M Tris buffer) for 15 min at room temperature. Further incubation is carried out in a complete DAB solution (1 drop of 0.3% hydrogen peroxide added to incomplete DAB solution) for 5 min in the dark at room temperature. These cells are postfixed with buffered 2% OsO_4 for 30 min at 4°C.

After dehydration and embedding, BEEM capsules are placed in an oven at 80°C for 1 day. Hardened blocks are then removed from the capsules. After examination of the embedded cell layers under an inverted microscope, appropriate areas are selected, and the blocks are trimmed with files and razor blades for sectioning.

Open-Face Embedding for Correlative Microscopy (Ruiter *et al.*, 1979)

The following procedure allows a comparison between the light microscopic characteristics of cells in smear preparations and their ultrastructural counterparts. The technique is useful when available cells are limited in number (e.g., from puncture specimen) and are present in small groups. The cells obtained are carefully smeared on a plastic sheet (Melinex, ICI) that is mounted on a glass slide with glue (cyanolit) and immediately fixed with 1.5% glutaraldehyde in 0.1 M cacodylate buffer (pH 7.2) for about 15 min at 4°C. After being rinsed in the same buffer, the smear preparations are stained according to Papanicolaou (without xylene), mounted in buffer, and examined by light microscopy. Areas with cells of interest are selected, photographed, and marked with a small circle drawn by a diamond pen on the back of the glass slide.

The cells are postfixed with 2% OsO_4 in buffer for 12 min at 4°C, dehydrated, and infiltrated with a mixture (1 : 1) of propylene oxide and Epon. After evaporation of the propylene oxide overnight, the Epon is thoroughly wiped from the plastic sheet except at the marked areas. A gelatin embedding capsule filled with Epon is placed face down over each marked area (Fig. 2.15A) and polymerized for 48 hr at 60°C. The marked area becomes adherent to the Epon block, thereby incorporating the selected cells (Fig. 2.15B). The glass slide is removed with a razor blade from the plastic sheet. The plastic sheet is cut away from the area of adhesion (Fig. 2.15C). The plastic sheet adherent to the Epon block is removed by gentle lifting with a pair of tweezers (Fig. 2.15D).

The Epon block is placed in a cardboard or wax mold adjusted in the central opening of the stage of a light microscope (Fig. 2.15E). The cells in the Epon

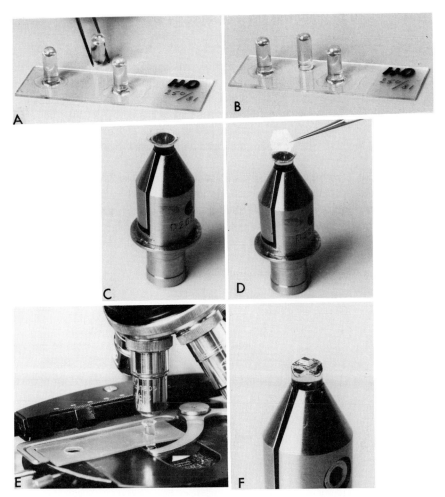

Fig. 2.15. Open-face embedding of a Papanicolaou-stained smear preparation on a Melinex film with dimensions of commonly used glass slides. See text for details. (From Ruiter *et al.*, 1979.)

block are located by light microscopic examination of the block face and comparison with photomicrographs. The selected area is indicated with ink dots around the cell group. The block is trimmed (Fig. 2.15F) and mounted on the ultramicrotome. The block is oriented exactly and approached very carefully so that the complete block surface is already cut after a few thin sections. The procedure can also be applied to paraffin sections (thickness 4 μm or greater) that are stretched on a plastic sheet for 48 hr at 60°C.

Resin Slide Embedding for Correlative Microscopy

Method 1

In the following technique, thick sections that have been observed under the light microscope can be resectioned for electron microscopy (Sigee, 1976). The resin slide functions both as an optical medium for light microscopy and as a permanent base for the sections. Resin sheets are prepared by pouring the requisite amount of Spurr resin into a shallow polypropylene dish and then polymerizing at 60°C for 12 hr. Resin slides are cut with a handsaw from sheets of resin 2 mm thick.

Sections of 2- to 6-μm thickness are cut with a dry glass knife from the block of embedded specimen. These sections are individually placed on drops of 0.05% gelatin solution on the resin slide (Fig. 2.16, step 2), which is placed on a slide warmer to expand the sections and dry the gelatin. The slide is left for at

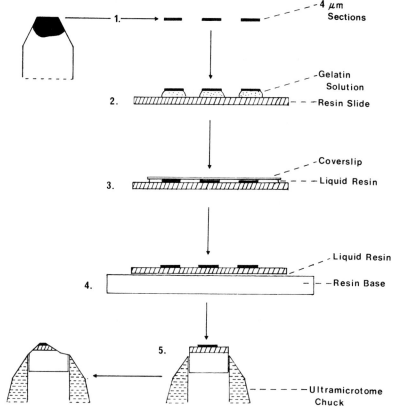

Fig. 2.16. Schematic representation of the resin slide technique. Numbers 1–5 refer to the major stages described in the text. (From Sigee, 1976.)

least 1–2 hr to dry completely. The position and number of each section are marked on the slide with a fine-tip pen. Drops of glycerol are placed on the sections with a coverslip on top (step 3), and the slide is then observed and photographed under the light microscope. The position of cells that are to be resectioned for electron microscopy is noted.

The coverslip is removed and the resin slide cleaned of all glycerol by rinsing in water. The slide is drained and dried in hot air to remove all the water. The resin slide is placed on a 3-mm-thick resin base coated with liquid Spurr resin (Fig. 2.16, step 4) and left on top of a slide warmer overnight to polymerize. At this stage excessive heat should be avoided to prevent resin slide distortion. (Alternatively, this step of mounting the resin slide on the thick resinbase can be omitted by simply using a very thick resin slide and trimming directly.) Segments bearing sections that are required for ultramicrotomy are cut out with a jeweler's saw and placed in a chuck (step 5). The segment is viewed under a binocular microscope and trimmed to a particular area, using the information obtained earlier. Ultrathin sections are carefully cut to ensure that the block face is in the correct plane relative to the knife's cutting edge.

Method 2

The following method was devised to facilitate the selection of desired areas in fixed and cytochemically stained Vibratome sections by light microscopy for semithin or ultrathin sectioning (Giammara and Hanker, 1982). This method is also useful for cultured cells.

The specimens are embedded in the form of an epoxy microscope slide (Cast-A-Slide). Vibratome sections (or cells) that have been fixed with glutaraldehyde, poststained with OsO_4, and cytochemically stained are dehydrated and infiltrated with epoxy by routine or rapid embedding methods. For the final embedment, the sections are placed in a small amount of epoxy within the rectangular slot of a polyethylene Cast-A-Slide mold (Fig. 2.17A) (Ted Pella, Inc., Tustin, California). The epoxy is spread to the edges of the recesses by flexing the mold and by teasing with a stick or pipette. Excess epoxy is removed. The Cast-A-Slide mold is then put into a level oven and heated at 60°C for 48 hr (or a shorter time for rapid-cure epoxies like Medcast) until polymerization is complete (Fig. 2.17B). After cooling to room temperature, the epoxy slides are readily removed by merely flexing the mold (Fig. 2.17C). Areas for ultrathin sectioning are selected with the light microscope (Fig. 2.17D), cut out, glued to a blank block with cyanoacrylate, and sectioned for electron microscopy.

Epoxy Resin Slides for Handling Unfixed Cryostat Sections

The following method allows the use of unfixed cryostat sections of uniform thickness for enzyme cytochemical studies (Altman and Barrnett, 1975; Bülow

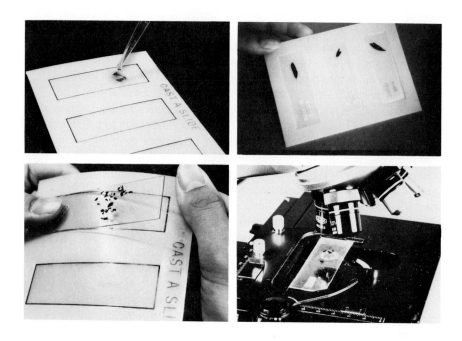

Fig. 2.17. One-step casting procedure that allows the embedment of stained cells or tissue sections in the form of an epoxy microscope slide. (A) The tissue is placed within the rectangular slot in the mold. (B) The mold is placed flat in an oven. (C) The epoxy slide is removed by flexing the mold. (D) The epoxy slide is viewed by a light microscope. (From Giammara and Hanker, 1982.)

and Høyer, 1983). Specimens are frozen and sectioned and then incubated. The conditions for optimal pretreatment, incubation, and posttreatment can be controlled. A mold is prepared of silicone rubber strips that are glued together, so that the inner dimensions of the mold are 76 × 25 × 1 mm. The mold is filled with resin and polymerized in an oven. The resin slide is removed from the mold, and the edges and corners are straightened and cleaned. Tissue blocks (2 × 2 mm) are frozen for 1 min at −70°C in *n*-hexane cooled by an ethanol–dry ice slurry. The sections are cut at 10 μm in a cryostat with a cabinet temperature of −30°C and the knife at −70°C. Sections are transferred from the knife to the slide in the cryostat chamber. Incubation, fixation, postfixation, and dehydration can be carried out on the resin slide.

BEEM capsules are filled partly with Epon mixture and are placed inverted on the slide still wet with 100% ethanol. The empty pyramidal top of the capsule is cut off to allow complete hardening of the entire specimen block. A heavy object is placed on top of it to keep the capsule pressed down and thus avoid excessive spill of Epon. The resin slide–capsule is placed in an oven at 60°C for 48 hr. After polymerization, the capsule containing the section is broken off at room temperature and the block face is trimmed for thin sectioning.

Wafer Embedding (Romanovicz and Hanker, 1977)

The following procedure allows light microscopic screening of embedded tissues prior to ultrathin sectioning. It is useful when specimens are obtained with an automatic tissue sectioner and treated cytochemically, yielding an electron-opaque reaction product.

Tissues are fixed with an aldehyde and then rinsed with 0.22 M sucrose in 0.1 M cacodylate buffer (pH 7.4). Either a Sorvall TC-2 or an Oxford Vibratome is used to obtain sections 10–30 μm thick. After incubation and postfixation with OsO_4, the tissue sections are dehydrated and infiltrated with a resin. The sections are transferred from the final 100% resin mixture to a small drop of fresh resin on a fluorocarbon-coated cover glass (Polysciences, Inc.) (Fig. 2.18A). Another coated cover glass is placed on top and pressed to spread the resin and sections. To support the cover glass–specimen sandwich without contact along the edges, a platform with disk-shaped pedestals ($\frac{5}{8}$ in. in diameter, $\frac{1}{8}$ in. high) is used (Polysciences, Inc.) (Fig. 2.18B). If embedments of minimum thickness are desired, a 50-g weight can be placed on the top cover glass during polymerization.

After polymerization, the cover glasses are carefully pried away from the embedded material. Alternatively, they can be separated from the polymerized resin by immersing the entire embedment in liquid nitrogen. The resulting thin embedment (wafer) (Fig. 2.18C) is mounted on a microscope slide with immersion oil and examined by light microscopy to select suitable areas for sectioning. Such areas are cut out of the wafer and glued with the desired orientation to blank epoxy blocks, using fast-curing cyanoacrylate cement. Sections of 4 μm thickness can be cut and mounted on glass slides for further light microscopic monitoring prior to ultrathin sectioning.

Identification of Areas of Interest in Human Breast Tissue before Embedding (Ferguson and Anderson, 1981)

Breast biopsies are sliced and areas of gray/white connective tissue (10 × 5 × 2 mm) are excised. The tissue is fixed with 3% glutaraldehyde in 0.1 M cacodylate buffer (pH 7.2) for 3 hr. The tissue is chopped into 60-μm-thick slices with a tissue sectioner. The slices are washed overnight in the buffer, postfixed with 1% OsO_4 in buffer for 1 hr, dehydrated, and infiltrated with a resin.

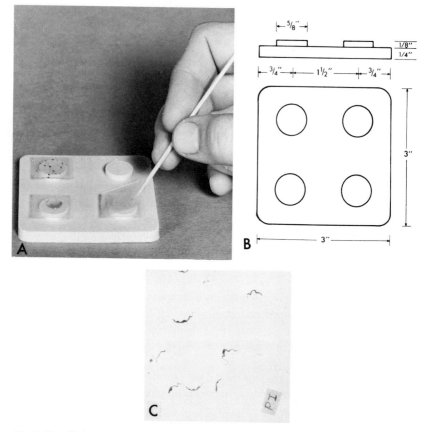

Fig. 2.18. Wafer embedding. (A) Transfer of epoxy-infiltrated tissue sections onto a fluorocarbon-coated coverglass supported by the embedding platform. (B) Dimensions of the embedding platform (side and top view). (C) A wafer containing cytochemically stained 25-μm sagittal sections of mouse plate. (From Romanovicz and Hanker, 1977.)

When the slices are examined with a dissecting microscope, parenchymal structures can be identified; areas containing normal ducts and lobules are separated from those with altered parenchymal architecture. A representative number of these slices are photographed to record the subgross appearance of the tissue before embedding it in the resin, using flat-bottomed BEEM capsules for subsequent ultrastructural examination. The connective tissue is uniformly stained light brown, while the epithelial cells of the ducts and lobules take up more OsO_4 and appear dark brown. The remaining slices are placed in 4 × 2-cm disposable plastic tissue-embedding molds and are then covered with 2 mm of the resin and polymerized at 60°C. Sections 1 μm thick are stained with toluidine blue, and a light microscope is used to select appropriate areas for thin sectioning.

Reembedding of Paraffin-Embedded Tissue in Resin

Reembedding of paraffin-embedded tissues in a resin allows electron microscopy in cases in which light microscopy has failed to provide a diagnosis. Additionally, electron microscopy can confirm light microscopic findings. The use of paraffin-embedded tissues allows better controlled sample selection, based on a light microscopic survey section. It is easy to randomly exclude a specific area in a very small tissue block used in electron microscopy without prior survey of a large tissue block embedded in paraffin. However, deparaffinization always results in some damage, especially loss of membrane clarity, to the tissue. Several methods are presented here.

Aldehyde-fixed and paraffin-embedded tissue blocks are cut into pieces no larger than 3 mm^3 and deparaffinized by placing them in a test tube containing 100% xylene and rotating them for 1 hr. The specimens are rehydrated in an ethanol series of descending concentrations (100, 95, 75, 50 and 15%), being allowed for 15 min each, after which there is a final rehydration in buffer overnight. The specimens are retrimmed to pieces no larger than 1 mm^3 and postfixed with OsO$_4$. These specimens are further processed according to standard procedures.

One-Step Methods

The following method eliminates rehydration and dehydration and combines deparaffinization and osmication in one step (Chien *et al.*, 1982). This simple procedure allows adequate diagnostic interpretation of tissue lesions. An area of interest is removed from the paraffin block and trimmed into cubes 1–2 mm in size. These cubes are immersed in a 2% solution of OsO$_4$ in xylene for 30–60 min with gentle stirring. They are then treated with xylene, infiltrated with mixtures of a resin and xylene, and finally embedded in a resin. When the tissues under study are embedded in paraffin-containing plastic polymers (Paraplast), toluene is preferred over xylene.

Another method, presented by Bergh Weerman and Dingemans (1984), is faster than that introduced by Chien *et al.* (1982). Paraffin-embedded tissue blocks (1–2 mm^3) are placed in xylene (or toluene) containing 1% (w/v) crystalline OsO$_4$ and the paraffin is dissolved at 40°C in an oven for 10 min; gentle agitation of the blocks expedites the dissolution. The specimens are infiltrated (using agitation) with 1 : 1 mixture of xylene and epoxy resin at 40°C for 10 min, and then with pure resin at 40°C for 10 min. Embedding is accomplished either overnight or for 1–2 hr at 100°C. Workers must be protected from the OsO_4 vapors emitted, especially in the first step.

Reembedding of Paraffin-Embedded Tissue Sections in Resin

Method 1

Thick sections of aldehyde-fixed and paraffin-embedded tissue are mounted on glass slides and then deparaffinized by placing the slides in a couplan jar containing 100% xylene. This procedure is followed by rehydration in a descending acetone series (100, 95, 75, and 15%), 10 min being allowed for each, and final rehydration in buffer. The sections on slides are postfixed with OsO_4 and dehydrated. One or two drops of a 1 : 1 mixture of acetone and resin are placed over the section, and after 20 min this mixture is replaced by a drop of the pure resin. A gelatin embedding capsule filled with resin is placed face down over the resin-covered thick section on the glass slide and polymerized in an oven at 60°C. The section adheres to the resin block, which is separated from the slide by immersion in liquid nitrogen or dry ice. Similar procedures can be used for embedding cells grown on coverslips. Before cells are grown on them, the coverslips should be lightly sprayed with Teflon and heated at 250°C for 20 min.

Method 2 (Kraft et al., 1983)

In the following method, heat is used to separate the section from the glass slide. Tissues are fixed with aldehydes, embedded in paraffin, and sectioned at 6–15 μm. The section is placed on a glass slide and stained for light microscopy. Localization of the region of interest on the slide is accomplished by using a dissecting microscope to make a drawing of the selected areas. After removal of the coverslip with an appropriate solvent, the section is rehydrated rapidly through graded alcohols into 0.12 M PBS (pH 7.2). The section is poststained with 0.5% OsO_4 in 0.12 M phosphate buffer for 10 min and then rinsed in buffer.

The section is stained with either 1% aqueous methylene blue or 2% toluidine blue and then rapidly dehydrated through graded alcohols to 100% alcohol. Next, n-butyl glycidyl ether (BGE) is applied drop by drop to the slide after it has been drained of excess alcohol. After 5 min, a 1 : 1 mixture of BGE and Quetol 651 mixture is added to the slide. At 5 min later the slide is drained, and the Quetol mixture is applied twice, 5 min each time. BEEM capsules (size 3), with cap and conical portion removed, are used for embedding. The top rim of the capsule is placed over the selected area on the slide. One hand holds the cap upside down with tweezers while the other rapidly pipettes the embedding mixture into the inverted capsule. When the meniscus is seen above the capsule rim, the cap is placed on it, thus excluding gross air bubbles.

Following three to four evacuation cycles in the vacuum oven, the slide (with capsule attached) is left in the oven under vacuum at 60°C for at least 24 hr. After the curing period, the slide is removed from the oven and maintained at 60°C, because separation becomes difficult, incomplete, or even impossible at lower temperatures. Placing the slide on a hot plate at 62°C ensures maintenance of the proper temperature. The capsule is grasped near its base with forceps, and the slide is held down with another instrument. Separation is accomplished by snapping off the capsule. The block is now ready for trimming and sectioning.

The Epon–Araldite resin gives equally good results and allows the use of BEEM capsule, because the viscosity of the Epon–Araldite is greater than that of the Quetol 651 resin. Although the quality of ultrastructural details observed in the sections will not be the same as that in tissues specifically processed for electron microscopy, good results are obtained with this method when specimens have been fixed with glutaraldehyde.

Method 3

Sometimes a paraffin section is the only tissue available for reembedding. The slide is oriented (section side down) onto a specially designed silicone-rubber embedding mold (Fig. 2.19) (Chien *et al.*, 1982) (Polysciences, Inc.), allowing the area of interest to be captured in an epoxy block. Any area of a section can be selected because the wells in this mold are asymmetrically distributed. After polymerization, the block is easily removed from the slide by heating it on a hot plate (100°C) for 15 sec.

Pop-Off Method for Reembedding (Bretschneider *et al.*, 1981)

The pop-off method allows a precise area in a paraffin-embedded section to be reembedded in a resin. The method is especially useful for cell smears and

Fig. 2.19. A silicone rubber mold for reembedding paraffin section from a glass slide. (From Chien *et al.*, 1982.)

monolayers because they may be the only specimens available for study. Although ultrastructural detail is often poor, an identical section can be studied under both light and electron microscopes. The area of interest in the stained paraffin-embedded section mounted on a glass slide is lightly circled (the same size as a BEEM capsule) with a diamond pencil on the undersurface of the slide before removing the coverslip. The slide is immersed in xylene to remove the coverslip and then rinsed with fresh xylene to ensure removal of all the mounting medium. The slide is dipped in equal parts of xylene and propylene oxide and then in propylene oxide.

The slide is placed in a 2 : 1 mixture of propylene oxide and resin for 2 min. After being transferred to a 1 : 1 mixture of propylene oxide and resin for 2 min, it is placed in a 1 : 2 mixture of propylene oxide and resin for 10 min. A labeled BEEM capsule is placed in a BEEM capsule holder (Fig. 2.20A). The holder is modified by cutting away a conical area at its base, thus allowing a capsule to be pushed down into the holder so that its top is level with the surface (Fig. 2.20B). The BEEM capsule is filled to overflowing with resin and placed in the holder. The section adhering to the glass slide is inverted gradually over the top of the resin-filled capsule (Fig. 2.20C). If a bubble forms, the slide is removed, 1–2 drops of resin are added, and the slide is inverted again.

Polymerization is carried out overnight in an oven at 76°C. After cooling to room temperature, the glass slide–BEEM capsule unit is removed from the holder by pushing the capsule up from the underside of the holder. The slide is placed on a hot plate (100°C) for 15 sec. After the unit has been removed from the hot plate, the capsule is rocked slightly until it pops the section off the slide. The pop-off section is at the wider surface of the BEEM capsule. The tip of the capsule is cut off as evenly as possible so that the reembedded section can be viewed under the light microscope by placing the block on top of the condenser

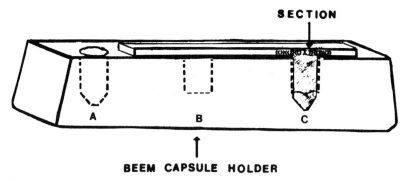

Fig. 2.20.   BEEM capsule holder for section reembedding: A, standard capsule well; B, bottom of well cutaway; C, resin-filled capsule in cutaway well. The section is inverted over the capsule. (From Bretschneider *et al.*, 1981.)

lens. The area of interest can be located and marked for ultrastructural studies. The cutting face of the capsule is trimmed into a mesa. Extreme caution is necessary while sectioning, because the 5- to 6-μm layer of cells (from the paraffin block) is located immediately at the surface of the block. Particular attention must be taken to ensure accurate orientation of the block face parallel to the knife edge prior to sectioning.

Reembedding of Tissue Culture Cells (Connelly, 1977)

The following method is used to reembed tissue culture cells. The cells are cultured in Falcon plastic Petri dishes (35 × 10 mm) in an appropriate medium to confluency or near confluency, depending on the experimental design. They are fixed with 3% glutaraldehyde in 0.1 M phosphate buffer (pH 7.3), rinsed in buffer, and postfixed with 1% OsO_4. Dehydration is carried out in 30, 50, 70, and 90% ethanol (propylene oxide should not be used) and then in 90 (two changes), 95, and 100% (two changes) HPMA. This process is followed by infiltration with 1:1, 1:2, and 1:3 mixtures of HPMA and Epon and then embedment in Epon. The polymerization is accomplished by placing the uncovered Petri dish overnight in an oven at 60°C. The embedded cells are removed from the dish by completely breaking off the sides of the dish. With the resin side down, the resin disk is flexed several times and the two layers are separated.

Small rectangular blocks are cut out of the resin layer with either a jeweler's saw or a straight-edge razor blade. These pieces are reembedded in a small amount of resin reserved from the batch used to originally embed the cells. The reembedment is accomplished by first half-filling a 4-mm well in a flat embedding mold with the resin. One block of embedded cells is then placed into the well (cell side up), and after the resin has flowed over the surface of the cells, a second block is placed (cell side down) directly over the block already in the well. Care must be taken to avoid air bubbles between the two blocks. Finally, the well is filled to the top with more resin, and the mold is placed in an oven at 60°C for 24–36 hr for polymerization to be completed. The block is trimmed (Fig. 2.21) and sectioned, with the lines of cells perpendicular to the knife edge to avoid the breakout of cells due to the possible different states of polymerization.

Reembedding of Thick Resin Sections

Method 1

Johnson (1976) introduced a rapid and simple method for reembedding thick sections under microscope control by using the rapidly setting cement cyanoacrylate. A thick section (1 μm) is placed on a drop of water on a no. 1 coverslip and

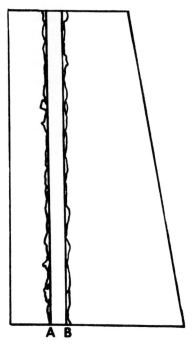

Fig. 2.21. Drawing showing typical section shape, which is asymmetrical in order to identify orientation of cell monolayers (A and B) when different populations are embedded together. (From Connelly, 1977.)

dried on a hot plate. The section is stained with toluidine blue in 1% borax or with any stain suitable for light microscopy. The coverslip is placed (section side down) on a glass slide without a mounting medium and viewed with the light microscope (a coverslip-corrected objective lens is used).

After a section or a structure for ultrastructural study has been selected, the coverslip with the section is gently removed from the slide and placed on a flat surface with the section side up. A drop of cyanoacrylate is applied to a blank epoxy block, which is quickly pressed against the section and hand-held for several seconds while it sets. After incubation of the section for 1 hr in an oven at 60°C, ice is applied to the coverslip. The coverslip separates from both the section and the block, leaving a smooth surface that may be thin-sectioned without further facing.

Method 2

The following method is useful for reembedding thick resin sections (mounted on glass slides) that have been examined with the light microscope (King *et al.,*

1982). An embedding capsule filled with a resin and containing a label is inverted over a selected thick resin section mounted on a glass slide. This slide, with the capsule, is placed in an oven at 60°C long enough for the resin to become firm but not yet completely polymerized; usually it takes 12–24 hr. The slide is transferred to an oven at 80–90°C for a few minutes and then slipped into a special slide holder (Fig. 2.22A) (Polysciences, Inc.). The slide, held in place by the rim of the slot that fits comfortably in the hand, is grasped firmly and uniformly around the edge, and the capsule is separated from the slide with gentle lateral force (Fig. 2.22C). The capsule should easily snap from the slide. This step should be carried out rapidly, without allowing the slide to cool significantly. The capsule containing the reembedded section is returned to the oven and allowed to complete polymerization for another 24 hr at 60°C. This block with the reembedded section can be sectioned for electron microscopy. Sturrock (1984) has designed a simpler and less expensive holder to help reembed thick resin sections.

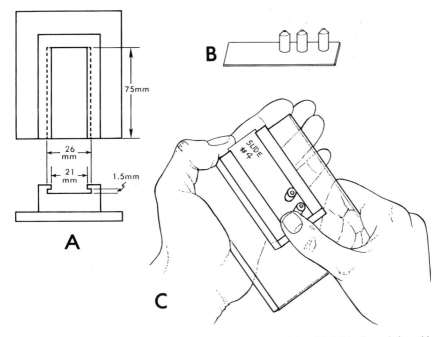

Fig. 2.22.   Reembedding thick resin sections. (A) Diagram of the slide holder for assisting with the removal of epoxy capsules containing reembedded sections from glass slides. The dimensions indicated are appropriate for holding standard slides (24 × 75 × 1 mm). (B) The placement of inverted BEEM capsules over selected sections on a slide. (C) The slide holder in use, illustrating a typical grip for applying force to separate capsules from the slide. (From King et al., 1982. © 1982 The Williams & Wilkins Co., Baltimore.)

Osmication and Flat Embedding of Large Tissue Sections (Schwartz, 1982)

Thick sections (40–300 μm), with or without prior embedding in agar, tend to roll up or curl during aldehyde fixation and buffer washes. Once osmicated, such sections cannot be flattened. The following procedure facilitates flat embedding, and hence the study of the entire cut face of thick sections, by producing either semithin or ultrathin sections. Thick sections are fixed with aldehydes and washed with buffer in glass vials (25 mm in diameter) having a slightly convex bottom. After removal of the last buffer wash, sections are unrolled and flattened on the crown of the convexity with a fine brush (Fig. 2.23A). A small amount of OsO_4 solution is introduced along the sides of the vial so that the crown remains above the fluid level. Thus sections are exposed only to the vapor. After 20 min in the closed vial, a slice 200 μm thick is completely blackened.

Additional OsO_4 solution is allowed to flow down the sides of the vial so that the sections float. After an additional hour in the OsO_4, the sections (brain) are rinsed in buffer and then stained *en bloc* with uranyl acetate. They are then dehydrated and infiltrated with resin. The pointed end of a BEEM capsule is cut off, and the cap is stuck to a piece of masking tape attached sticky side up to a glass plate. A drop of resin is placed in the cap and the section is laid over it. A small piece of nylon net is placed over the section and fastened in place by snapping the body of the BEEM capsule into the cap (Fig. 2.23B and C). The capsule is filled with resin and polymerized. Attaching the capsule to a tape-covered support allows the capsules to be separated easily from the excess resin that leaks out. The tape and excess resin can be discarded and the support reused. The block face should be trimmed so that no strands of nylon are present in the face that is to be cut.

Reembedding of Autoradiograms of Semithin Sections for Correlative Microscopy (Barajas *et al.*, 1981)

The following method allows both light and electron microscopy of the same section. It is well suited for studying structures that can be autoradiographically labeled but are small and relatively infrequent, because it permits their search by light microscopy and further analysis by electron microscopy. This technique has been used in the identification of renal neuroeffector junctions and could also be applied to other junctions. Electron microscopic autoradiography of the ultrathin sections is also possible. A similar approach might be useful in methods using markers such as horseradish peroxidase.

Serial sections (1–2 μm) of resin-embedded tissue fixed for electron micros-copy and containing the radioactive label are mounted on gelatin-coated slides,

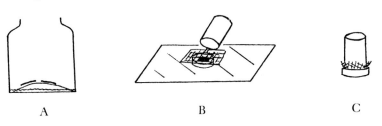

A   B   C

Fig. 2.23. Flat embedding of large tissue sections. (A) Sections brushed flat on the convex bottom surface of the vial and exposed to osmium vapors. (B) BEEM capsule cap on sticky surface with section and nylon mesh in place. (C) Capsule body snapped into cap, sandwiching section flat between cap and mesh. (From Schwartz, 1982.)

and light microscopic autoradiograms are prepared. The sections are stained with 1% toluidine blue or 0.1% crystal violet. The autoradiograms are examined, and the areas containing accumulations of grains are photographed. A low-magnification micrograph of each section is prepared to locate the area in the reembedded section for subsequent thin sectioning. The stain is then removed by immersion of the autoradiograms in acid–alcohol for those stained with toluidine blue or in 70% alcohol for those stained with crystal violet. Immediately before reembedding, the autoradiogram is completely dried by passing it through an ascending series of alcohols and allowed to dry in a slide box containing Drierite.

A BEEM capsule (gelatin capsules are not recommended) with the tip cut off is placed base down on top of the section, filled with enough Epon to just cover the section, and prepolymerized in the oven at 80°C for 1 hr. The capsule is then filled with Epon and polymerized for 48 hr at 60°C. After polymerization, the glass slide is scored with a diamond pencil around the section with the Epon-filled capsule on top, and excess glass is removed. The glass beneath the section is scored deeply. Immersion in liquid nitrogen shatters the glass attached to the capsule. The glass is easily removed, exposing the semithin section embedded flat at the capsule base.

Retrieval of Poorly Resin-Embedded Tissue

Poor embedment of tissues is caused mainly by hydrated embedding component(s) and/or insufficient dehydration. These tissues can be rescued for reembedding. Sections 10 μm thick of the poorly embedded tissue are placed in a small, disposable aluminum weighing dish and covered with distilled water. The dish is heated on a hot plate until the water has evaporated, leaving the sections adhering securely to the bottom. Then the dish is filled with a resin mixture, placed in a vacuum desiccator, and transferred to an oven at 60°C for 48 hr to complete polymerization and reembedding.

3

Sectioning

INTRODUCTION

There are two main reasons for ultramicrotomy (ultrathin sectioning): (1) the electron beam of the transmission electron microscope operating at a conventional accelerating voltage (50–60 kV) will not penetrate sections much thicker than 100 nm; and (2) cell components will overlap in thick sections, confusing the image. Sections are generally cut at a thickness ranging from 70 to 100 nm, showing silver–gold interference color. Thinner sections are needed for high resolution studies. Although thicker sections (100 nm) yield higher contrast, they allow lower resolution even at a higher accelerating voltage (80–100 kV). The use of higher voltages, however, permits increased penetration by the electron beam and reduces chromatic aberration.

It is difficult to obtain ultrathin sections of good quality, and considerable experience is needed to master the skill of sectioning. The most important factor for success is patience, for both technical and artistic abilities are a prerequisite for successful sectioning. At the learning stage, one must find a sufficient amount of time for sectioning. Sectioning has numerous attendant problems; these and their remedies are listed on pages 151–155. There is no need to be discouraged. Thousands of high quality electron micrographs have been published, which proves that ultrathin sectioning can be accomplished. For a detailed discussion on sectioning, see Hayat (1981a).

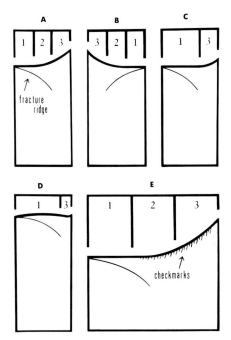

Fig. 3.1. Diagram showing the relationship between various configurations of the cutting edge of glass knives and the length of the usable part in each type. Longer spikes in (A) and (B) are accompanied by a short usable cutting edge.

EXAMINATION OF GLASS KNIVES

A considerable amount of time is wasted unintentionally when a defective glass knife is used. Some of the imperfections in the thin section can be traced directly back to defects in the knife's cutting edge. Therefore it is imperative that the knife be carefully examined for major or minor defects before being used. Major defects can be seen with the naked eye by slowly rotating the knife under an ordinary light source until light is reflected off its edge. With careful viewing, a fracture ridge (stress line) following a curved path can be seen on the front face of the knife (Fig. 3.1A). This fracture ridge runs into one corner of the cutting edge, and the best part of the edge extends from this corner, for it is here that the break starts and the speed of fracture is at its lowest.

Sawtooth striations (check marks) extend for a variable distance from the opposite corner of the edge, rendering this part of the edge unfit for sectioning (Fig. 3.1E). The dimension of striations increases in the part where the fracture speed is higher. A short spike of glass may be visible at this corner of the edge.

The spike is formed because of a slight displacement of the score line from the exact diagonal of the 25-mm glass square. The length of the striation part increases with the length of the spike. The spike is not visible on a good knife. However, the presence of a short spike does not make the knife useless. A long spike is accompanied by a very short usable cutting edge (Fig. 3.1A and B).

Detailed examination of the knife can be carried out under a stereomicroscope (dissecting microscope) at a magnification of ×20–30 by using incident light illumination from above. A beam of light from an appropriate source is focused onto the cutting edge at such an angle that the edge appears as a narrow band of bright light against a dark background. Against such a background the presence of striations, usually on one part of the edge nearer the spike, is easy to observe because they scatter the light (Fig. 3.2). As noted earlier, the part free from striations is used for sectioning. The knife edge can be divided roughly into three parts, depending on the quality of the cutting edge. Some workers indicate the three parts by marking on the face of the knife (away from the cutting edge) with a fine pen, because once the knife is mounted on an ultramicrotome, the three parts are difficult to see. The part nearest the spike is useless, the middle third is used for preliminary sectioning, and the part farthest from the spike is used for good sectioning. It should be noted that the entire cutting edge shows striations under the scanning electron microscope (SEM), but the striations on the useful part of the edge are too small to be visible under an optical microscope (Hayat, 1981a).

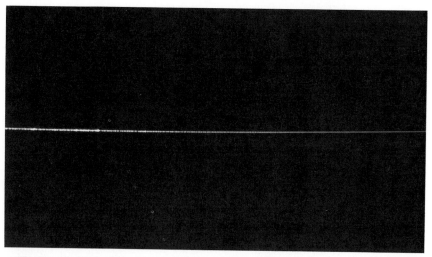

Fig. 3.2. Image of a glass knife edge seen with light scattering. The imperfections are seen to the right (nearer the spike) as irregular spots of light. The almost flawless part of the edge appears to the left as a thin, even, light line. (From Ward, 1977.)

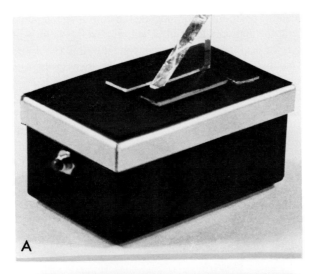

Fig. 3.3. Light-box for selecting the best part of a glass knife. (A) General view of the light-box with a glass knife in position over the slot in the lid. Pieces of 2-mm-thick black card are attached to the lid to hold the knife in the correct relationship with the lamp. The press-switch can be seen at left. (B) Plan view of the box with the aluminum lid removed. (From Franchi and Symons, 1980.)

Another approach for selecting the best part of the knife is to use a light-box designed by Franchi and Symons (1980). This device exploits internal reflection of light within the knife, which is illuminated from below. The box incorporates a lens-ended flashlight bulb powered by AA (E91) penlight batteries and is operated by a small press switch. The glass knife is supported in the optimum position for inspection over a slot cut in the lid (Fig. 3.3). The great advantage of the device is its ability to be used with any convenient microscope without requiring other elaborate lighting equipment.

During the construction of the box it is essential to achieve the correct relationship between the positions of the slot and the bulb so that light is directed into the glass knife at the appropriate point. The knife itself is held in place between supports with its vertical face 1 mm or so beyond the slot. The desired effect is to emit light from the cutting edge only, which is then visible against a dark background, as discussed earlier. The dimensions of the slot can be made adjustable for knives 5–12.5 mm thick by using movable knife supports secured with a proprietary stationer's putty. The device with assembled knife is positioned under a dissecting microscope (×50 or higher as desired), and the edge is illuminated by pressing the switch. An updated version of the Franchi–Symons prototype knife evaluator is commercially available (Marivac Ltd., Halifax, Nova Scotia). A glass-knife inspection stand containing a control knob that adjusts the knife angle and position under the stereomicroscope is also available (Ted Pella, Inc., Tustin, California).

DIAMOND KNIVES

Diamond knives are far superior to glass knives for sectioning either soft or hard specimens. Diamond knives are almost indispensable for sectioning hard specimens such as collagenous or calcified tissue and plant materials. Glass knives usually produce less-than-satisfactory sections of hard tissues. Diamond knives are ideal for obtaining serial sections and extremely thin sections (20 nm) for high resolution electron microscopy; however, their routine use is limited because of the high cost and vulnerability to damage. A diamond knife can be serviceable for several years if used carefully; knives with minor nicks can be resharpened by the manufacturer.

The length of the cutting edge of diamond knives ranges between 1 and 6 mm, and the included angle varies between 40 and 60°. Relatively large included angles are preferred for sectioning hard materials. The optimal clearance angle of a new knife with its holder is determined by test sectioning. This procedure is carried out by first using a low clearance angle and then increasing it 1° at a time until the optimal angle is achieved. A knife angle ranging from 2 to 5° is commonly used. It should be noted that the angle of the knife varies along the cutting edge. Therefore every time the knife is moved laterally during sectioning, it will require realignment with the face of the specimen block.

Unlike a glass surface, the surface of a diamond knife is difficult to wet with water. New diamond knives are more hydrophobic than old ones. If the meniscus is repelled by the diamond surface, the cutting edge may have to be wetted by drawing the water up, using a hair moistened with saliva or an appropriate wetting agent. The trough fluid can also be drawn up to the cutting edge with the flat end of a soft wooden stick carrying a bit of saliva. Additional details concerning the use of knives have been presented elsewhere (Hayat, 1981a).

It is important to remember that, at the beginning of sectioning session, the specimen block should be mounted before the knife is mounted on the ultra-microtome. Diamond knives should not be used for block trimming or cutting sections thicker than 1 μm (greenish-blue interference color). The cutting edge should be cleaned immediately after use while it is still wet; otherwise, leftover sections or their fragments and other forms of debris may tenaciously adhere to the cutting edge. Such adhered debris is very difficult to remove.

The cutting edge can be cleaned with a stick of soft orangewood, which is sharpened to a thin chisel edge (spatula-shaped). Before being used, the stick should be thoroughly soaked in a nonpolar solvent such as acetone to extract natural wood resins and then rinsed in water. While the knife edge is viewed under a dissecting microscope, the flat end of the stick is gently pressed against the cutting edge as though to split the stick (Fig. 3.4). The stick movement must only be made *parallel* to the cutting edge; any other orientation, whether oblique or at right angles to it, must not be attempted. The stick should always travel the entire length of the cutting edge on each cleaning stroke. If one follows the basic precautions of sectioning well-cured specimen blocks and removing section fragments from the cutting edge by thorough rinsing with water after each sectioning session, it is not necessary to use a stick more than once or twice a week. Wooden sticks have the potential for producing fine nicks in the cutting edge.

Other methods of cleaning the cutting edge have been suggested. Gorycki and Oberc (1978) recommend the preparation of a cleaning solution (diluted seven parts in distilled water), which is then filtered through a 0.8-μm Nalgene filter. After the cutting edge is submerged in this solution for 10 min, the edge is

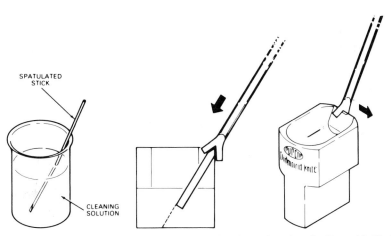

Fig. 3.4. Proper use of the wooden stick for cleaning the cutting edge of a diamond knife. The stick must be moved gently, parallel to the entire length of the cutting edge. (Courtesy of E. I. du Pont de Nemours & Co.)

cleaned with a squeegee of Tygon tubing (R-3603, Fisher Scientific Co.); this tubing has an inner diameter and wall thickness of 7.9 and 1.6 mm, respectively. The squeegee is prepared by cutting a 2-mm-long tubing into four equal segments, one of which is impaled along its length on a needle in a holder (Fig. 3.5). The cutting edge is cleaned with a gentle unidirectional stroking motion of the wet squeegee parallel to the edge. This cleaning is done under observation through a binocular microscope.

Another method of cleaning the diamond knife involves soaking it for 30 min or more in a 0.1–1.0% nonionic detergent such as Triton X-100, or Freon 113; common laboratory detergents (e.g., Alconox) should be avoided. The knife is then thoroughly rinsed with distilled water. This treatment also tends to improve the wettability of the cutting edge. A small blast of compressed air (e.g., Freon) can also be employed to clean the cutting edge.

Another alternative is ultrasonic cleaning by treatment with a detergent (e.g., Triton) (Wallstrom and Iseri, 1972). The knife is immersed in 7% aqueous solution of a detergent in a 50-ml Tri-pour polyethylene beaker. Because of the internal geometry of this type of beaker, the knife edge does not come in contact with the beaker wall. The beaker is placed in water in an ultrasonic cleaner (40 W, 60 Hz) for several minutes. The knife is removed and rinsed thoroughly in distilled water. Evaporative drying of the cutting edge should be prevented. The knife is immersed in distilled water in a second 50-ml beaker and subjected to ultrasonic radiation for several minutes. It is then rinsed again in distilled water. This rinsing is necessary irrespective of the type of cleaning method used. Sonication should be used only when debris cannot be removed by any other means.

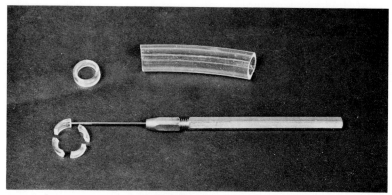

Fig. 3.5. A 2-mm-long piece of Tygon tubing cut into equal segments (lower left), one of which is impaled on a needle to form a squeege used to clean the surfaces and edges of diamond knives. The cutting edge is cleaned with a gentle unidirectional stroking motion of the squeege parallel to the knife edge. (From Gorycki and Oberc, 1978.)

The presence of a glass chip in the resin block can seriously damage a diamond knife. Potential sources of such chips are glass knives that are used for block trimming, Pasteur glass pipettes employed during dehydration and/or infiltration, and ampoules containing OsO_4 or other reagents. If available, an "old" diamond knife can be used for trimming. Plastic pipettes are superior to glass pipettes. The reagents prepared in a bottle containing the broken ampoule should be pipetted into another bottle before using them. Since a single-edge razor blade, when used for trimming, may leave a steel chip in the trimmed block, a discarded diamond knife should be used instead. Sectioning with a glass knife can also result in the introduction of a glass chip into the block face. This possibility should be kept in mind when sectioning with a glass knife is followed by sectioning of the same block with a diamond knife.

GRIDS

Grids with an extensive variety of designs are commercially available (Fig. 3.6). The choice of a grid is determined by the size of the section and the objective of the study. The most commonly used grids are made of copper having an outer diameter of 3.05 or 2.3 mm and are available in a wide range of mesh sizes ranging from 50 to 1000 bars/in. The 300–400 mesh grids are convenient for routine studies because they have relatively small openings, which can support even individual sections without a support film. Grids of other materials such as gold, silver, nickel, stainless steel, aluminum, molybdenum, titanium, and copper/palladium are also available. Because they are more robust, nickel grids are superior to copper grids. The difference in the price of copper and nickel grids is negligible. However, since nickel is a ferromagnetic metal, these grids may deflect the electron beam away from the grid bars, resulting in astigmatism. Also, these grids may adhere to steel forceps; this problem can be eliminated by using platinum-tipped forceps. Although grids with square holes are the most popular, grids with openings of other shapes (e.g., hexagonal) are available. Grids with hexagonal-shaped holes provide the maximum open area and give better support to very thin sections.

The open area that a grid provides for viewing depends on the size and shape of the opening as well as the width of the bar (Fig. 3.7). For example, the standard 200-mesh grid with square openings provides an open area of about 60%; the widths of the hole and bar are 94 and 28 μm, respectively (Fig. 3.7A). The 200-mesh grid with square openings but slim bars has an open area of about 84%; the widths of the opening and bar are 115 and 10 μm, respectively (Fig. 3.7B).

Grids with single or multiple holes or slots of various dimensions, as well as parallel bars, are available (Fig. 3.6H and I). Various types of reference points in

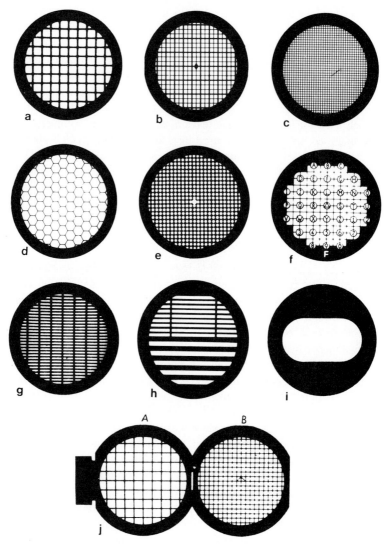

Fig. 3.6. Various grid designs. (a) 100-mesh/in. grid. (b) 200-mesh/in. grid with central mark. (c) 400-mesh/in. grid. (d) Hexagonal-type grid. (e) Round-hole grid with central mark. (f) Finder grid. (g) Rectangular-mesh grid (75 × 300/in.). (h) Multiple-slot grid. (i) Single-slot grid. (j) Folding grid (side A is 100 mesh/in. and side B is 200 mesh/in.). (From Baumeister and Hahn, 1978.)

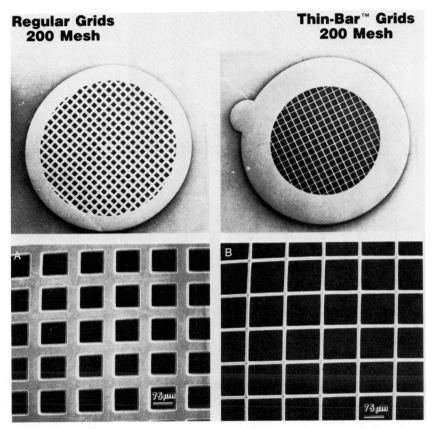

Fig. 3.7. Comparison between conventional grids and grids with slim bars. Top, ×45; bottom, ×270. (Courtesy of SPI Supplies.)

the grid help one locate and relocate specific specimen areas. Nylon-mesh, carbon-coated grids are ideal for X-ray microanalysis to reduce background noise. Other types include finder grids and folding grids (Fig. 3.6F and J). The latter has a curved securing tab that folds to the curvature of the grid.

The practice of bending the edge of a grid so that it can be lowered flat over the floating sections may damage the grid. This problem can be avoided by using tabbed grids (Fig. 3.8). Sections are picked up by bending the tab (handle) of the grid. The bent tab forms an angle of 45° to the grid bars, making it easy to retrieve sections at a natural angle. Sections can be collected from the top or from below by a scooping motion. Forceps are not moistened while picking up the sections. The tab is also useful for transferring the grids during poststaining and rinsing. It is cut off before the grid is placed in the specimen holder in the TEM.

Fig. 3.8. A tabbed grid. (Courtesy of SPI Supplies.)

The grids most commonly used are of the Athene type, which usually has a rim and is characterized by both sides being flat, with one side polished and the other dull (matte). Sections should be collected on the dull side because they get a better contact and are easily seen against a dull background with the naked eye or a light microscope.

Carbon–Polymer Support Grids for X-Ray Microanalysis

In X-ray microanalysis, the generation of extraneous X rays and their detection should be drastically reduced if not eliminated. Therefore materials of low atomic number (e.g., beryllium, carbon, or aluminum) are used to support the specimens. Grids made of these materials are commercially available. However, such grids have disadvantages. Beryllium grids are expensive, and their cleaning for reuse is hazardous. Aluminum grids produce relatively more X rays under the electron beam in the TEM. Nylon grids are not electrically conductive and show poor adhesion to the specimens. Electrical nonconductivity causes charging. Although carbon–polymer grids are conductive, their price is prohibitive. Further, these grids become brittle with aging, and reuse is impractical.

Smits *et al.* (1983) have introduced a simple and rapid procedure for the production of single- or multihole carbon–polymer grids. These grids are electrically conductive, do not become brittle even 2 months after production, have both a smooth and a rough surface, and are flat. The X-ray background generated by these grids is equivalent to that produced in a beryllium grid, and less than that generated in an aluminum grid. If needed, these grids can be covered with a support film.

A 4 : 1 mixture of spectroscopically pure carbon powder and 2.5% (w/w) solution of collodion in amyl acetate is prepared at room temperature. One milliliter of this mixture is spread on a glass slide with a wooden stick or a glass rod. After the mixture has dried, a layer of about 50 μm thick is obtained on the

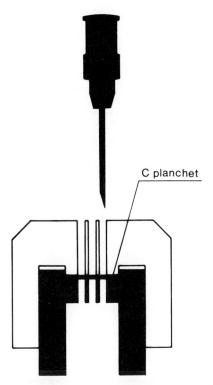

C planchet

Fig. 3.9. Schematic drawing of the needle guide used for preforming carbon–polymer planchets. (From Smits *et al.*, 1983.)

slide. This layer (foil) comes off the slide easily. Disks (planchets) of about 3 mm are punched with a cork drill or a grid punch into this foil. One or more holes are drilled in each disk with a hypodermic needle (0.4–1.0 mm in diameter). A simple tool for needle guidance can be constructed (Fig. 3.9) to obtain reproducible holes. Figure 3.10A shows an SEM image of a support grid prepared with this tool. The smooth and rough sides of the grid can be distinguished from each other by the naked eye (Fig. 3.10B and C).

SPECIMEN BLOCK TRIMMING

Resin blocks containing the embedded specimens invariably require trimming prior to sectioning. Trimming is needed irrespective of whether the specimens are embedded in capsules or flat molds. The main purpose of trimming is to remove the excess resin surrounding the specimen at the tip of the block (Fig.

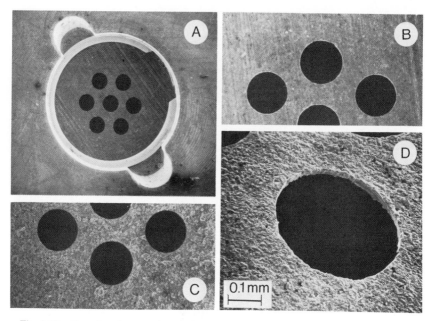

Fig. 3.10. Scanning electron micrographs of a carbon-polymer grid with 0.4 mm drilled holes. (A) Whole grid, fastened in the SEM grid holder by a snap ring. (B) Smooth grid surface, originally attached to the glass slide. (C) Rough back surface of the same grid. (D) Tilted view showing the sharp edges and the small grid thickness. (From Smits *et al.*, 1983.)

3.11). The result is an extremely small block face and sections devoid of plain resin surrounding the specimen. For conventional studies the block face should be as small as possible (about 0.3 mm^2 or less) because the smaller the block face, the better the quality of the sections. However, the objective of the study determines the size of the block face and thus the size of the section. It should be kept in mind that during sectioning, the block face is enlarged. A section compared with the previous section is slightly larger in size, provided the pyramid is obtuse near the block tip. Thus the block face will need retrimming after a certain number of sections have been cut.

The specimen blocks are coarse-trimmed and then fine-trimmed. This step requires a certain degree of patience; hasty trimming tends to damage the specimen block. The block is mounted in the chuck or vise-type holder, which is then mounted on a special trimming block. This assembly is placed on the stage of a stereomicroscope (dissecting microscope). If the trimming block is unavailable, the specimen block can be mounted on an ultramicrotome for rough trimming, but this alternative is not preferred.

First, a horizontal cut is made across the tip of the block, provided the

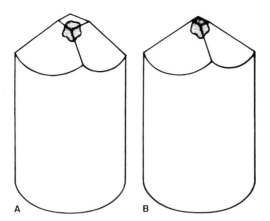

Fig. 3.11. Diagrammatic representation of untrimmed (A) and trimmed (B) specimen blocks. (A) shows the resin around the specimen at the tip of the block. The block face is large in size. In (B), excess resin has been trimmed and the block face is of a smaller size and contains only the specimen.

specimen is more than about 0.5 mm deep. Under low- to mid-range magnifications, excess resin is then removed from the sides of the block tip by short shallow cuts made with a single-edged razor blade. Conventional shaving razor blades are excessively flexible and should not be used. The razor blade should be cleaned with acetone before use and be discarded at the first sign of dullness. A damaged razor blade will produce scratches on the sides of the block, resulting in corrugated edges of the block face that may hinder ribbon formation and cause other problems. During trimming, the upper and lower sides of the block face must be made parallel to each other so as to facilitate the formation of a straight ribbon. The other two sides are sloped. The purpose is to form a shallow angled pyramid topped by a trapezoid (Fig. 3.12) that contains only the specimen and not the surrounding plain resin. The angle of the sloping sides should be sufficiently large to provide adequate rigidity to the block during sectioning. A block

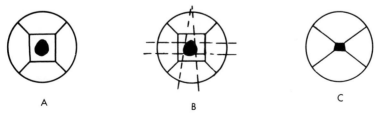

Fig. 3.12. Steps used for trimming the specimen block to obtain a truncated trapezoidal block face (top view). (A) Untrimmed block face. (B) Cutaway lines. (C) Final trapezoidal face (<0.1 mm).

face with a rectangular or square shape is acceptable when a ribbon is not required.

Rough trimming can also be carried out with the Butler block trimmer (Fig. 3.13), which supports and guides the razor blade during trimming. By using this trimmer, possible damage to the specimen during uncontrolled freehand trimming can be avoided. In addition, the angle, location, and thickness of the slices can be controlled. The trimmer can be used with any stereomicroscope that is provided with adequate epi-illumination. The specimen block (dowel-shaped) (G in Fig. 3.13) is inserted into the chuck (B). The micrometer screw (D) is rotated several full turns to back the knife guide (C) away from the chuck. A flat object is laid across the tip of the specimen block, which is pressed against the ejector spring (E) into the chuck until the block tip is even with the angled surface of the knife guide. The retaining screw is tightened with a screwdriver. The knife guide is oriented with respect to the specimen, and the micrometer screw is adjusted to

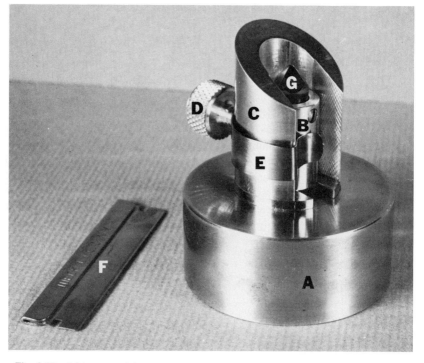

Fig. 3.13. Main parts of the precision hand trimmer: A, base; B, specimen chuck; C, knife guide; D, micrometer screw; E, restoring spring; F, hair-shaper blade; G, specimen block. This trimmer permits delicate control of the pyramid shaping of a tissue block face and can be used with any stereomicroscope. (From Butler, 1974.)

position the guide so that a thin slice will be removed from the block when the blade, resting on the surface of the guide, is stroked downward.

The screw is readjusted, and additional slices are removed from the block until one face of the pyramid is formed. The entire trimmer is turned 90°, and the knife guide is rotated about the axis of the chuck to establish the plane of the second pyramid face. Slices are removed to form the second face, after which the operations just described are repeated to produce the two remaining pyramid surfaces.

The coarse-trimmed block can be fine-trimmed on an ultramicrotome by using a glass knife or a discarded diamond knife. Even the best trimming by hand does not produce upper and lower edges of the block face exactly parallel to each other. Moreover, the sides of the pyramid are not smooth. Fine trimming produces the smooth sides needed to obtain a straight ribbon. The knife stage is rotated about 30° to the right and thick sections are cut. It is then rotated 30° to the left and again the block is trimmed. Thick sections are cut at both times until the desired area of the specimen is exposed. This operation will result in the two edges of the block face being exactly parallel to each other. The block is rotated about 90° so that the other two sides come in position to be trimmed. These two sides are trimmed in the same manner as the first two sides. My preference is to use a glass knife to accomplish both coarse and fine trimming on the ultramicrotome. It is desirable to gently clean the block face with a moist filter paper to remove any debris before sectioning.

At least three precision machines for specimen block trimming are commercially available. The Cambridge Block Trimmer utilizes a steel blade as the trimming edge. The specimen is illuminated from underneath. The Reichert TM-60 specimen trimmer uses a diamond milling cutter that revolves at high speed. A fixed focused light from above is used for illumination. The LKB Pyramitome employs a glass knife for trimming and advances the specimen by a hand wheel. A focused light can be rotated about the cutting point. This machine can be fitted with the Target Marker, which enables the image of a stained survey section on a glass slide to be superimposed on the block face. Thus the desired area can be accurately located on the block face for retrimming. The details of the operation of these three machines can be found in the manufacturers' manual.

Most of the machines just described require that the same specimen area seen in the thick section be relocated on the block face before selective trimming can begin. This procedure is tedious and time-consuming. A simple, selective trimming method is described by Troyer and Wollert (1982). The only special equipment required is a calibrated ocular micrometer for the light microscope. Selected areas seen in thick sections need not be found correspondingly in the block face. This method is not, however, very useful when specimens require orientation before sectioning. Another possibility, the mesa technique of specimen trimming, has been discussed elsewhere (Hayat, 1981a).

Excessively hard resin blocks (e.g., Spurr mixture and Epon) can be softened for easy trimming by immersing them overnight in 95% ethanol (Pfeiffer, 1981). They are then rinsed in distilled water and trimmed to the desired shape with a razor blade. After a block has been trimmed, it is allowed to dry overnight at room temperature or 6–8 hr in an oven at 60°C. The ethanol penetrates and softens the block at 0.3–0.5 mm/hr for the first 3 mm, after which the rate is considerably reduced. Depending on the duration of treatment, any concentration from 70 to 100% softens the block sufficiently.

PREPARATION OF TROUGHS

To prevent adhesion of ultrathin sections to a dry cutting edge of the knife, a leak-proof trough containing the flotation fluid is built around the knife edge. Diamond knives are sold with a trough incorporated, but glass knives must be provided with one. Troughs are made of several types of materials. Metal troughs of various sizes are commercially available. This type of trough, which is inert, is positioned on the knife in such a way that the upper edge of the trough is level with the cutting edge. This placement will subsequently be of help in achieving a satisfactory meniscus. The trough can be sealed by placing a piece of dental wax (paraffin is not recommended) inside the heel of the trough and melting it by pointed heat application, using an alcohol hand torch (McKinney, 1969) or some other appropriate heat source. The trough sides are heated, and capillary action draws the melted wax up the sides. The wax solidifies and seals the trough in a few minutes. These troughs can be reused after being cleaned with acetone or xylene or after being heated.

Tape troughs are disposable and relatively small, although perfectly satisfactory in size. The black color of the tape prevents the formation of reflections that may otherwsie interfere with viewing the sections on the water surface. A piece (0.7 × 3 cm) of black electrical adhesive tape (Scotch Brand no. 3) is wrapped from one side to the other of the cutting edge of the glass knife, as shown in Fig. 3.14. The tape should not be too wide; otherwise the knife clamp may press on the tape as it is tightened and thus break the seal. While the tape is being wrapped, its lower edge should preferably be parallel to the lower edge of the knife; the upper edge of the tape must be level with the cutting edge. The tape can be placed at a 2–3°angle relative to the lower edge of the knife to compensate for tilting the knife during sectioning. This positioning results in a nearly straight trough that is tilted neither toward nor away from the cutting edge. A straight trough facilitates obtaining an optimal meniscus level, which is necessary because it enables the sections to glide away smoothly from the cutting edge and to correctly show interference colors.

The tape is held by both ends and placed on the knife as described and then

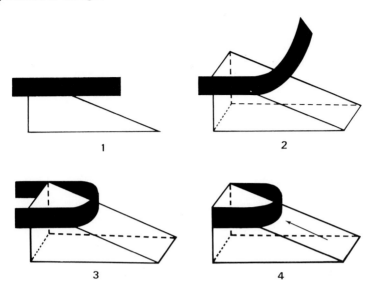

Fig. 3.14. Diagram illustrating the four stages in the making of a tape trough; arrow indicates the heel of the trough, which is sealed with dental wax or nail polish.

pressed against the glass; both upper (in level with the cutting edge) and lower edges of the tape should be pressed. It is then wrapped around the angled portion of the knife. Excess tape extending from both sides of the knife face is removed by a single slashing cut with a clean razor blade. The blade should cut the tape at an angle of about 45° so that it does not touch the back surface of the cutting edge. If some tape is left protruding beyond the cutting edge, it may touch the specimen block as it passes the knife, resulting in breakage of the seal. The trough thus obtained is about 0.8 cm long. Troughs that are too big should be avoided because of the difficulty in handling the floating sections properly in such a large space.

To make the trough waterproof, the region of the heel is sealed with molten dental wax, which is kept melted in a small dish on a hot plate or in any paraffin oven. Molten wax is applied on the outside of the trough to form the seal, using a small brush or cotton swab. Alternatively, nail polish can be used as a sealant. Although nail polish is easy to apply, it takes a little longer to dry. I prefer dental wax over nail polish. Since the inside of the tape is sticky, it must be kept clean; otherwise flotation fluid may become contaminated.

Another material used to prepare troughs is plastic microslides (Preiser Scientific, Louisville, Kentucky) (Gulati and Akers, 1977). Trapezoidal pieces are cut out with a razor blade (Fig. 3.15). Each piece is scored twice in the middle and bent 90° toward the scored side at each score line. The distance between the two

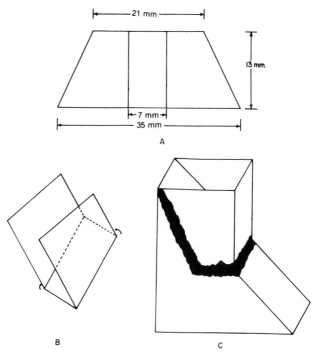

Fig. 3.15. (A) Diagram showing approximate dimensions of the plastic trapezoid used to make a trough. (B) Folding of the trough. (C) Plastic trough sealed onto the glass knife. (From Gulati and Akers, 1977.)

scores must be equal to the thickness of the glass used to prepare the knife. The trough obtained after bending the sides is sealed to the knife with Scotch superstrength adhesive or with nail polish. These troughs resist distortion and do not react with water.

For quick examination of one or two sections, a simple "trough" can be obtained by drawing a hydrophobic line across the front surface of the glass knife, using a wax pencil and making the line 6–8 mm distant from and parallel to the cutting edge. This line will hold a drop of water extending to the cutting edge. Sections floating on this drop are difficult to see because of the convex shape of the water surface. Alternatively, a wedge of bakelite can be cemented several millimeters below the cutting edge (Westfall, 1961).

MOUNTING THE SPECIMEN BLOCK AND KNIFE

The trimmed specimen block should be mounted in the specimen holder before mounting the knife, although exact orientation of the block will be accomplished later. This precaution avoids damaging the knife. The specimen block should be pushed into the collet jaw until a few millimeters extend outward. If the block is too long, it should be removed and shortened with a file. The shorter the block, the less the vibration during sectioning. Before the knife is mounted, the knife stage is withdrawn a sufficient distance from the mounted specimen block so that the cutting edge will not hit the specimen accidentally. This distance is also needed to allow the necessary orientation of the specimen block in relation to the mounted knife.

The holder with its securely fastened knife is mounted on the ultramicrotome, and the knife angle is set to the desired sectioning angle. The angle should be sufficiently large so that the specimen block clears the back of the knife after striking its edge. At the same time, this angle should be small enough to maintain knife sharpness. The clearance angle between the knife and the specimen block, as set on the ultramicrotome, is commonly in the range of 2 to 5°; knife holders have an engraved scale for determining this angle. The clearance angle is determined primarily by the properties of the specimen block and the shape of the cutting edge. Glass knives prepared from rhombi, for instance, inherently have a larger angle. Clearance angles larger than 5° rapidly dull the cutting edge.

The height of the knife should be level with the midpoint of the specimen arm movement. If the knife height is not optimal, the tangential relationship between the cutting edge and the radius of the arm movement will change, disrupting the movement of the arm through an arc. Ultramicrotomes are equipped with a reference mark for setting the knife height. The knife stage is moved laterally so that the useful part of the cutting edge comes in line with the face of the specimen block. The knife, the knife holder, and the knife stage should be clamped firmly. Many problems encountered during sectioning are caused by loose connections of these parts. Even the smallest vibrations resulting from a loose connection between the knife and its holder and between this holder and the ultramicrotome will adversely affect the quality of sections, and may lead to permanent damage to the knife.

The cutting edge should be parallel to the block face (Fig. 3.16B), so that the whole of the specimen block face is cut; in other words, it should be perpendicular to the specimen arm. If the cutting edge is positioned obliquely to the specimen arm, lateral distortions will occur in the sections. To obtain a straight ribbon, both the upper and lower edges of the block face should be made parallel to the cutting edge by viewing the block face under the binocular microscope and rotating the specimen holder. The shorter of the two parallel edges of the trapezoid block face should be facing down. This position will reduce compression because the shorter edge will strike the cutting edge first. If the block face is rectangular, the longer side should come in contact with the cutting edge first. The advantage of orienting the block face obliquely to the cutting edge to section hard laminated fibrous tissues is given on p. 159.

At the start of sectioning, the cutting edge is brought very close to the block

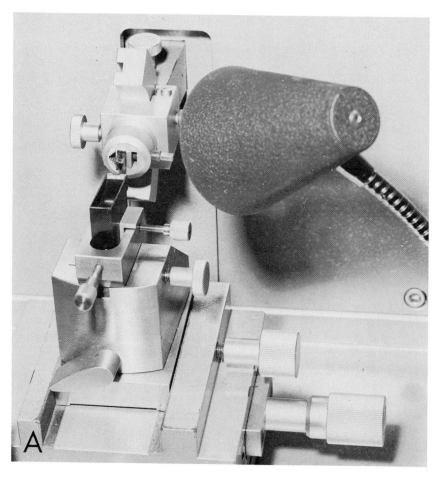

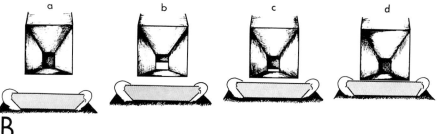

Fig. 3.16. (A) Photograph of high-intensity lamp positioned so as to cast a reflection of the diamond knife on the face of a specimen block mounted in chuck. This positioning facilitates the safe approach of knife edge to the block, as explained here and in the text. (B) Schematic representation of

face. This operation is risky and time-consuming and is carried out with the aid of a stereomicroscope at relatively high magnifications ($\times 30$–40) under the illuminating system of the ultramicrotome. The light is adjusted to obtain a bright white reflection from the block face. The knife is advanced first by using the coarse advance until it reaches as close as about 1 mm to the cutting edge. This distance indicates the gap between the cutting edge and the block face. With the fine advance, the knife is carefully advanced about 2 μm per microtome cycle until contact is made between the cutting edge and the block face. If any scratches are visible on the sections, a different area of the cutting edge should be used.

The aforementioned alignment of the knife with the block face is difficult. This difficulty can be minimized by the use of concentrated illumination from a small, high-intensity lamp positioned laterally adjacent to the microtome (Fig. 3.16A) (Nelson and Flaxman, 1973). When the light is directed on the knife, a reflection of the knife edge is seen on the block face as a bright band, the upper edge of which may be used for alignment. This procedure is useful for either glass or diamond knives and provides good visibility at magnifications up to $\times 40$. The reflection disappears when the knife edge has advanced to within 10 μm of the block face. At this stage the lamp is withdrawn and the block is allowed to cool for at least 1 min before the advance of the knife is resumed. This pause counteracts the thermal expansion of the block caused by the heat radiated from the lamp. This procedure is simpler than those using reflectors.

SECTIONING PROCEDURE

The ultramicrotome should be protected from external vibrations, draft, changes in temperature, and sunlight, and should be housed in a room without distractions. It is essential to use a chair with an adjustable seat height, because a comfortable seating position contributes to obtaining satisfactory sections. Before the sectioning, the worktable and other surfaces should be wiped free from dust with a wet sponge, and the forceps, pipette, and other instruments should be absolutely clean. The points of the forceps must be in perfect condition; they should be periodically cleaned with a solvent and should be protected with a tip shield when not in use. Only freshly prepared solutions and fresh water should be used.

The materials are as follows: trimmed specimen block, several glass knives

---

steps during approach of the knife to the block face when using the lamp: a, no reflection of the knife is visible on the block face when the knife and block face are separated by a distance greater than 200 μm; b, bright reflection appears on the lower portion of the block face as the knife approaches to within 100 μm of the block; c, reflection on the block face diminishes in height as the knife moves closer to the block; d, no reflection is visible on the block face when the knife advanced to within 10 μm of the block. After this last step, the knife should be advanced extremely carefully. (From Nelson and Flaxman, 1973.)

with a trough attached (or a diamond knife and a cleaning stick), fresh distilled water in a small bottle, acetone-rinsed grids in a Petri dish with its bottom mantled by layers of filter paper, a glass fiber drawn out from a micropipette into a fine thread (or a camel's hair brush with a single hair or an eyelash glued to one end of a stick), a small-bore syringe or pipette, chloroform in a small bottle with a stopper or a heat pen, cotton swabs, a pair of fine forceps with a locking device, lens tissue, filter paper, a Petri dish (with two layers of filter paper at the bottom) for transferring grids after picking up the sections, and a wire transfer loop for transferring ribbons of sections directly to the stains.

It is assumed that the worker is familiar with the operation of the microtome. The following step-by-step procedure should be helpful in obtaining ultrathin sections. Although this procedure has worked well in my laboratory, other procedures may of course work equally well or even better.

1. Reset the advance mechanism of the microtome.

2. Retract both the coarse and fine stage advances. This is an essential precaution especially when using a diamond knife to avoid bumping the knife and the specimen block into each other.

3. Mount the trimmed specimen block in its holder and screw tightly; make certain that the long axis of the block face is parallel to the cutting edge.

4. Gently wipe off any resin debris present on or near the block face with a strip of lightly moistened lens paper.

5. Mount a glass knife or diamond knife in its holder and screw tightly. The height of the knife should be adjusted.

6. Set the knife angle between 2 and 5°, depending on the type of the specimen. Knife holders are provided with a scale to determine this angle.

7. Cycle the microtome manually and stop when the face of the block is slightly above the height of the cutting edge.

8. Position the binocular microscope and the illumination system so that the block face can be viewed through the microscope.

9. Bring the mounted knife forward slowly and carefully until the cutting edge and the block face can be viewed through the microscope.

10. Using a pipette, add distilled water in the trough until the cutting edge is wet. With a proper angle of illumination, a uniformly bright surface of the water normally indicates a slightly negative meniscus level.

11. Adjust the angle of illumination so that some light is reflected against the block face.

12. Move the knife by laterally adjusting the knife holder so that sections will be cut by the best portion of the cutting edge.

13. While viewing through the microscope, very slowly advance the knife toward the block face, by using first coarse adjustment and then fine adjustment. Advance the knife in increments of 1 μm between each stroke, until the first

contact between the cutting edge and the block face is made. The importance of this first contact cannot be overemphasized, since a very thick section can damage both the knife and the block face.

14. Adjust the desired cutting speed (2–3 mm/sec can be used as a starting point) and section thickness (80 nm, pale gold in color). If the need arises, the thickness should be changed only during the return part of the cycle, not during the cutting part.

15. Adjust the meniscus level so that the interference color of the sections can be seen clearly.

16. Make sure that the ribbon is straight and that no knife marks are seen on the sections.

17. After the desired number of sections have been cut, gently paddle them away from the cutting edge as well as from the sides of the knife and group them together in the middle of the trough by using an extremely fine glass fiber.

18. Expand the sections by holding a cotton swab dipped in chloroform close to but not in contact with the sections for 6 sec.

19. Lower the grid onto the sections and collect them on the dull side of the grid. Alternatively, bring the grid up under the floating sections; I prefer the former.

20. Place the grid on a filter paper (section side up) to dry, and cover the Petri dish.

Throughout the cutting process, the quality of sections should be closely watched, and the correct meniscus level should be maintained by adding or removing small amounts of water with a syringe. Alternatively, one can use a commercially available device for precisely controlling the water level in a knife trough (E. F. Fullam, Inc., Schenectady, New York). The hand control allows microamounts of water to be pumped in or out of the trough to maintain the proper meniscus. Advanced ultramicrotomes have a built-in device for controlling the meniscus level. Only enough sections should be cut at one time to be mounted on one grid. The same portion of the glass cutting edge should not be used for cutting more than one or two ribbons. Periodic checks ought to be made to ensure that the knife, specimen block, and knife stage assembly are tightened firmly. After such checks, the block face will require realignment in relation to the cutting edge. Once the first section has been cut, every stroke must produce a section; otherwise the sections will be either too thick or too thin, or the block face will rub against the clearance facet of the knife. For a detailed discussion on sectioning, see Hayat (1981a).

The problem of static electricity shown by grids kept in a plastic Petri dish can be solved by spraying the lids (inside up) with an antistatic spray such as Hansa (Agar Aids). The lids are sprayed lightly from a distance of 1 ft; this treatment lasts for the life of the dish.

Sectioning speeds of 1.5–2 mm/sec for diamond knives and 1.5–3.5 mm/sec for glass knives are recommended. Diamond knives should not be used at cutting speeds exceeding 2 mm/sec. Sections of a large block face are easier to cut at slower speeds. To obtain a fairly constant section thickness for serial sectioning, sections should be cut at a slow speed (about 0.5 mm/sec). When difficulties arise, lowering the sectioning speed helps. A pause in sectioning for several minutes is also helpful.

The durability of glass knives can be improved by coating the cutting edge with a film of evaporated tungsten metal (Roberts, 1975). Coated knives are thought to section hard specimens better than standard glass knives. Freshly prepared knives are attached to the base plate of a vacuum evaporator with adhesive tape. They are hinged to allow tilting toward the evaporation source. The film is deposited on both faces of the knife simultaneously. The evaporation source is a 25-mm long and 0.5-mm thick tungsten wire (V-shape filament), which is mounted at a distance of 8–10 cm from the cutting edge. A pressure of 5 $\times$ 10^{-5} torr for 3–5 min is required to provide a sufficiently thick film. A glass knife with an angle of 50° seems to be more durable than that with an angle of 45°; the former also has a longer cutting edge.

Glass knives are thought to section hard tissues embedded in epoxy resins with less difficulty when 1% Dow Corning 200 fluid silicone plastic additive (Dow Corning, Midland, Michigan) is added to the embedding medium (Langenberg, 1982b). It has been claimed that glass knives last 5–15 times longer when cutting modified rather than unmodified resin. The silicone additive is added to the resin mixture while stirred and just before addition of the accelerator.

The thickness of a section is commonly estimated by observing the interference color of light reflected from the section while it is floating on the surface of a liquid in the trough. Since the interference colors form a continuous spectrum instead of clearly separate colors, the following color scale provides only a guide to the actual thickness:

Interference color	Approximate thickness (nm)
Gray	60
Silver	90
Gold	150
Purple	190
Blue	240

DEFECTS APPEARING DURING SECTIONING

The defects that can occur during sectioning, their possible causes, and their remedies are described in Table 3.1.

TABLE 3.1.

Defects Appearing during Sectioning

Defects	Possible causes	Remedies
I. Sections are not cut	1. Face of block rubbing on back of knife	1. Increase clearance angle
	2. Knife, knife holder, or block loose	2. Tighten them up
	3. Vibrations and/or change in temperature	3. Check possible causes of vibration or change in temperature such as draft, lights, and air conditioner
	4. Advance mechanism reached to the end	4. Reset advance mechanism
	5. Blunt knife edge	5. Replace knife
	6. Face of block away from knife edge	6. Bring face of the block close to knife by cycling microtome while still advancing by fine adjustment
II. Sections vary in thickness	1. Knife, knife holder, or block loose	1. Clamp them tightly
	2. Face of block too large	2. Reduce size by trimming
	3. Knife tilted too much forward or backward	3. Adjust knife angle (2–5°)
	4. Cutting speed too fast	4. Reduce speed
	5. Blunt knife edge	5. Try another portion of knife edge; get knife resharpened; replace knife
	6. Advance setting too much reduced to get very thin sections	6. Increase advance setting and after some time try again to reduce it
	7. Interruption in microtome cycling	7. Cycle microtome rhythmically
	8. Mechanical vibration or thermal draft	8. Check possible causes including draft and heaters
	9. Too soft block	9. Heat block in oven at 60–80°C for 24 hr or change block
	10. Defective mechanism or installation of microtome	10. Check and reinstall or return microtome to maker

(*continued*)

TABLE 3.1. (*Continued*)

Defects	Possible causes	Remedies
III. Variations in thickness within a section	1. Dull knife 2. Uneven consistency of specimen block 3. Vibrations 4. Hydrated hardener	1. Change knife 2. Retrim block and remove plain resin 3. Remove cause of vibration 4. Use fresh supply
IV. Block lifts section	1. Meniscus too high 2. Too small clearance angle 3. Upper edge of block face or knife is dirty 4. Fluid drop on block face 5. Fluid drop on back of knife 6. Too soft block 7. Face of block electrified 8. Knife, knife holder, or block loose	1. Remove fluid from trough just enough to achieve bright reflection under binocular microscope near knife edge 2. Increase clearance angle by tilting knife backward 3. Replace glass knife; clean diamond knife with a stick; clean tissue block face with lens paper 4. Clean block face with lens paper 5. Replace glass knife; dry diamond knife with lens paper 6. Heat block in oven at 60–80°C for 24 hr or change block 7. Increase room humidity; ionize air with high frequency charge 8. Tighten them up
V. Block face gets wet	1. Drop of fluid on vertical back of cutting edge 2. Incomplete infiltration 3. Too much fluid in trough 4. Too slow cutting speed 5. Debris on back of cutting edge	1. Remove drop with a piece of lens paper 2. Change specimen block 3. Lower fluid level 4. Increase cutting speed 5. Clean by wiping with a hair or lens paper

Problem	Cause	Remedy
VI. Ribbon is not formed	1. Too high meniscus causing section to leave knife edge as soon as it is cut	1. Lower meniscus level
	2. Upper and lower edges of block face are not parallel	2. Retrim block face; coat sides of block with a thin layer of Tackiwax
	3. Upper and lower edges of block face are not straight	3. Retrim block face
	4. Cutting speed too slow	4. Increase cutting speed
	5. Block face or knife area electrified	5. Touch block face with moist lens paper to remove charge; increase humidity so that charge may leak out into moist air
	6. Unsteady or break in cycling rhythm of microtome	6. Cycle microtome steadily and without interruption
	7. Upper edge of block face not parallel to cutting edge	7. Reorient or retrim block face
	8. Leading end of ribbon is obstructed by the side of trough or debris in trough fluid	8. Center ribbon in trough and remove debris
	9. Debris adhered to cutting edge	9. Clean cutting edge
VII. Ribbon is curved	1. Upper and lower edges of block face not parallel	1. Retrim block face
	2. Knife not uniformly sharp causing differential compression across section	2. Try another portion of knife; get knife resharpened; replace knife
	3. Upper and lower edges of block face not parallel to knife edge	3. Reorient block face
	4. Edges of block parallel to knife edge not straight	4. Retrim block face
VIII. Sections are difficult to see	1. Too much fluid in trough	1. Lower fluid level
	2. Incorrect angle of illumination	2. Adjust angle of illumination

(continued)

TABLE 3.1. (*Continued*)

Defects	Possible causes	Remedies
IX. Sections show scratches	1. Fine nick in knife edge	1. Try another portion of knife edge; get knife resharpened; replace knife
	2. Dirty knife edge	2. Replace glass knife; clean edge of diamond knife
	3. Hard material in specimen	3. Replace specimen or use diamond knife
X. Sections crumple or stick to knife edge	1. Low meniscus level causes dry cutting edge	1. Raise meniscus level
	2. Dirty knife edge	2. Replace glass knife; clean edge of diamond knife
	3. Blunt knife edge	3. Replace glass knife; get diamond knife resharpened
	4. Too small knife angle	4. Increase knife angle
XI. Sections show wrinkles	1. Poor embedding	1. Heat block in oven at 60–80°C for 24 hr; get another block
	2. Dull or uneven sharpness of knife edge	2. Use different portion of knife edge; get knife resharpened; replace knife
	3. Too large block face	3. Reduce size of block face
	4. Knife, knife holder, or block loose	4. Tighten them up
	5. Normal compression	5. Use vapors of organic solvent or heat
XII. Sections show tiny holes	1. Tiny air bubbles in block face	1. Eliminate bubble by retrimming block face

2. Debris on knife edge		2. Try another portion of knife edge; replace knife; remove debris with moist lens paper
3. A piece of hard material in specimen		3. Bring up another specimen using a different preparatory procedure; use a harder block corresponding to hardest part in specimen
XIII. Specimen crumples and drops out of section	1. Imperfect embedding	1. Extend duration of infiltration
XIV Sections show chatter	1. Dull knife	1. Change knife
	2. Very hard resin	2. Use softer resin mixture
	3. Too fast or unsteady cutting	3. Cycle microtome slowly & rhythmically
	4. Too large knife tilt	4. Reduce angle
	5. Chatter localized on section	5. Change area of knife or portion of block face
XV. Block too brittle	1. Excess hardener	1. Reduce the proportion of hardener; include dibutyl phthalate as a plasticizer
	2. Wrong ingredient mix	2. Prepare fresh embedding mixture
XVI Blocks discolored	1. Old catalyst	1. Use fresh supply
XVII. Blocks too soft	1. Incomplete polymerization	1. Repolymerize; increase polymerization temperature
	2. Insufficient catalyst	2. Increase the proportion of catalyst
	3. Wrong ingredient mix	3. Prepare fresh embedding mixture

SERIAL SECTIONING

For most studies the two-dimensional information yielded by study of individual sections is adequate. A third dimension may be added to individual sections by stereomicroscopy or be inferred from stereologic methods. However, to analyze the form and structural relationships of cells or their components in greater depth, their structural details can be reconstructed in three dimensions from examination of serial sections. Serial sections can also aid in locating specific constituents within cells and their relative numbers. The following method has been presented by Wells (1974).

A glass microscope slide is polished with fine alumina powder, dipped into a solution of 0.3% Formvar in dichloroethane, and drained vertically. The film is cast onto the water surface (Fig. 3.17A) and plastic rings (6 mm diameter, 1 mm thick, and 2 mm high) are carefully placed onto the film (Fig. 3.17B). The filmed rings are removed with a piece of coarse paper towel (Fig. 3.17C) and coated in an evaporator with a thin layer of carbon (Fig. 3.17D). A ribbon of sections is maneuvered within the single slot of the grid (Fig. 3.17E and F), which is then placed on a large drop of water on a suitable hydrophobic surface (Fig. 3.17G).

After the sections are stained and washed, the grid is placed on a ring containing the Formvar film (Fig. 3.17H); the filmed ring has been attached to a filter paper with double-sided Scotch tape for preventing surface tension from lifting the ring toward the grid. After air drying in a Petri dish, the grid, along with the supported ribbon, is removed from the ring by inverting the assembly over a peg of the same diameter as the grid (Fig. 3.17I and J). The surface of the peg and the support film do not come into contact. This step is carried out using a dissecting microscope.

Several methods are in use for visualizing and reconstructing three-dimensional structures from a series of electron micrographs, and they have been reviewed by Hayat (1981a). According to one method, images from the negatives are enlarged onto sheet film, which is developed as a positive print (Brown and Arnott, 1971). The positive image on the film is, in effect, a transparency. The transparencies are stacked between glass or Plexiglas plates that approach a thickness of 90 nm on the magnified scale. Usually 6–10 transparencies can be analyzed simultaneously. In order to use this technique, it is essential that the original negatives are of high contrast.

Street and Mize (1983) have developed a computer system which digitizes serial sections through cells, aligns the sectional profiles, and displays the completed reconstructions at different rotations. Apparently serial sections need to be cut, collected on grids, stained, and photographed with the TEM. This system works for both light and electron microscope materials and can be used with electron micrographs, slides, or back-projected film strips. The role of the operator is necessary in data entry and analysis during reconstruction.

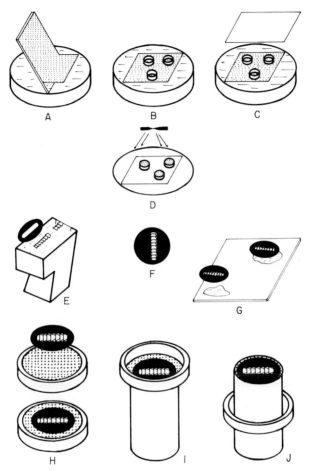

Fig. 3.17. Method for collecting ultrathin serial sections. See text for details. (From Wells, 1974.)

SPECIFIC METHODS

En Face Sectioning

En face sections facilitate a three-dimensional reconstruction of the fine structure of urothelium and other types of epithelial tissues. Fixed tissue pieces (1 mm²) are placed lumenal side down (each in a drop of Spurr's resin) on a 7-mm-thick Mylar film supported on a flat surface (Fig. 3.18A) (Walton *et al.*, 1982). The upper surface of the tissue is covered with another piece of Mylar, and the tissue is flattened with a brass weight (5 × 8 × 1 cm). After polymerization in an

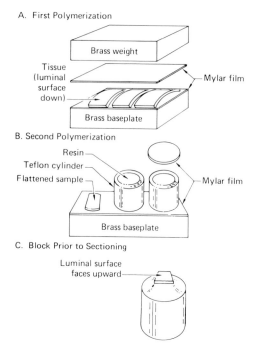

A. First Polymerization

Brass weight

Tissue (luminal surface down)

Mylar film

Brass baseplate

B. Second Polymerization

Resin

Teflon cylinder

Flattened sample

Mylar film

Brass baseplate

C. Block Prior to Sectioning

Luminal surface faces upward

Fig. 3.18. *En face* sectioning technique. (A) A sample of bladder tissue is placed lumenal side down onto Mylar film in a small drop of epoxy resin, and then the resin is allowed to polymerize. (B) Additional resin is added to the upper surface of the tissue and allowed to polymerize. (C) After the Mylar film and Teflon cylinder are removed from the block, it is trimmed and aligned for immediate collection of thin sections. (From Walton *et al.*, 1982.)

oven at 60°C for 24 hr, the brass weight is removed, and the upper Mylar film is peeled off.

A 0.5-cm cylinder of silicone rubber (Teflon) (internal diameter 6 mm) is placed over the tissue. This well is filled with epoxy resin, overlaid with a small piece of Mylar to prevent leakage, and the resin is polymerized at 60°C (Fig. 3.18B). Both pieces of Mylar and the Teflon cylinder are removed from the polymerized resin block containing the tissue piece (Fig. 3.18C). Excess resin is trimmed from around the specimen. The block face is carefully aligned to the knife edge prior to sectioning.

Sectioning Hard Mineral Fibers in Tissues

During conventional ultrathin sectioning of tissues containing hard mineral fibers (e.g., crocidolite and amosite), the fibers tend to tear and fall out of the sections. Johnson and Ibe (1981) have introduced a method of sectioning that permits the retention of fibers *in situ*. Semithin sections (0.3–0.5 μm) are cut at a speed of 1 mm/sec (cutting angle 8°) with a diamond knife. The sections are picked up on folding grids and poststained by immersion in uranyl acetate for 15 min and then in lead citrate for 8 min. The sections are viewed in the scanning transmission electron microscope at an accelerating voltage of 200 kV. Fibers can be present within cells both in vacuoles and free in the cytoplasm.

Sectioning Hard Laminated Fibrous Tissues

Laminated fibrous tissues (e.g., mineralized dermal tissues of vertebrates) may be composed of several superimposed layers of collagen or chitin. The orientation of these fibers varies from one layer to another. A mineral phase within the fibrous organic matrix may also exist in such tissues. This heterogeneous structure is difficult to section. If fibers in the tissue are oriented parallel to the knife edge, good sectioning becomes almost impossible. It is difficult to orient all the fibers perpendicular to the knife edge because the orientation of fibers varies from layer to layer in a single tissue block.

Allizard and Zylberberg (1982) have presented a method of orienting the specimen block face that greatly improves sectioning of heterogeneous tissues such as fish scales. Figure 3.19A shows the heterogeneous composition of a fish scale; the block face is oriented so that the two longest edges are parallel to the cutting edge. During sectioning, the unmineralized part of the tissue comes in contact first with the knife edge. In Fig. 3.19B, on the other hand, the block face

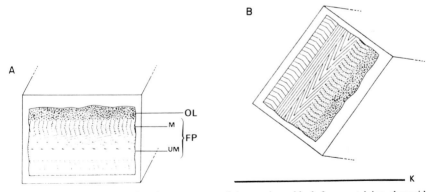

Fig. 3.19. Diagram showing the appearance of the specimen block face containing elasmoid scales of the goldfish. (A) This orientation will result in sections perpendicular to the surface of the scale. The osseous layer (OL) constitutes the upper part of the scale and is always mineralized. The fibrillary plate (FP) is organized in a partially mineralized plywoodlike structure; M, mineralized part of the fibrillary plate; UM, unmineralized part of the fibrillary plate. (B) The block face is oriented obliquely; K, knife edge. (From Allizard and Zylberberg, 1982. © 1982 The Williams & Wilkins Co., Baltimore.)

is oriented obliquely to the cutting edge so that the mineralized part of the scale is cut before the unmineralized one. Moreover, the first impact with the cutting edge is kept to a minimum, resulting in better sections.

Vertical Sections of Cultivated Anchorage-Dependent Cells

The following method is useful for preparing cultivated monolayers of cells for vertical sectioning, which allows observation of the relationships between the cells and the growth surface (Pentz *et al.*, 1981, 1983). A tissue culture slide chamber (TCSC-1) (Heraeus GmbH, D-6450 Hanau, W. Germany) is connected to the surface on which cells are grown (Fig. 3.20). All procedures are carried out in this chamber. Cells are grown on a polyester foil which, after polymerization, can be easily taken off, and a second resin layer is poured over the embedded cells. This bilayer of the resin permits vertical sectioning.

Cells are grown in nutrition medium, HAM F IO (see Appendix), containing 10% fetal calf serum on a polyester foil (23 μm; ICI) in the chamber. This foil is resistant to organic solvents. The cover from the PTFE-adaptor of the chamber is removed and the medium is pipetted off. The cells are carefully washed twice with the medium (37°C) without serum and fixed with 2% glutaraldehyde in 0.1 M cacodylate buffer (pH 7.2) for 1 hr at room temperature. After being washed twice in buffer, cells are postfixed with 1% OsO_4 in buffer for 1 hr at room temperature. They are then dehydrated, infiltrated with ERL, and polymerized at 65°C for 48 hr. During polymerization, the cover of the adaptor is displaced. The adaptor is unscrewed and the underlying foil can be pulled off. The resin block can be tapped out with the help of a hollow cylinder and a plunger (Fig. 3.20). A thin fiber of resin is layered onto the resin block so that the resin is present on all sides of the cells. Embedded cells can be observed with an inverted, phase-contrast microscope, and desired cells can be marked with a needle.

In Situ Thin-Section Microscopy of Cell Cultures (Handley *et al.*, 1981a)

The following method is useful for preserving topographic cell–substrate relationships. In this method tapered rotary beveling is employed to reduce the thickness of the Petri dish to a dimension suitable for direct thin sectioning. Cells grown in Dulbecco's Modified Eagle Medium with 5% fetal calf serum are fixed with 2% glutaraldehyde in buffer for 30 min at room temperature. To visualize extracellular matrix, cells are postfixed with 1% OsO_4 containing 0.1% ruthenium red for 2 hr. The cells are treated with 2% uranyl acetate, dehydrated, and embedded in Spurr's resin, using ethanol as transitional fluid to avoid dish solvation.

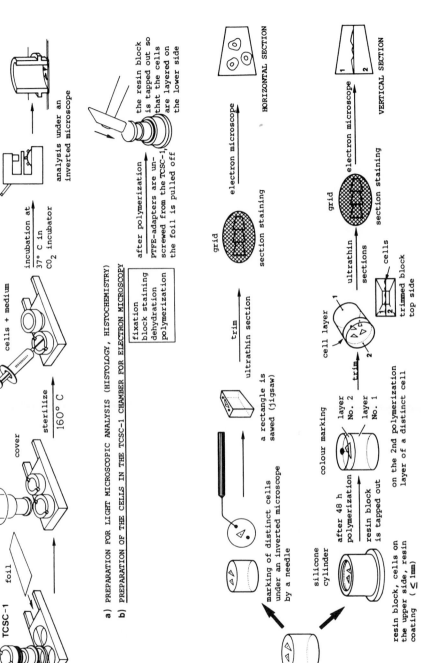

TCSC-1

foil

cover

sterilize
160° C

cells + medium

incubation at
37° C in
CO$_2$ incubator

analysis under an
inverted microscope

a) PREPARATION FOR LIGHT MICROSCOPIC ANALYSIS (HISTOLOGY, HISTOCHEMISTRY)

b) PREPARATION OF THE CELLS IN THE TCSC-1 CHAMBER FOR ELECTRON MICROSCOPY

fixation
block staining
dehydration
polymerization

after polymerization
PTFE-adapters are un-
screwed from the TCSC-1,
the foil is pulled off

the resin block
is tapped out so
that the cells
are layered on
the lower side

grid

electron microscope

HORIZONTAL SECTION

section staining

marking of distinct cells
under an inverted microscope
by a needle

a rectangle is
sawed (jigsaw)

trim

ultrathin section

silicone
cylinder

colour marking

after 48 h
polymerization

resin block
is tapped out

layer
No. 2

layer
No. 1

on the 2nd polymerization
layer of a distinct cell

cell layer

trim

1

2

trim

2

1

trimmed block
top side

cells

ultrathin
sections

grid

electron microscope

VERTICAL SECTION

section staining

resin block, cells on
the upper side, resin
coating (≤ 1mm)

Fig. 3.20. Preparation of cultivated anchorage-dependent cells. (From Pentz *et al.*, 1983.)

The bottom of the resin-filled dish (Fig. 3.21, 1) is beveled by using a lathe at a 4–8° angle in successive cuts until a circumferential cell-free zone is formed (Fig. 3.21, 2). Standard blocks (5 × 10 mm) are cut to include a portion of the cell-free zone (Fig. 3.21, 3A) and rough-sectioned with a glass knife until a 50–200-μm leading edge of the Petri dish is obtained (Fig. 3.21, 3B). Both trimming into a rectangular face and vertical thin-sectioning (Fig. 3.21, 3C) reduce the chatter distortions normally encountered with samples having dissimilar cutting properties.

Resectioning of Semithin Sections (Campbell and Hermans, 1972)

The following method is useful for resectioning semithin resin sections that have been stained for light microscopy, mounted on slides, and examined under oil immersion. Sections of 0.5–4 μm thickness are picked up from a water surface with a fine brush or a wire loop. They are placed on a drop of water on a glass slide, and allowed to dry at 25–100°C. They are then stained with 1% toluidine blue in distilled water or 1% aqueous borax solution and stored without coverslips. For high magnification study, they are covered with a drop of immersion oil.

After examination with the light microscope, the sections are cleansed of immersion oil by flushing them with a stream of xylene. Individual sections are lifted from the slide by placing a drop of water beside each section and sliding a sharp needle under it. As the section is freed from the slide surface, the drop of water raises it. When the entire section is loosened and is floating on the liquid surface, it is transferred (Fig. 3.22A) to the surface of a water droplet on a preshaped Epon block (explained below). The section can be transferred on the

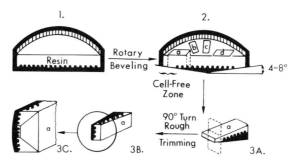

Fig. 3.21. Cross-sectional view of *in situ* embedding before (1) and after (2) rotary beveling. Cells are denoted as closed circles. Regular-size blocks are easily cut (3A) and rough-trimmed (3B) before vertical thin sectioning (3C). (From Handley *et al.*, 1981a.)

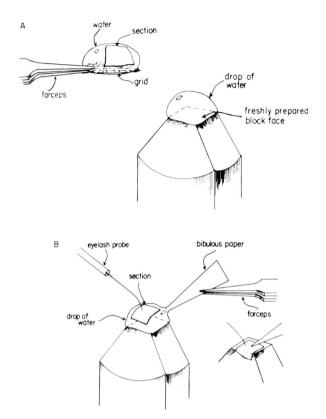

Fig. 3.22. (A) The section, suspended on the water surface, is transferred to a drop of water on a smooth, freshly sectioned block face. (B) The section is held in place with an eyelash probe while water is removed with a fine wedge of bibulous paper. Inset at right: after most of the water has been withdrawn, the surface of the section is stroked with the moist bibulous paper to smooth and flatten it to the surface of block. (From Campbell and Hermans, 1972. © 1972 The Williams & Wilkins Co., Baltimore.)

surface of liquid held in a wire loop, on the tip of a needle, on a razor blade fragment, or on a grid.

An Epon block is trimmed into a pyramid and then smoothly truncated on the microtome. The surface is made smooth by finally producing ultrathin sections from it. The block face should be as large as the semithin section to be mounted. A drop of water is placed on the surface and the section floated on it (Fig. 3.22B). The water drop is carefully absorbed with a filter paper while the section is held in place with an eyelash probe (Fig. 3.22B). The surface of the section is stroked smooth with the paper wedge to ensure that the section does not wrinkle as it dries on the block surface (Fig. 3.22B).

SEMITHIN SECTIONING

Introduction

The use of semithin resin sections is firmly established in histology and clinical diagnostic studies. Because embedding in resins preserves cellular components better than embedding in paraffin, the importance of semithin (thick) resin sections is obvious. During hardening, resins shrink less than does paraffin. Semithin sections of resin-embedded specimens provide greater definition of cellular details in light microscopy than that obtained in sections of paraffin-embedded tissues. Resin sections (1 μm thick) have only about 20% of the image blurring that occurs in paraffin sections (5 μm thick). Semithin sections provide a bridge between the images obtained with a light microscope and those obtained with an electron microscope. These sections (1 μm thick) are about five times thinner than an average paraffin section, and many times thicker than an ultrathin section. The image of a semithin section resembles a low-magnification electron micrograph. It therefore provides the identification of corresponding cellular regions in photomicrographs and electron micrographs.

Semithin sections can be examined under a phase-contrast microscope or, after staining, under a standard light microscope. The specificity of staining achieved in semithin sections by using chromatic stains has not been obtained in ultrathin sections. Complex intracellular as well as extracellular structures are more easily and accurately studied by using semithin sections than by using serial ultrathin sections. Semithin sections are more easily cut and handled and are physically more stable. Finally, spatial relationships between cellular components are better preserved, without significant overlapping of structures.

By determining precisely the area of interest in a semithin section prior to ultrathin sectioning, one can enhance the accuracy of the study and save a considerable amount of time. One procedure is to locate the area of interest in a semithin survey section (stained or unstained) under the light microscope and then trim the block face so that it contains only the region of interest or mainly this region. This block is used for ultrathin sectioning. A more accurate approach involves the use of the Target Marker that is attached to the LKB Pyramitome. This allows the image of a stained semithin section to be superimposed over the face of the specimen block. Since cellular details are clearly seen in the semithin section, the block can be trimmed down precisely to the desired area. Another advantage is that the same semithin section can be used for both light and electron microscopy, allowing correlative studies.

A disadvantage of using semithin resin sections rather than paraffin sections is the small size of the specimen that can be viewed. Paraffin sections measuring about 600 mm² can be cut, whereas the size of semithin sections cut by an ultramicrotome is usually less than 1 mm². However, this difficulty can be circumvented by using a long-edged Ralph knife (discussed later) instead of a Latta & Hartmann triangular glass knife. The latter's cutting edge extends only up to 10 mm, whereas the former has a much longer cutting edge. Moreover, microtomes designed specifically for large-area sectioning of resins have been introduced: Sorvall JB-4A (Du Pont Instruments), HistoRange (LKB), and Autocut (Reichert Jung). With these microtomes, specimens embedded in resin blocks can be cut as large as 12 × 16 mm. These microtomes can cut block faces about 200 times larger than those used in electron microscopy, and a section thickness ranging from 0.25 to 10 μm can be obtained. They can also cut paraffin blocks up to 40 mm long and accept glass or steel knives. Large-size semithin sections of glycol methacrylate resin can be easily cut.

Ralph Knife

Bennett *et al.* (1976) have published a method for making long-edged glass knives, which they called the Ralph knives in honor of the late Dr. Paul Ralph, who invented the method. The width (not the thickness) of glass strips determines the length of the cutting edge. The glass strips commonly used have a width of about 25 mm and a thickness of about 6.5 mm (the same width is used for the Latta & Hartmann type of knife), although, theoretically, glass knives of unlimited edge length can be produced; an edge length of 4 cm has been produced. The choice of the length of the cutting edge is based on the principle that the longest length of a block face to be sectioned must be shorter than the length of the knife edge. Approximate lengths of the cutting edges of the Latta & Hartmann and the Ralph knives are compared in Fig. 3.23.

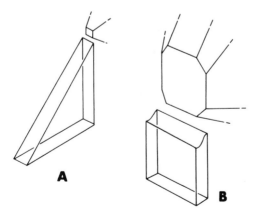

Fig. 3.23. (A) Latta & Hartmann glass knife aligned with the specimen block face. (B) Ralph glass knife aligned with the specimen block face; the larger block face and longer knife edge are apparent.

Both ultrathin and semithin sections undergo compression during the cutting process. The compression of semithin resin sections ranges between 4 and 19% (Helander, 1984); nuclei show less compression than that of the cytoplasm. The degree of compression depends on manifold factors including the angle of the cutting edge, section thickness, and ambient temperature. Generally, the larger the edge angle, the higher the compression. Thus the use of knives with small edge angles is recommended. However, resin sections are difficult to cut when the edge angle is smaller than 35°. Figure 3.24 shows the best angle (shape) of the cutting edge of the Ralph knife. The edge angle of these knives varies between 12 and 58° (Helander, 1984). The right portion of the cutting edge is superior to the left portion; the latter has a large number of ridges that can be visualized with the SEM.

The Ralph knives can be made by hand (Bennett *et al.*, 1976) or with one of the commercially available instruments: Histo KnifeMaker (LKB) and Longknife Maker (Polaron Instruments). Histo KnifeMaker provides glass knives 25 or 36 mm long. Glass knives are superior to steel knives even for cutting large sections. Satisfactory sections thinner than 2 μm are difficult to cut with steel knives. The Ralph knife can be used immediately after it has been made by attaching it to a glass knife holder (Fig. 3.25) (Szczesny, 1978); this assembly is then placed in a Sorvall JB-4 microtome steel knife holder for cutting thick sections.

The preparation of this glass knife holder is as follows. A piece of steel (4 × 1

Fig. 3.24. Photomicrographs of the Ralph knives. Different types of knives are shown with different angles. The knife marked with an asterisk in (A) and (B) has an angle which is optimal for majority of specimens. The knives shown to the left are too extreme in their angles and those to the right are excessively flat. (From Johansson, 1983.)

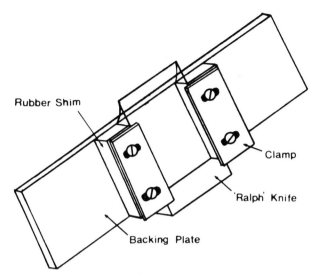

Rubber Shim

Clamp

Ralph Knife

Backing Plate

Fig. 3.25. Holder for the Ralph glass knives, with the side of the assembly facing the block to be sectioned. Note the curvature of the edge of the knife and the position of the overhanging clamps. (From Szczesny, 1978.)

$\times \frac{1}{8}$ in.) is used as a backing plate. A total of four holes, tapped for $\frac{1}{8}$-in. machine screws, are placed in $1\frac{1}{4}$ in. from each end of the backing plate and $\frac{1}{4}$ in. from each longer edge. Two pieces of hard rubber ($1 \times \frac{7}{16} \times \frac{1}{4}$ in.) are used as shims. Two centrally located horizontally elongated $\frac{1}{8}$-in. holes, $\frac{1}{4}$ in. from each shorter edge, are made in the face of each piece. Two pieces of stiff metal ($1 \times \frac{1}{2} \times \frac{1}{10}$ in.) are used as clamps. Each of these has two $\frac{1}{8}$-in. horizontally elongated holes, centered $\frac{1}{4}$ in. from each shorter edge and $\frac{7}{32}$ in. from one of the longer edges, cut in the face of each piece to match the holes in the rubber blocks. Four $\frac{1}{8}$-in. machine screws are used.

The rubber shims are glued to the clamps with an epoxy cement, ensuring that the elongated holes are lined up and the clamp has a $\frac{1}{16}$-in. overhang on one side of the rubber shim. This $\frac{1}{16}$-in. overhang is sufficient to secure the knife, which is about 1 in. wide and $\frac{1}{4}$ in. thick. These units are attached to the backing plate with the machine screws, and a knife is placed between the units with the cutting edge facing the clamps (Fig. 3.25). The shim-clamp units are pushed into contact with the edge of the knife and their screws tightened. The assembly is now ready for use.

A modified knife holder for use with the Sorvall Porter-Blum MT-2 ultramicrotome has been described by Gorycki and Sohm (1979), while details of the construction of the Ralph knife holder (Fig. 3.26) for use with MT-1 are given by Smith and Wren (1983).

Instead of using razor blades with the Vibratome as is generally done, one can

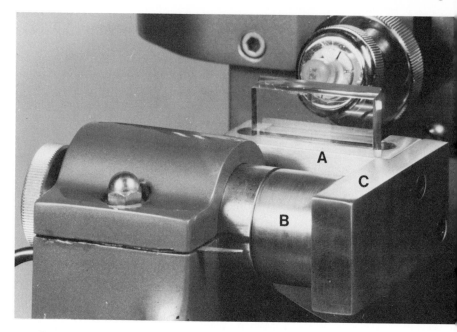

Fig. 3.26. The Ralph knife holder in position on a "Porter-Blum" MT-1 ultramicrotome: A, knife holder; B, locating boss; C, attachment arm. (From Smith and Wren, 1983.)

use Ralph knives by attaching them to this instrument with a special holder (Electron Microscopy Sciences, Fort Washington, Pennsylvania) (Johansson, 1983). The sectioning is carried out at a vibration rate of 7 scale units, a feeding of 2 scale units, and a thickness of 5–50 μm.

Glass knives are better than razor blades for freehand trimming of the large glycol methacrylate blocks. The former are especially useful for obtaining smooth block faces. Butler (1980) has described the procedure for specimen block trimming. A disadvantage of the Ralph knife is that it tends to dull rather rapidly. The durability of this knife can be increased by reducing the clearance angle. Hard specimens such as undecalcified bone are sectioned better with tungsten carbide knives.

Sectioning

Tissue blocks are trimmed and mounted so that on each cutting stroke the knife cuts through about 1 mm of resin before encountering the tissue. Semithin sections (0.5–2.0 μm) are cut by using a slow sectioning speed on a dry glass knife. As the section begins to be cut, the edge of the section is lifted from the knife with a brush (or forceps) to minimize folding. Superficial folds can be removed by warming the section with one's breath. Deeper folds can be eliminated by floating the section on water at room temperature. If this fails, folds can be pulled apart with brushes or forceps while the section is on a glass slide wet with water. Alternatively, the section can be immersed in 100% ethanol for a few minutes and then floated on water. It is then mounted on a glass slide freshly cleaned with ethanol. Mounting is accomplished by inserting the slide at an angle into the water and bringing it up to the surface under the section. The slide with the section is removed from the water, and the section is positioned with a brush and allowed to dry on a hot plate at 60°C.

The resistance of the knife to the resin block can be minimized by sectioning onto a water-filled trough, as is done in conventional ultramicrotomy. A trough of suitable size can be made from a discarded plastic coverslip container (Fig. 3.27A and B) (Stephenson, 1982). A larger trough (40 × 76 mm) is especially useful for collecting serial sections. The transparent lid is marked by offering the glass knife to its longest side and scribing along the touching glass edges. Cuts are made along the scores with a jeweler's saw. The recess should be the same thickness as the knife. The waste plastic is chipped out after being scribed with a scalpel blade, and if necessary, the sides are filed to fit the knife closely. The trough can be attached to the knife mounted in its holder on the microtome; the plastic–glass contact is then sealed with dental wax (Fig. 3.27C). When the trough is filled with water, the sections can be lifted onto a corner of a glass slide. After the sections are positioned on the slide, the wet sections are expanded by applying gentle heat.

A simple method for transferring a large number of semithin sections to a glass slide has been introduced by Leknes (1985). This method is particularly useful for transferring 0.5- to 1.0-μm sections and when serial mounting is not required. When an appropriate number of sections have been cut, additional water is carefully added into the trough with a syringe until the water surface is strongly convex (Fig. 3.28A). The knife is removed from the microtome and is then turned very slowly 180° around a horizontal axis (Fig. 3.28B) and then subsequently 30–40° counterclockwise around an axis perpendicular to the paper (Fig. 3.28C). In the second step the water surface "membrane" is cut by the knife edge, and the water along with the sections is dropped onto the slide. After expanding the sections, the water is removed with a syringe and dried. Staining procedures for semithin sections are given on page 226.

Serial Sectioning

Since semithin sections do not easily form ribbons, a brief discussion of various procedures that facilitate ribbon formation is in order. Thicker sections form stronger ribbons. Tackiwax applied to the specimen block edges promotes ribbon formation (Rohde, 1965). When this sticky wax is used, sections should

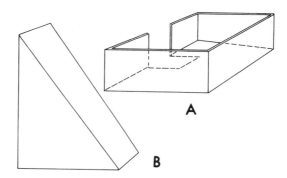

A

B

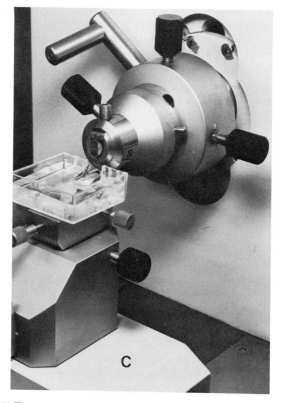

C

Fig. 3.27. (A) The transparent plastic trough ready for attachment to the triangular glass knife (B) made from glass 6–12 mm thick. (C) Photograph of the knife and trough mounted on a Reichert-Jung Autocut. (From Stephenson, 1982.)

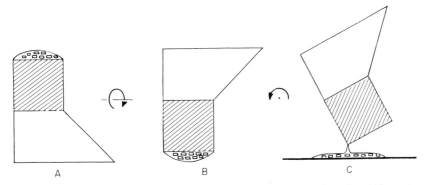

Fig. 3.28. Diagram showing the simultaneous transfer of a large number of semithin sections onto a glass slide. For further explanation, see the text. (From Leknes, 1985. © 1985 The Williams & Wilkins Co., Baltimore.)

not be expanded by heat. Obtaining ribbons by supplementary embedding of resin blocks in a paraffin–plastic matrix (Greany and Rubin, 1971) is time-consuming. Another method is to coat the top and bottom edges of a trimmed rectangular face of the resin block with contact cement (Weldwood) (Henry, 1977). Before being used, the cement is diluted with an equal volume of toluene and allowed to dry for a few minutes. It does not contaminate the knife. All these methods have proven useful for methacrylates, but not for epoxy blocks.

Richards (1980) double-embedded the epoxy blocks to obtain ribbons (Fig. 3.29). In this method, polyethylene glycol 6000 (PEG) is melted at 55°C. A mixture of Araldite CY 212 (10 parts), DDSA (10 parts), and DMP 30 (0.4 parts) is heated to 55°C. The final secondary embedding medium is prepared by mixing PEG (3 parts) with the epoxy mixture (1 part), stirring well, and heating in an oven at 55°C for 1 hr before use. The epoxy block to be double-embedded is first trimmed, forming a square or rectangular face. It is then placed face downward in an embedding mold. Hot secondary medium is poured into the mold above the level of this block and allowed to cool for 30 min before sectioning. Since the secondary medium contains unpolymerized resin components, care should be taken to avoid skin contact.

Probably the simplest method for obtaining continuous series of uniform ribbons was introduced by Childress and McIver (1983). Resin blocks (containing the specimens) are trimmed to form a mesa, and Tackiwax (Central Scientific Co., Chicago) or red dental wax is applied to the top and bottom sides (Fig. 3.30). An advantage of the mesa is that sections do not increase in size during sectioning. The tendency of ribbons to break is greatly reduced by the mesa-shaped blocks as well as by the use of Tackiwax. Rectangular pieces of glass coverslips (7 × 22 mm) are broken to fit the diamond knife trough and carefully submerged in the trough before sectioning. During sectioning, the ribbon may be

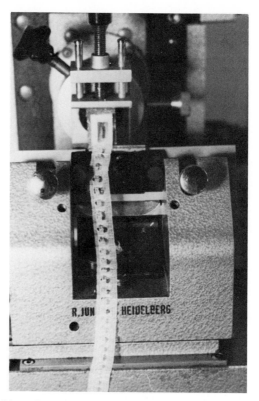

Fig. 3.29. A ribbon of semithin sections (3 μm) obtained after double embedding, using a fluorocarbon-coated Ralph glass knife. Ribbons are cut dry. Epoxy or methacrylates can be used. (From Richards, 1980.)

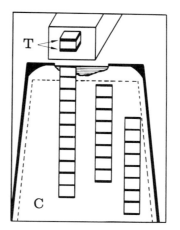

Fig. 3.30. Diagrammatic representation of cutting a series of ribbons of thick sections. T, Tackiwax; C, coverslip piece. (From Childress and McIver, 1983.)

parted every 20–25 sections with an eyelash mounted on an applicator stick. When enough ribbons have been produced, they are aligned over the coverslip piece and expanded with ethylene dichloride; the water level in the trough is lowered by using a capillary tube. Coverslip pieces are placed in an oven at 60°C until dry to expand fully and fix sections to the glass.

Campbell (1981) combined some of the preceding methods with the use of a stainless steel trough. A thin layer of contact cement (Wilhold) is applied to the leading and trailing edges of the face of a trimmed epoxy block. Sectioning can be started after several minutes. A hydrophobic line is drawn with a wax pencil across the front surface of the glass knife, 6–8 mm distant from and parallel to the cutting edge. The glass surface should be absolutely clean before drawing the wax line. This line will hold a droplet of water extending to the cutting edge (Fig. 3.31A), causing the droplet to form a curved rather than horizontal fluid surface.

A trough is prepared by grinding a portion of a length of stainless-steel tubing (Small Parts, Inc., Miami) and beveling one end. All edges should be polished.

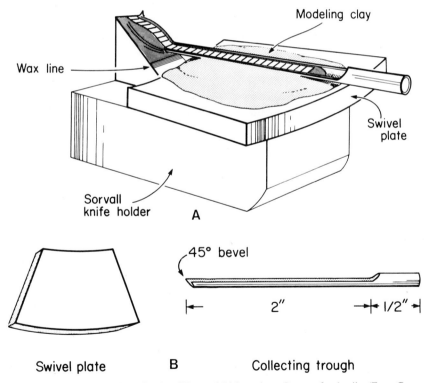

Fig. 3.31. Apparatus for collecting ribbons of thick sections. See text for details. (From Campbell, 1981. © 1981 The Williams and Wilkins Co., Baltimore.)

Figure 3.31B shows the dimensions and shape of a convenient trough, which is held in place with modeling clay on an aluminum swivel plate. This plate serves both to adjust the height of the trough to that of the cutting edge and to allow the trough to be angled sideways in case the ribbon is not straight. The trough is pushed into position so that its beveled tip is in the water droplet. Sectioning can be started after the trough has been filled with water, using a syringe. Once a ribbon of desired length has been obtained, the trough is removed and the ribbon is transferred onto a large drop of water on a glass slide, which is then heated on a hot plate to expand the sections.

CRYOFIXATION AND CRYOULTRAMICROTOMY

Chemical fixation, dehydration, and embedding cause distortion of biologic activity, lead to extraction or dislocation of many soluble components, and limit to a great extent chemical and immunologic reactions. Cryoultramicrotomy was developed to minimize these adverse effects. This technique is almost indispensable in a meaningful application of X-ray microanalysis, because most electrolytes are either lost or displaced during conventional processing.

Frozen ultrathin sections of unfixed specimens are the most desirable material for the study of the localization of diffusible substances for X-ray microanalysis, autoradiography, and immunocytochemistry. However, chemically fixed and frozen sections have provided significant information in the area of immunocytochemistry, particularly for the demonstration of intracellular antigens. Prefixation and cryoprotection prior to freezing can be allowed to localize intracellular antigens. It is desirable to stabilize the cryosections before exposing them to aqueous solutions of immunoreagents. On the other hand, aqueous fixation is undesirable in X-ray microanalysis.

Vapor fixation of freeze-dried cryosections seems to be the best approach. Frederik *et al.* (1984) have tested several organic and inorganic vapor fixatives on freeze-dried sections to preserve antigenicity and ultrastructure. Their studies demonstrate that vapors of OsO_4 and dry formaldehyde yield satisfactory preservation of the ultrastructure, while formaldehyde vapors preserve the antigenic sites. This method of vapor fixation of cryosections of fresh-frozen specimens appears to be ideal for the localization of antigenic sites as well as element distribution. The morphological details obtained with vapor fixation are less clear compared with those achieved with conventional fixation. However, in certain studies chemical information is more important than high-quality preservation of ultrastructural morphology.

Whether or not specimens have been vapor-fixed, they are subjected to rapid freezing to minimize the size of ice crystals. Visible ice crystals will be absent when the specimen is smaller than 0.3 mm (Bachmann and Schmitt, 1971).

However, it is not always practical to obtain unfixed tissue blocks smaller than 0.3 mm in diameter without extensive mechanical damage. The smaller the size of the tissue block, the larger the proportion of the block showing the mechanical damage. Irrespective of the size of the tissue block, excision will inevitably cause structural damage to cells at the cut surface. The depth of the zone of mechanical damage can be minimized by dicing the animal tissues in a concentrated buffered hydrophilic polymer solution (e.g., polyvinyl pyrrolidone) (Barnard, 1980). However, the artifacts introduced by cryoprotectants cannot be disregarded. Cells in culture and very small organs can be frozen without mechanical damage.

Ice crystal damage induced during freezing may be compounded through the subsequent processing; for example, ice crystals may grow in size. Although it has been suggested that melting occurs during sectioning, recent studies indicate that the energy liberated during cutting does not lead to any significant change in the structure of ice crystals (Frederik and Busing, 1981; Karp *et al.*, 1982). Thus there is neither a section melting nor a growth of ice crystals during cryosectioning. Exposure of frozen thin sections to OsO_4 vapor seems to prevent the formation of certain artifacts (Frederik and Busing, 1980). When tissue freezing and cryosectioning are carried out under optimized conditions, diffusion of ions, heat-labile enzymes, and soluble compounds during sectioning at about $-80°C$ seems to be less significant at the resolving power currently available.

Specimen Collection

The mode of specimen acquisition or excision from solid organs is the first step that invites artifacts for chemical determination. Therefore this step should be carefully planned. Devices recommended for specimen collection include spring-powered freezing clamps, plyers, and a projected biopsy needle cooled by cold nitrogen gas (cryogun).

Encapsulation

If needed, specimens can be encapsulated (supported) by bovine serum albumin, gelatin, fibrin, or polyethylene glycol. This treatment differs from embedding in that the media do not penetrate the cells or tissues. Tissues such as muscle do not require encapsulation, whereas cell suspensions do. This step is not used for chemical studies.

Fixation and Cryoprotection

The treatment given to a specimen before freezing is determined by the objective of the study. There are basically three alternatives: (1) the specimen is frozen without any prior treatment; (2) it is frozen after a brief fixation (5–15 min) with glutaraldehyde; and (3) it is frozen after fixation and antifreeze treatment. Treatment with a cryoprotectant agent may reduce the formation of artifacts caused by freezing, especially in the case of specimens with higher water content. The agents that have been used for cryoprotection include glycerol, DMSO, PVP, and hydroxyethyl starch. Specimens can also be soaked in ice-cold saturated sucrose or 1.5% methyl cellulose. About 20–25% glycerol in a suitable buffer is most commonly used. However, cryoprotectants may not only interfere with the elemental distribution in a specimen but also affect the ultrastructure. When rapid freezing is used, cryoprotectants do not protect the cells; they may be even harmful. Franks (1977) and Skaer et al. (1977) have discussed the advantages and limitations of various cryoprotectants.

Coolants

The choice of a suitable cryogenic liquid is a compromise between safety for the worker and speed of cooling. Extremely high cooling rates ($-180°C$) are required to prevent the formation of ice crystals larger than 10–15 nm. Visible ice crystals will be absent when the specimens less than 0.3 mm are frozen. In practice, the specimen must be frozen fast enough to prevent the formation of ice crystals large enough to interfere with the resolution of the cellular structure under study. Although Freon and liquid nitrogen are not the most efficient coolants, they are commonly used because of their relative safety. Propane has greater cooling efficiency, but it is highly inflammable. Ethanol is more efficient than propane between 0 and $-50°C$ (Silvester et al., 1982). Ethanol is relatively safe and inexpensive, and is useful in freeze-substitution (Marchese-Ragona, 1984).

Freezing

Specimens can be frozen by contact with solid or liquid coolants. A high cooling velocity can be obtained by freezing on a cold metal surface, spray freezing, high pressure freezing, or propane-jet freezing. The choice of the method for rapid freezing depends on the objective of the study and whether or not the specimen has been fixed and cryoprotected. The conditions needed to freeze unfixed and unprotected specimens are more critical.

Fixed and cryoprotected materials mounted on standard specimen holders are frozen by plunging them into a cooling liquid. Even fresh specimens can be frozen directly in a cooling liquid provided they are thin and mounted on low-mass specimen holders (a thin metal sheet of good thermal conductivity) and the renewal of the coolant around the specimen is sufficiently rapid to prevent the formation of a gaseous layer which would retard heat dissipation. Rapid renewing of the coolant can be achieved by high immersion velocity with a spring-assisted device (Handley *et al.*, 1981b). Alternatively, rapid immersion can be obtained by exposing the specimen to propane jets (Müller *et al.*, 1980).

Freezing by immersion can be accomplished by plunging small pieces of the tissue (0.5–1.0 mm) into Freon kept at its melting point by cooling with liquid nitrogen. The specimens are placed on top of metal specimen pins which are then lowered into the coolant. Alternatively, the specimens can be frozen in solid/liquid nitrogen slush. Very rapid quenching is obtained with liquid nitrogen maintained near its melting point of −210°C at a pressure in excess of the critical value of 33.5 atm; this pressure prevents excessive vapor formation. Another method is to freeze the specimens in ethanol supercooled to −125°C in liquid nitrogen.

Free cells adhered to the specimen support can be cryofixed by a jet of liquid propane (−190°C), obtained by condensing camping propane gas with liquid nitrogen. The liquid propane is shot by a nitrogen pressure ranging between 3 and 5 bar from both sides on the specimen support. This is a relatively simple and inexpensive method for freezing unpretreated cells. The spray freezing method has been used for suspended cells. Small specimen droplets (10 μm) are sprayed into liquid propane (−190°C) (Bachmann and Schmitt, 1971). The propane is removed by evaporation, and the frozen droplets are mixed with butylbenzene (−85°C). This mixture is transferred to a specimen holder and solidified by dropping it into liquid nitrogen. The cooling rate is estimated to be on the order of 10^5 °C/sec.

Impact freezing can be obtained on an ultrapure copper block cooled to −269°C with a spray of liquid helium (Heuser *et al.*, 1979). The high purity of the copper block enhances significantly its thermal conductivity at −258°C, enabling a rapid heat diffusion. This method yields cooling rates sufficiently high to freeze without any cryoprotectant only the superficial layers of cells of a tissue block; these layers are devoid of visible ice crystal damage. However, it requires an expensive and complex instrument. Another approach is to freeze the tissue by a punch biopsy method using a specially designed cryogun with a highly thermal conductive specimen holder (Chang *et al.*, 1980). This method allows simultaneous quick removal of small tissue blocks from the animal and their rapid contact freezing at −196°C.

Frozen specimens are stored in liquid nitrogen until needed for sectioning. The temperature of the frozen specimens during storage should not be allowed to rise above the recrystallization temperature (−133°C). Williamson (1984) has introduced a simple holder for storing frozen specimens under liquid nitrogen. During transfer of the specimens to the cryochamber for sectioning, they should be kept covered by liquid nitrogen to prevent condensation of moisture or momentary thawing. The usual precautions must be observed while using cryogenic liquids; the use of goggles, hand gloves, and fume hood is necessary.

Freezing Instrument

Many types of instruments used for rapid freezing of specimens are complex, expensive, and devised in different laboratories. A simple instrument made of inexpensive and readily available materials, using liquid nitrogen, was introduced by Verna (1983). This instrument is made of a cylindrical copper block that is partly immersed in liquid nitrogen contained in a styrofoam box (Fig. 3.32A). The copper (1) is of ordinary purity (electrolytical grade) and its surface is mirror polished. The styrofoam box (2) is covered by a bakelite plate (3) (5 mm thick). A large hole in this plate (4) allows the filling of the box with liquid nitrogen and may be closed by a rubber stopper (5). To avoid any condensation, a small hole (6) in the freezing chamber wall, located just above the copper surface, allows liquid nitrogen vapor to flow over the copper block.

The object-holder (7) which is disposed on a lateral striker arm (8) is a cylindrical aluminum stick (4 mm in diameter), supporting a small piece of foam

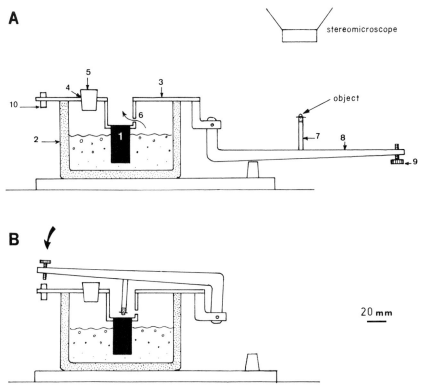

Fig. 3.32.   Diagram of the freezing device. See text for details. (A) Specimen orientation. (B) Freezing. (From Verna, 1983.)

rubber (5 mm thick). A 180° rotation of the striker arm around its axis applies the specimen against the copper block. The interval between the foam rubber and the copper surface may be adjusted by a stop-screw (9) to take into account the specimen thickness or to modify the pressure with which the specimen is apposed against the copper block. The interval screw (steel) comes into contact with a magnet (10) which avoids any rebound of the object-holder after the initial contact of the specimen with the block.

Operation

The apparatus (Fig. 3.32B) is filled with liquid nitrogen and the fill-in orifice is closed by a rubber stopper. After 5 min, the copper block is at liquid nitrogen temperature; more liquid nitrogen is added to reestablish a correct level, that is, 1 cm below the copper surface. The device is now ready for use. A small piece of aluminum foil is glued by a droplet of water onto the rubber of the object-holder. The specimen is placed on the foil, and, if necessary, it is oriented under a stereomicroscope. Excess water around the specimen is removed with a filter paper. The striker arm is quickly rocked by hand to apply the specimen against the cold copper block. Liquid nitrogen is immediately poured into the freezing chamber to avoid any rewarming of the specimen during subsequent operations. The piece of aluminum foil supporting the specimen is detached from the holder with precooled forceps and stored in liquid nitrogen.

Sectioning

Ultrathin cryosectioning is either cutting or fracturing, depending on the conditions such as ambient temperature and speed of sectioning. At a speed of 90 mm/sec and at temperatures lower than −70°C, the surface of the specimen block face displays a fracture image similar to that seen with the standard freeze-fracture method (Leunissen *et al.*, 1984). At a speed of 0.1 mm/sec the fracture image is encountered only at temperatures below −120°C. Although the under-surface of the section is complementary to the block face at the moment of separation, both surfaces of the section do not show fracture patterns. The reason may be that just after the initial fracturing, the ice in the section is temporarily melted, resulting in the rearrangement of lipid monolayers of split membranes (Leunissen *et al.*, 1984).

Cryosectioning can be carried out in cryostats or cryokits: Reichert FC4 cryomicrotome or LKB Ultratome III equipped with a Cryokit. In the cryostat the microtome is housed within a refrigerated box, which is cooled by compressors. Temperatures as low as −80°C can be attained in the atmosphere of the cryo-chamber. A cryokit is an insulated box that is attached to an ultramicrotome, so that it completely surrounds the specimen and knife areas. It is cooled by liquid nitrogen or cold nitrogen gas.

The temperature in the cryochamber should remain stable because this temperature effects the surface temperatures of the specimen and the knife. The temperature of the cryochamber should be monitored during cryosectioning. Regardless of the types of cryoultramicrotome and knife used, the latter is used dry. The technology of cryosectioning is continuously changing, and each group has -been modifying commercially available instruments to achieve better cryosectioning. Cryoultramicrotomy is not a widely used technique because of its complexity, high cost, and very stringent requirements to obtain satisfactory results. It should be attempted only when necessary.

The ideal sectioning temperature is a subject of considerable controversy. There appears to be an inverse relationship between section thickness and the lowest possible sectioning temperature (Barnard, 1982). The production of cryosections less than 0.5 μm thickness is thought to be free of apparent diffusion provided temperatures are maintained below −100°C. According to one approach, the specimen is maintained at −11°C, the knife at −100°C, and cryochamber at −130°C. At temperatures below −80°C, inspite of the contrary reports, no through section melting occurs during cutting. Cutting sections at very low temperatures is particularly difficult because of increased tissue brittleness, but the use of knives with very small cutting angles may minimize this difficulty. When glass knives are used, they should have a scoring angle of 40° and a clearance angle of 6–9°.

The pin with frozen tissue is mounted in a cryoultramicrotome, and ultrathin sections are cut dry by using a glass knife without trough fluid. A cutting speed of 2–5 mm/sec is recommended. Dry sections tend to roll up during sectioning and may curl over the cutting edge. According to one method, a vacuum pipette is used to suck the sections out horizontally from the knife edge, forming a ribbon that slides onto a grid placed on the flat surface of the underlying knife. Alternatively, the section can be flattened by pressing it onto the grid with a chilled, polished copper rod. Another technique involves an antiroll plate that consists of a piece of a glass coverslip attached close to and parallel to the glass cutting edge. Sections can also be transferred to the grid with a hollow plastic straw (5 mm in diameter) plugged with cotton wool, to the pointed end of which is glued a white dog hair (Barnard, 1982). An eyelash probe precooled with liquid nitrogen can also be used to transfer the sections.

The simplest approach is to transfer the section with the aid of an eyelash probe onto a precooled grid that has been coated with either Formvar or carbon or both and then placed on the knife face about 1 mm from the cutting edge. A shelf on which to rest the grid can be made by placing a piece of electrical tape on the knife face about 5 mm from the cutting edge. If sections are repelled immediately before contacting the eyelash probe due to their electrical charge, a commercially available discharge device can be used to achieve easier manipulation. After sections are transferred to the carbon-coated grid, a second empty grid (carbon-coated) is laid on top of it so that the sections are sandwiched between two layers of carbon (Zierold, 1984). The grids are gently pressed together by a cold polished surface of a metal rod. Then the grids are separated, and each grid bearing sections can be examined either without staining or after positive or negative staining with vapors of OsO_4 or PTA. The low contrast in frozen-hydrated sections is caused by the very small difference in the mass thickness of ice, carbon, and organic material, resulting in similar electron scattering.

Frozen ultrathin sections can be examined in the electron microscope at room temperature, or they can be viewed in the frozen state. Among the advantages of viewing the sections in the frozen state is the reduced movement of ions in the presence of the frozen vitreous water in the tissue. Devices have been developed to transfer frozen sections from the cryoultramicrotome to the microscope without depositing ice from atmospheric vapor (see, e.g., Ross *et al.*, 1981). Apparently, cryotransfer systems are essential for electron microscopy of frozen-hydrated sections.

4

Positive Staining

INTRODUCTION

Positive staining increases the electron scattering power of biologic specimens. This increase is necessary for two reasons: (1) biologic specimens are
composed mainly of elements of low atomic number (carbon, hydrogen, oxygen,
and nitrogen), which possess a relatively low electron scattering power; and (2)
to obtain thin sections, specimens are usually embedded in an epoxy resin, which
is of biologic origin and therefore possesses the same atoms as those constituting
cells and tissues. As there is little difference between specimen and resin in
electron scattering power, the result is a transmission image of inadequate contrast. The relative electron scattering power of a specimen is increased by introducing atoms of heavy metals (osmium, uranium, and lead) into either the tissue
or the section or both. Therefore the image obtained represents essentially the
intracellular distribution of bound heavy metals. Specimen contrast can also be
increased by using low accelerating voltages, a smaller objective aperture, and
thicker sections.

The areas of the specimen that allow electron penetration appear dark on the
film, and the areas that prevent it record as light. These density values are
reversed on the print; that is, dark areas on the print represent the absence of
electrons, while light areas record the presence of electrons. Since the heavy
metal atoms introduced during specimen processing prevent electron penetration,
areas of the tissue section impregnated with these atoms appear light on the film
but dark on the print.

Specimens can be stained during fixation (e.g., with OsO_4), before or during dehydration (e.g., with uranyl acetate *en bloc*), after embedding but before sectioning (e.g., with phosphotungstic acid, PTA), and/or after ultrathin sectioning (e.g., with uranyl acetate and lead citrate). Staining of a specimen is accomplished usually in most of the aforementioned steps. Staining after embedding but before sectioning can be carried out by trimming the resin block to expose the tissue and then immersing it in 2% uranyl acetate or 2% PTA in a sealed vial that is placed in an oven (60°C) for 12–24 hr (Locke and Krishnan, 1971). Both staining solutions are prepared in 95% ethanol. The stain penetrates about 15 μm into the block face, allowing hundreds of stained ultrathin sections. This method is also useful for thick sections for high-voltage electron microscopy. Double staining of ultrathin sections with uranyl acetate and lead citrate is the most commonly used method (see pp. 214–216).

The binding of heavy metal ions depends on both the spatial arrangement of the macromolecules within the specimen and the chemical interaction between the metal and biologic substance. The chemistry of staining has been presented by Hayat (1975, 1981a). To obtain adequate staining, a certain minimum amount of cellular substance must be present in an ultrathin section. Most stains penetrate best when ultrathin sections are not allowed to dry after removal from the knife trough prior to staining.

Major problems in staining are low specificity, contaminated staining solutions, and incomplete rinsing. Specificity can be improved by carefully controlling the stain concentration, the pH, and the duration of staining. Selective extraction of certain cell components can also increase the specificity. Freshly prepared staining solutions are preferred. If they are not available, the solutions should be filtered before use, especially in the case of lead stains. All the staining solution should be removed by thorough rinsing with distilled water; otherwise it will dry on the section surface, resulting in the formation of precipitates of contamination. A drop of staining solution trapped between the tines of the forceps must be washed off before the stained grid is finally picked up from distilled water.

STAINS

Acridine Orange

Acridine orange selectively binds to cell surface glycos- and galactosaminoglycan compounds in the presence of Na^+ in low concentrations (Timár *et al.*, 1979). Cultured cells are fixed *in situ* with 2.5% glutaraldehyde in 0.1 M buffer for 30 min at 4°C. Cells are rinsed with the same buffer and then stained with 1 mM acridine orange in the same buffer for 48 hr in the dark at 4°C. After being rinsed in the buffer, cells are postfixed overnight with 2% OsO_4 at 4°C.

Alcian Blue

Alcian blue is used for staining cell surface materials and cartilage components (Schofield *et al.*, 1975). Specimens are fixed with buffered glutaraldehyde containing 1–2% alcian blue. After a rinse with the buffer containing 0.5% alcian blue, the specimens are postfixed with OsO_4 and poststained with uranyl acetate and lead citrate.

Bismuth

Bismuth specifically stains Golgi beads and interchromatin granules after fixation with glutaraldehyde, whereas it stains nucleoli, basic proteins, and biogenic amines after formaldehyde fixation. This specificity can be achieved by *en bloc* staining under controlled conditions of fixation. Bismuth staining of sections is relatively nonspecific: it can stain glycogen, lysosomes, ribosomes, polysaccharides, ferritin, and DNA. This staining stems in part from bismuth's ability to enhance uranyl and lead staining and from its interaction with reduced osmium.

Specific En Bloc Staining (Locke and Huie, 1977)

Solution A: 400 mg of sodium tartrate is dissolved in 10 ml of 1 *N* NaOH. While this mixture is being stirred, 200 mg of bismuth oxynitrate ($BiONO_3 \cdot H_2O$) (bismuth subnitrate or basic nitrate) is added.

Solution B

0.2 *M* triethanolamine-HCl buffer (pH 7.0)

Staining solution: One part of solution A is added to two parts of solution B, and the pH is adjusted to 7.0 with 1 *N* HCl.

Aldehyde-fixed tissue blocks are rinsed in the above buffer and stained *en bloc* with bismuth for 1 hr at room temperature on a gentle rotary shaker. Postfixation is carried out with 1% OsO_4 in 0.05 *M* cacodylate buffer (pH 7.2). Poststaining is not necessary. For some studies, instead of *en bloc* staining, thin sections of aldehyde-fixed specimens can be stained with bismuth for 30–60 min.

Enhancing Contrast in General (Riva, 1974)

Alkaline bismuth subnitrate is used after uranyl acetate.

Sodium tartrate	400 mg
NaOH (2 *N*)	10 ml
Bismuth subnitrate	200 mg

Sodium tartrate is dissolved in NaOH. Drops of this solution are stirred into the bismuth subnitrate. After the addition of 6–8 ml of this solution, the mixture clears, and after the addition of all of this solution, all the bismuth is chelated. Conventionally fixed tissue specimens are stained *en bloc* with a saturated solution of uranyl acetate in water for 20 min. Thin sections on grids are stained under cover for 3 min by immersion in the bismuth subnitrate solution and then rinsed with distilled water.

Chromosome Staining (Albersheim and Killias, 1963)

Bismuth metal	20 mg
Nitric acid (concentration 5.4 M)	0.2 ml
Distilled water to make	80 ml
Citric acid	4 ml
NaOH (1 N) to adjust pH to 7.0	
Distilled water to make	100 ml

After the nitric acid has been added, the mixture should be kept covered to avoid oxidation of the metal. Conventionally fixed tissue specimens are rinsed once with 0.1 N HNO_3 solution in sodium citrate buffer (pH 7.0) and then stained *en bloc* with the bismuth solution for 90 min at 4°C.

Staining of Mucosubstances and Polysaccharides (Ainsworth et al., 1972)

Solution A

Picric acid (H_5IO_6)	0.8 g
Ethanol	70 ml
Sodium acetate (0.2 M)	10 ml
Distilled water	20 ml

Solution B

Sodium tartrate	400 mg
NaOH (2 N)	10 ml
Bismuth subnitrate	200 mg

Solution B is prepared by adding drops of the sodium tartrate and NaOH mixture to the bismuth subnitrate. After 6–8 ml has been added, the solution will clear. Thin sections of conventionally fixed tissues are mounted on grids (copper) and floated on solution A for 10–30 min at room temperature. After being thoroughly washed 20–30 times for 5–10 min with distilled water to remove periodate, the grids are floated on solution B for 30–60 min at room temperature. They are then again thoroughly washed with distilled water.

Cobalt

The cobalt technique (Pitman *et al.*, 1972) is used to demonstrate neural geometries in vertebrates and invertebrates and to locate the structure of nerve cells at both the light and the electron microscopy level.

Cobalt Sulfide–Silver Intensification (Hausen and Wolburg-Buchholz, 1980)

The method described here has five steps: filling of neurons, sulfide precipitation, fixation, postfixation, and intensification. Neurons in the brain and the optic lobes are filled with cobalt ions by using an extracellular diffusion technique. The animal (e.g., blowfly) is immobilized by cooling (4°C, 10–20 min) and glued for dissection. The back of the head capsule is opened, the airsacs are removed, and the brain is exposed. A micropipette (tip diameter 5–10 μm) filled with cobalt chloride (5%) is inserted into selected areas of the protocerebrum, and cobalt ions are allowed to diffuse for 1–3 hr into the tissue at room temperature. The brain is rinsed in a modified Millonig's phosphate buffer (pH 7.35), which is composed of 83 ml of 2.26% NaH_2PO_4, 17 ml of 2.5% NaOH, 1.2 g of glucose, and 1 ml of 1% $CaCl_2 \cdot 2H_2O$.

The brain is prefixed for 5 min with a mixture of 2.5% glutaraldehyde and 2% formaldehyde in the buffer. For precipitation, the brain is immersed in a solution of fresh ammonium sulfide, $(NH_4)_2S$ (5 drops of 20% ammonium sulfide in 10 ml of buffer), until the injection site turns black in 1–2 min. After being rinsed 3 times (1 min each) in the buffer, the brain is fixed with the aldehyde mixture for 2 hr at 4°C and then thoroughly rinsed in the buffer. Postfixation is accomplished with 1% OsO_4 for 1 hr at 4°C.

Silver intensification is accomplished by holding thin sections in water-filled plastic loops and transferring them onto the surface of a developer bath (2 ml of 5% hydroquinone, 2 ml of 10% citric acid, and 1 ml of 1% $AgNO_3$ added to a solution of 10 ml of 4% gum arabic in distilled water) for 10–30 min at 35°C in the dark. The progress of intensification is monitored by placing the loops onto a water-filled depression slide and examining the sections under the light microscope. Sufficiently intensified profiles in the section show a light yellow color. Sections are mounted on a grid and poststained with lead citrate for 10 min.

Concanavalin A

Concanavalin A (con A), a plant lectin, is useful for the demonstration of glycosyl residues in cell surface materials. Two methods of staining with con A are presented here.

Two-Step Lectin–Peroxidase (Bernhard and Avrameas, 1971)

Cells in Eagle's medium (without glucose) are incubated with 100 μg/ml of con A for 15 min at 4°C and rinsed three times with the medium, using slow-speed centrifugation at 4°C. They are then incubated in excess medium containing 50 μg/ml peroxidase for 15 min at 4°C and rinsed three times as before. Cells are fixed with 4% formaldehyde in 0.2 M cacodylate buffer (pH 7.4) for 30 min at 22°C.

To develop peroxidase activity, cells are treated for 15–30 min in the dark with 0.5 mg/ml solution of diaminobenzidine (DAB) in 0.1 M Tris-HCl buffer (pH 7.4) containing 1 drop of 30% H_2O_2 per 5 ml of incubation medium. Cells are then rinsed twice in the medium and postfixed with 2% OsO_4 in phosphate buffer. Controls are run in the presence of 0.5 M inhibitory saccharide.

One-Step Lectin–Peroxidase (Huet and Garrido, 1972)

In this method, 1.5 mg/ml of con A and 3 mg/ml of peroxidase are mixed in 0.1 M phosphate-buffered saline (pH 6.8). Then 0.03 ml of 1% purified glutaraldehyde is added to the mixture, which is allowed to react for 1–3 hr at 20°C. The mixture is dialyzed against the same buffer but containing 2 mg/ml glycine. After two more dialysis steps at 4°C, the conjugate is centrifuged at 36,000 g for 20 min at 4°C. The conjugate concentrations of 2.5–1200 μg/ml have been used in this method.

Cells are labeled for 15 min, rinsed, and fixed in 2.5% glutaraldehyde. Peroxidase activity is developed by treating the cells for 15–30 min in the dark with 0.5 mg/ml solution of DAB in 0.1 M Tris-HCl buffer containing 1 drop of 30% H_2O_2 per 5 ml of incubation medium. Cells are then rinsed twice in the medium and postfixed with 2% OsO_4 in the buffer. Controls contain 0.1 M of an appropriate saccharide inhibitor in labeling and rinsing solutions.

Diaminobenzidine–Osmium Tetroxide

In addition to its use in the staining of peroxidase, mitochondria, and microbodies, DAB is effective in staining sulfated mucopolysaccharides (Monga *et al.*, 1972). Thin sections of the tissue fixed with aldehydes and embedded in glycol methacrylate are collected on gold grids and rinsed in 5% (saturated) boric acid in distilled water for 5 min. The grids are floated on a freshly prepared 1% solution of DAB in 5% boric acid for 20–30 min. After two rinses in 5% boric acid (1–2 min each), the grids are floated on a 2% OsO_4 in distilled water for 10–20 min and rinsed in distilled water.

Gold

Gold is used for localizing intracellular antigens, polysaccharides, plant lectins (Horisberger and Vonlanthen, 1977, 1980), vitamin A (Wake, 1974), and nuclear pore complexes (Feldherr, 1974).

Preparation of Colloidal Gold

To prepare colloidal gold, 100 ml of aqueous $H(AuCl_4)$ solution (0.01%) is heated to boiling, 2.5 ml of 2% aqueous sodium citrate is added, and the mixture is kept boiling for an additional 10–15 min. The color of the solution changes from yellow to deep red. The colloidal gold particles produced are 20 nm in diameter; the size can be increased by decreasing the quantity of the citrate. The size increase is indicated by a shift of the color of the gold suspension from deep red (16–20 nm) to purple.

Gold–Protein Complex (Schwab and Thoenen, 1978)

To produce a gold–protein complex, tetanus toxin is dialyzed for 2 hr against water and 100 μg is added to 10 ml of the gold suspension. Since the gold particles are kept in suspension by their negative surface charge and electric repulsion, buffer salts added together with the proteins can cause precipitation and therefore have to be removed. If proteins with a basic isoelectric point are coupled, precipitation may also occur. In this case, an increase of the pH of the gold suspension by 0.2 N K_2CO_3 is often beneficial. The mixture is stirred for 1 min and then stabilized by the addition of 0.1 ml of 1% carbowax. The suspension is centrifuged for 30 min at 10,000 g. The pellet is then resuspended for rinsing and centrifuged again. The final pellet (100 μl) is diluted with 200 μl of water and injected.

Injection and Fixation

Once the animal (e.g., rat, 200 g body weight) has been anesthetized with ether, its anterior eye chamber is injected with 10 μl of toxin–gold complex, using a Hamilton syringe. After intervals of 1, 2, 4, or 6 hr, the animal is reanesthetized and perfused through the heart with a brief prerinse of Ringer's solution containing 1000 U of heparin and 0.1% procaine, followed by a mixture of 2.5% glutaraldehyde and 1% formaldehyde in 0.1 M phosphate buffer (pH 7.4) containing 5% sucrose. This perfusion lasts 10–15 min. Small blocks are cut from the dilator region of the iris, additionally fixed for 2 hr, postfixed with 1% OsO_4, and stained *en bloc* with uranyl acetate. The results show a selective binding of tetanus toxin–gold to the autonomic nerve terminals, followed by internalization and retrograde axonal transport.

Indium

Indium trichloride is effective in staining nucleic acid components of cell nuclei (Watson and Aldridge, 1964; Coleman and Moses, 1964; Eddy and Ito, 1971). Glutaraldehyde-fixed tissue blocks are gradually transferred from absolute acetone (at the end of dehydration) to pure pyridine over a period of 15 min.

After being rinsed three times (10 min each) in pyridine at 4°C, specimens are exposed for 2 hr at 4°C to pyridine saturated with lithium borohydride. Following three rinses in pyridine at 4°C, specimens are immersed overnight at room temperature in pyridine containing 40% acetic anhydride saturated with a trace of sodium acetate. Specimens are rinsed three times (10 min each) in pyridine and then rinsed in pyridine and acetone mixture (1 : 1) for 5 min. Specimens are rinsed three times (10 min each) in acetone and then stained *en bloc* for 2 hr at 4°C with a solution containing 25 mg of anhydrous indium trichloride in 1 ml of acetone.

Iodide–Osmium Tetroxide

An iodide–osmium tetroxide mixture is used as a fixative and stain for demonstrating synaptic vesicles, Ca^{2+} binding sites (Gilloteaux and Naud, 1979), and other cell structures (for review, see Hayat, 1975).

Sodium Iodide–Osmium Tetroxide

Osmium tetroxide   1 part
Sodium iodide (3%)   3 parts
Tissue specimens are treated for 24 hr at room temperature. No additional staining is required.

Zinc Iodide–Osmium Tetroxide

Zinc iodide solution
Metallic iodine   5 g
Metallic zinc   10–15 g
Distilled water   200 ml
The combined powders are slowly added to distilled water in a beaker, and after 5 min the solution is filtered. Caution should be observed while dissolving the powder, because this is an exothermic reaction.
Final solution
Zinc iodide solution   8 ml
Osmium tetroxide (2% unbuffered)   2 ml
The solution should have a pH of 5.0 and an osmolality of 218 mosmols. It should be kept in the dark before and during use. Tissue specimens are placed in this mixture in the dark for 24 hr at room temperature. Uranyl acetate is capable of dissolving the zinc–osmium precipitate.

Iodine

Iodine or iodinated compounds have been used to stain starch in *Chlorella* and glycogen in rat liver (Flood, 1970). Thin sections exhibiting low contrast after conventional staining may be exposed (after lead citrate staining) to iodine vapor for 30 sec in a number 00 gelatin capsule in a fume hood. The increase in contrast is shown in parts of the section not already exposed to the electron beam. This method was used by Williams and Adrian (1977) to stain microtubular subunits in the basal bodies found in *Tetrahymena* and rabbit oviduct.

Tetraiodophthalic Acid

When the reagent tetraiodophthalic acid is applied as a free acid, it interacts with molecules (proteins, nucleotides) containing basic groups. The staining solution is prepared by hydrolyzing 2 g of tetraiodophthalic acid in 100 ml of 1 M NaOH for 2 hr at 100°C (Allen *et al.*, 1955). This solution is acidified with concentrated HCl; the resultant needles (yellow in color) are purified by twofold precipitation from an alkaline solution with concentrated HCl. Thin sections are stained with a half-saturated solution of the stain in 50% ethanol for 20 min at room temperature.

Iron

Acidic mucosubstances (e.g., sialomucins and sulfomucins) are specially stained with iron solutions at pH 1.2–2.4. Staining can be carried out either by *en bloc* treatment or by staining of thin sections. Several types of iron solutions are in use.

Colloidal Ammonium Ferric Glycerate

Ferric chloride (anhydrous)	30 g
Distilled water	100 ml
Glycerol	40 ml
Ammonium hydroxide (28% NH_3)	22 ml

Glycerol is added after the $FeCl_3$ is completely dissolved in water. Ammonium hydroxide is added gradually in small amounts at a time and is stirred vigorously each time to dissolve the precipitate before adding more ammonium hydroxide; this procedure takes about 30 min. When the solution is an unclouded, deep reddish brown, it is poured into a cellophane tubing. The tubing is sealed with only scant space allowed for dilution to take place, and then is dialyzed against distilled water. The completion of dialysis requires 8–10 changes of distilled water over a period of 72 hr. This stock solution can be stored at 4°C for several months. Just before the stock solution is used, sufficient glacial acetic acid is added to it to obtain a pH of 1.8–2.0. The ratio of acetic acid to the stock solution is 1 : 3 to 1 : 4.

Negative Colloidal Ferric Hydroxide

A solution of negative colloidal ferric hydroxide is prepared by dissolving 6.53 g of $FeCl_3$ in 100 ml of distilled water and then adding this mixture fairly rapidly (almost in a stream) to 1 l of boiling water. This solution should have a pH of 1.8 and contain 1.7 g of iron/l. Before the solution is used, the pH is lowered to 0.8 by adding HCl.

Positive Ferric Oxide

The preparation of positive ferric oxide begins by dissolving 6.75 g of $FeCL_3 \cdot 6H_2O$ in 50 ml of distilled water. Then 50 ml of this solution is added in a slow stream to 600 ml of boiling distilled water. This solution is dialyzed against 10 volumes of distilled water for 5–6 days with two changes of water each day. Distilled water is added to the iron solution to obtain a final concentration of 1.2 g of iron/l. Finally, 100 ml of this solution (pH 5–6) is mixed with 100 ml of 0.0012 N HCl. This mixture (pH 3.5) is the stock solution. The final staining solution is:

Stock solution   10 ml
Glacial acetic acid   10 ml
Distilled water   20 ml

This solution has a pH of 1.8 and contains 0.15 g iron/l.

Negative Ferric Oxide

A solution of negative ferric oxide is prepared by recharging the positive ferric oxide solution with ferrocyanide ions. First, 100 ml of 8 mM potassium ferro-cyanide is dissolved in boiling distilled water with constant stirring and poured into a beaker. Then 100 ml of the stock solution of the positive ferric oxide is slowly poured into the beaker. The 1 : 1 ratio of the two chemicals results in the maximum stable negative charge on the Fe_2O_3 colloid. This stable stock solution contains 0.3 g of iron/l and can be stored for 1–2 weeks without coagulation. The final staining solution is prepared by combining 20 ml of this stock solution with an equal volume of boiled distilled water. The pH of this mixture is 6.0. A pH of 3.4–4.2 can be obtained by mixing 20 ml of the negative solution with 20 ml of 0.01 M acetate buffer (pH 3.4). These solutions become unstable at a pH lower than 6.0; the degree of instability is related to the increase in hydrogen ions. These unstable solutions show a green or greenish blue color, which indicates partial conversion of the negative colloid into Prussian blue or other iron blues.

Staining Procedures

En Bloc Staining

After a thorough rinse in cold 7% sucrose solution, fixed tissue blocks are frozen in dry ice and cut into cryostat sections 40 μm thick. These sections are immersed in the staining solution for 2–10 hr, dehydrated, and embedded in a resin. Poststaining is not required.

Staining of Thin Sections

Thin sections of fixed specimens are floated free on a solution prepared imme-diately before use by mixing 5 ml of the colloidal iron stock solution with 9 ml of distilled water and 6 ml of glacial acetic acid. The pH should be 1.8; the staining time is 1 hr. Sections are rinsed twice in 10% acetic acid, then twice in distilled water. Sections are mounted on grids. Poststaining is not required.

Lanthanum

Lanthanum is useful as a marker for extracellular and intercellular materials such as mucopolysaccharides and glycoproteins. It should be used at room temperature. Postfixation with OsO_4 is necessary to retain lanthanum. Phosphate buffer is undesirable.

Lanthanum Hydroxide

Tissue specimens are fixed with 2% glutaraldehyde for 1–2 hr and then postfixed with 1% OsO_4 for 1–2 hr; both fixatives contain 1% lanthanum hydroxide. Lanthanum hydroxide is prepared by bringing a 3% solution of $La(NO_3)\cdot6H_2O$ to pH 7.6 by gradually adding 0.01 N NaOH. During this titration, the solution should be stirred vigorously. At pH 7.8, lanthanum hydroxide becomes insoluble, the result being flocculation.

Lanthanum Nitrate

Tissue specimens are fixed with buffered glutaraldehyde containing 1% $La(NO_3)\cdot6H_2O$ for 2 hr. After being rinsed for 30 min with the buffer containing 1% $La(NO_3)\cdot6H_2O$, specimens are postfixed for 30 min with 1% OsO_4 in the buffer containing 1% $La(NO_3)\cdot6H_2O$.

Lead

Lead salts stain many cellular components and are used as routine stains, either alone or after staining with uranyl acetate. Lead staining is enhanced when it is preceded by staining with uranyl acetate. The reasons for this phenomenon seem to be that lead binds primarily to uranyl acetate and/or that certain cellular substances are lost during lead staining if they have not previously been preserved with uranyl salts. It is known that prolonged staining with lead results in the destaining of thin sections except glycogen. Lead solutions react with even the smallest trace of CO_2 to form $PbCO_3$ precipitate. Thus special precautions must be taken to exclude CO_2 from the staining and rinsing solutions.

Lead Acetate

Method 1 (Dalton and Zeigel, 1960). A saturated solution of lead acetate is prepared in boiled distilled water and kept in a glass stoppered bottle containing some undissolved crystals of the stain. The bottle is kept completely filled; whenever a few drops of the stain are removed, the bottle is refilled to the top with boiled distilled water. Sections on the grid are stained for 3–20 min at pH 5.9 and then immediately given a thorough rinse in boiled distilled water. The grid is blotted dry and then exposed for 5 sec to 1–5% ammonium hydroxide vapor. Lead acetate imparts less contrast than does lead hydroxide.

Method 2 (Björkman and Hellström, 1965). The staining is done at nearly neutral pH, and the precipitation of $PbCO_3$ is almost eliminated. First, a saturated solution of lead acetate is prepared by dissolving 39 g of lead acetate in 100 ml of distilled water. Then 100 ml of this solution is mixed with 18.5 g of ammonium acetate. The ammoniated concentrated lead solution remains clear for several weeks. The staining solution loses its stability on adding as little as 2% water. The solution is always kept saturated with respect to lead acetate. Tissue blocks are stained by immersion in the solution for 45 min.

Method 3 (Kushida, 1966). The staining solution is prepared by adding lead acetate in excess to 100% ethanol. The mixture is gently stirred for 5 min and then filtered. This is the staining solution for thin sections.

En bloc staining can be carried out with the following modified staining solution. An equal volume of 100% acetone is added to the staining solution just described. The mixture is stirred for 5 min and then filtered and used immediately. Tissue blocks are stained for 1–2 hr and then rinsed in two changes of 1 : 1 mixture of 100% alcohol and 100% acetone for 10–20 min.

Lead Aspartate (Walton, 1979)

A stain containing lead aspartate is used for *en bloc* staining at an acidic pH. Lead nitrate and aspartic acid are present in the staining solution in the ratio of 0.02 *M* and 0.03 *M*, respectively. The pH is adjusted to 5.5 with 1 *N* KOH. Tissue blocks in a stoppered vial are covered completely with the staining solution and incubated for 30–60 min at 60°C.

Lead Citrate

Method 1 (Reynolds, 1963)

Lead nitrate, $Pb(NO_3)_2$	1.33 g
Sodium citrate, $Na_3(C_6H_5O_7)\cdot 2H_2O$	1.76 g
Distilled water (CO_2-free)	30 ml

The mixture is shaken vigorously at intervals for 30 min in a 50-ml volumetric flask. The completion of conversion of lead nitrate to lead citrate is indicated by the appearance of a uniform milky suspension. To this suspension is added 8 ml of 1 *N* NaOH (4%). It is diluted with distilled water to 50 ml and then mixed by inversion until lead citrate dissolves and the suspension clears up completely (pH 12.0). For staining, this solution is used either as is or diluted 5–1000 times with 0.01 *N* NaOH (0.04%), depending on the contrast desired.

Distilled water used for preparing the staining solution should be free of CO_2, which can be removed by boiling the water for about 8 min. It is then covered and allowed to cool before use. Since this procedure is time-consuming, CO_2 can be removed by using a vacuum desiccator. The NaOH used should be fresh and carbonate-free. Stock and staining solutions will keep for several weeks if kept tightly stoppered; if precipitates appear, the solution should be discarded. Lead solutions should be stored in plastic rather than glass containers and can be centrifuged before use.

Method 2 (Venable and Coggeshall, 1965). In this method, 0.01–0.04 g of lead citrate and 0.1 ml of 10 N CO_2-free NaOH are added to 10 ml of distilled water in a screw-topped vial. The vial is closed tightly, and the mixture is shaken vigorously until the lead citrate is dissolved. The solution can be centrifuged before use. A staining time of 1 min is adequate.

Method 3 (Fahmy, 1967). Previously boiled, cooled, double-distilled, and Millipore-filtered water (which can be stored in the refrigerator) is used. One pellet of NaOH (0.1–0.2 g) is dissolved in 50 ml of this water. After about 0.25 g of lead citrate is added, the solution is ready for use (pH 12.0). Sections are stained for 3–15 min. Freshly prepared staining solution is used each time.

Method 4 (Sato, 1967)

Lead nitrate	1 g
Lead acetate	1 g
Lead citrate	1 g
Sodium citrate	2 g
Distilled water	82 ml

This mixture is stirred at room temperature for 1 min; the solution appears milky. After 18 ml of 4% sodium hydroxide is added to it, the solution is stirred for 2–3 min. The staining solution can be used in its concentrated form or after dilution 1 : 7 with distilled water. The grids are stained by immersion in this solution for 7–10 min, followed by a thorough rinse in distilled water.

Lead Hydroxide

Method 1 (Watson, 1958). About 8.26 g of lead acetate is dissolved in 15 ml of distilled water. Then 3.2 ml of 40% NaOH solution is added, the solution is vigorously stirred, and the precipitated PbOH is centrifuged into a pellet. The supernatant is discarded after its volume is measured. The pellet is suspended in twice this volume of distilled water by vigorous stirring. The suspension is centrifuged and the supernatant is discarded. The previous two steps to wash the precipitated PbOH are repeated. The final supernatant is stored in a sealed bottle containing a small quantity of the washed precipitate. This is the staining solution, which is centrifuged before use. Usually, excessive contamination of sections occurs with this staining solution. However, the solution yields a strong contrast, and the pH is lower than that of alkaline lead solutions.

Method 2 (Karnovsky, 1961). First 270 mg of PbO (lead monoxide) is gently boiled in 20 ml of 1 N NaOH in a flask for 2 min with continuous stirring and then rapidly cooled to room temperature. The mixture is filtered through Whatman no. 1 filter paper. This is the concentrated stock solution (pH 11.2), which is diluted 1 : 20 to 1 : 50 with distilled water and centrifuged before use. Both the stock solution and the diluted staining solution are stable.

Method 3 (Karnovsky, 1961). Lead monoxide is added in excess to 15 ml of 10% Na-cacodylate solution in a flask. The flask is stoppered and the mixture is thoroughly stirred for 10–15 min (a magnetic stirrer is preferred) and filtered. The filtrate is diluted 2 : 8 with 10% Na-cacodylate solution with continuous stirring, and then 1 *N* NaOH is added drop by drop. A faint cloudiness may form, but the addition of more alkali drops will clear the solution. If the cloudiness does not appear, a maximum of 6 drops are required. The clear solution is the staining solution, which can be stored for months.

Lead Tartrate

Method 1 (Millonig, 1961)

NaOH	12.5 g
K-Na-tartrate	5 g
Distilled water	50 ml

First 0.5 ml of this stock solution is diluted to 100 ml with distilled water and heated; then 1 g of PbOH is added. After cooling, the solution is filtered and should stay clear (pH 12.3). Staining time is 5–15 min.

Method 2 (Millonig, 1961)

NaOH	20 g
K-Na-tartrate	1 g
Distilled water to make	50 ml

Of this stock solution, 1 ml is added to 5 ml of a 20% lead acetate, $Pb(CH_3COO)_2 \cdot 3H_2O$, solution. The solution is stirred, diluted 5–10 times with distilled water, and filtered. This is the staining solution. Staining time is 5–15 min.

Molybdenum Blues

Solutions of molybdenum oxides are colloidally dispersed mixtures of the oxide hydroxides of hexavalent and quinquevalent molybdenum. They are prepared from aqueous phosphomolybdic acid solutions and bind to polyhydroxylic components. These solutions have been used for staining mast cell granules, the secretion content of goblet cells, cytoplasmic granules in salivary glands, and chromatin (Ferrer *et al.*, 1984). The stain is prepared by mixing a highly concentrated $H_3PO_4 \cdot 12MoO_3 \cdot 24H_2O$ solution (10 g in 50 ml of distilled water) with zinc powder (5 g) under constant agitation. After 5 min, zinc is removed by centrifugation and the deep blue solution is evaporated at 40°C. The dried material is dissolved in the same volume of methanol. After removal of the white insoluble sediment, the blue methanolic solution is allowed to dry at room temperature. Purified samples of the stain are dissolved in 50 ml of distilled water and used as the staining reagent. Thin sections of specimens fixed with glutaraldehyde and OsO_4 are stained by immersion in the solution for 5–6 hr at room temperature.

Osmium Amine

This method is useful for staining chromatin and extracellular polysaccharides (Gautier *et al.*, 1973). It seems to possess specificity and intensity in both Feulgen-type and PAS-type reactions. Specific staining of chromatin can be obtained by floating the section on a drop of 5 *N* HCl solution for 15 min at 20°C (acid hydrolysis), followed by staining with 1% aqueous osmium amine solution. Before staining, osmium amine solution is treated with bubbling SO_2 for 10 min. For a PAS reaction, the section is floated on a drop of 1% periodic acid solution for 15–45 min at 20°C. The staining procedure is the same as the preceding one. Osmium amine is a black crystalline powder that is easy to prepare (Cogliati and Gautier, 1973).

Osmium Tetroxide

Prolonged impregnation with OsO_4 intensely stains Golgi complex and multivesicular bodies (Friend, 1969). Membranous structures such as endoplasmic reticulum are also stained but less intensely. Small tissue blocks are fixed with a conventional buffered OsO_4 solution. Specimens are immersed in 2% aqueous OsO_4 (unbuffered) in vials wrapped with aluminum foil to exclude light for 36–48 hr at room temperature or at 37°C. The solution is decanted after 24 hr and replaced by fresh solution. Specimens are treated *en bloc* with 0.5% aqueous uranyl acetate in 0.05 *M* maleate buffer (pH 5.3) for 90 min before dehydration. Specimens become friable and should be handled with extra care.

Osmium Tetroxide–Acrylic Acid (Herbert et al., 1972)

This procedure selectively stains unsaturated residues in cellulose, allowing determination of fibrillar size. About 10 g of a cellulose-containing compound or tissue is immersed for 20 hr at 25°C, with continuous stirring, in 200 ml of benzene containing 0.06 *M* acrylic acid and 0.15 *M* trifluoroacetic acid. Specimens are then rinsed in sequence with benzene, methanol, and distilled water. Finally, specimens (e.g., strands of cellulose or cell wall fragments) are stained with 2% OsO_4 in 0.1 *M* phosphate buffer (pH 6.8) for 20 hr at room temperature.

Osmium Tetroxide–Dimethylenediamine (Os-DMEDA)

This method is useful for staining polyanionic macromolecules (nucleic acid and acid polysaccharides) (Castejón and Castejón, 1972), basophilic material in the matrix of cartilage (chondroitin sulfate) (Seligman *et al.*, 1968), ribosomes, and chromatin. The solution does not stain membranes.

Thin sections of an aldehyde-fixed tissue are mounted on nickel grids and etched by immersion in dimethyl formamide for 20 min. The grids are immersed in aqueous 1% Os-DMEDA solution for 16 hr at room temperature and then rinsed in distilled water. Sections are poststained with uranyl acetate followed by lead citrate. The staining quality is inconsistent.

Osmium Tetroxide–p-Phenylenediamine

The intensity of overall staining produced by OsO_4 can be enhanced by *p*-phenylenediamine. The extracellular material also stains densely. Fixed tissue blocks are posttreated with 1% *p*-phenylenediamine in 70% ethanol for 15–25 min; sections do not require additional poststaining (Ledingham and Simpson, 1972). Semithin sections of the tissue thus treated can be viewed with the light microscope without additional staining. Caution is warranted in handling *p*-phenylenediamine, because it can cause contact dermatitis and bronchial asthma.

Osmium Tetroxide–Potassium Ferricyanide or Ferrocyanide

This method is useful for tracing certain membrane systems such as rough endoplasmic reticulum. It demonstrates morphologic variations in the smooth endoplasmic reticulum and its association with the mitotic apparatus of plant cells. The method also improves the preservation and staining of certain cell components such as the sarcotubular network in muscle cells and mucilaginous coats of bacteria. It is not a specific staining procedure for calcium-sequestering membrane systems. Prefixation with an aldehyde fixative containing di- or tri-valent cations (Ca^{2+}, Mg^{2+}, Mn^{2+}, or La^{3+}) is a prerequisite for this staining to occur. Extraction of calsequestrin from isolated vesicles of the skeletal muscle sarcoplasmic reticulum prevents their staining with this method.

Tissues are fixed with buffered glutaraldehyde containing 5 mM $CaCl_2$ for 1 hr at room temperature. After being rinsed in the same buffer, specimens are postfixed with a mixture of 1% OsO_4 and 0.8% potassium ferricyanide [$K_3Fe(CN)_6$] in 0.1 M cacodylate buffer (pH 7.4) for 1 hr at 4°C. After another rinse in the initial buffer, specimens are stained *en bloc* with 0.5% aqueous uranyl acetate for 30–60 min.

Osmium Tetroxide Buffered with Imidazole

The following method is effective in staining tissue lipids, containing unsaturated fatty acids, as well as lipoprotein particles (Angermüller and Fahimi, 1982). The lipid droplets appear well circumscribed with no evidence of diffusion. This method results in better staining of lipids than that yielded by conventional postfixation with buffered or unbuffered OsO_4.

Rat liver is fixed by vascular perfusion with 1.5% glutaraldehyde in 0.1 M

cacodylate buffer (pH 7.4) containing 4% polyvinyl pyrrolidone and 0.05% $CaCl_2$. Sections about 50 μm thick are cut with a Smith–Farquhar TC-2 chopper and placed in 0.2 M imidazole in which 4% aqueous OsO_4 is added; the final concentration of OsO_4 is 2% and the pH is 7.5. Note that the tissue must first be placed in imidazole buffer; then the OsO_4 is added. This postfixation is carried out for 30 min at room temperature. Osmium tetroxide–imidazole penetrates 50-μm-thick sections and 1-mm³ tissue blocks in 5 and 30 min, respectively.

Oxalate–Glutaraldehyde

The oxalate precipitation method is effective in demonstrating Ca^{2+}-accumulating sites in cells and tissues (Constantin et al., 1964). Immediately after removal from the animal, muscle blocks are immersed in a medium containing 40 mM potassium oxalate and 140 mM potassium chloride (pH 7.4, adjusted with KOH) for 10 min at room temperature (Popescu et al., 1974). The presence of a high potassium concentration is required to increase membrane permeability to the oxalate. Superficial bundles of fibers are dissected and fixed with glutaraldehyde for 2 hr at 4°C. After a rinse in the buffer, specimens are postfixed with OsO_4 for 1 hr. All solutions contain 40 mM potassium oxalate. Poststaining is not required, because uranyl acetate may dissolve calcium oxalate precipitates.

Phosphotungstic Acid (PTA)

Phosphotungstic acid (PTA) is useful primarily for staining basic proteins, glycoproteins, and polysaccharides, depending on the pH and concentration at which it is used. Aqueous PTA solutions at strongly acidic pH values show a strong affinity for the carbohydrate moiety of glycoproteins. Thin sections of tissues fixed with an aldehyde followed by OsO_4 and embedded in a resin are collected on polyethylene rings. They are floated on aqueous 10% solution of H_2O_2 for 1 hr and then rinsed in two to three changes of distilled water. Sections are stained on a drop of a 10% aqueous solution of PTA for 5 min. A direct contact between copper grids and PTA should be avoided. If PTA precipitate develops, treatment with 1% PTA solution before drying the sections is helpful (Farragiana and Marinozzi, 1979).

Ethanolic Phosphotungstic Acid (EPTA)

This method allows visualization of synapses for both qualitative and quantitative studies (Bloom and Aghajanian, 1968). Following 95% dehydration, aldehyde-fixed tissue specimens are stained en bloc with 1% EPTA for 1 hr; the final solution contains 2–10 drops of 95% ethanol per 10 ml of PTA solution. Specimens are rinsed in 100% ethanol.

Chromatin and nucleoli can also be specifically stained with EPTA. Glutaraldehyde-fixed tissue specimens are dehydrated in ethanol. After the first bath of

100% ethanol, the specimens are immersed in 1% PTA in precooled 100% ethanol for 12 hr at 4°C. After a thorough rinsing in 100% ethanol followed by propylene oxide, specimens are embedded. Poststaining of thin sections is not required.

Phosphotungstic Acid–Acriflavine (Chan-Curtis et al., 1970)

One component of this mixture is fluorescent; the other has adequate electron density. It is useful for localizing nucleic acids with the fluorescence microscopy or electron microscopy. When stained with this mixture, aldehyde-fixed tissues show chromatin as intensely black and nucleoli and ribosomes as dark gray. Poststaining with uranyl acetate accentuates the electron density of ribonucleo-proteins and deoxyribonucleoproteins. This mixture is available from Polysciences, Inc., Warrington, Pennsylvania.

Glutaraldehyde-fixed tissues are washed for 20 min with 0.1 M phosphate buffer (pH 7.3) containing 5% sucrose. During dehydration, specimens are treated with a 1 : 1 mixture of PTA and acriflavine and 50% ethanol for 2 hr at 4°C.

Phosphotungstic Acid–Chromic Acid

This method results in selective staining of plasma membranes of plant cells (Roland *et al.*, 1972) and mammalian sperm (Yunghans *et al.*, 1978). Specimens are fixed with an aldehyde followed by OsO_4. Thin sections are transferred to a 1% aqueous solution of periodic acid (HIO_4) for 30 min. After a thorough rinse in distilled water, sections are stained wtih 1% PTA in 10% chromic acid (CrO_3) for 5 min and then rinsed thoroughly with distilled water.

Phosphotungstic Acid–Hematoxylin

This mixture has been employed to stain basic protein in central nervous system (CNS) and peripheral nerve myelin (Adams *et al.*, 1971). The staining solution is prepared by dissolving 1 g of hematoxylin and 20 g of PTA separately, each in 200 ml of distilled water. The former solution is heated until the hematoxylin dissolves. After cooling, it is added to the PTA solution, and the mixture is diluted to 1000 ml with distilled water. To this mixture is added 0.177 g of $KMnO_4$ to facilitate "ripening," a process that takes about a month to complete. The pH of 50 ml of the stock solution is adjusted to 5.0 with 0.5 N NaOH.

Small segments of the tissue are fixed with 3% glutaraldehyde for 4 hr and washed overnight with 2 M sucrose in 0.2 M phosphate buffer (pH 7.4). Specimens are stained *en bloc* with PTAH (pH 5.0) for 3 hr at 37°C.

Platinum

Platinum–pyrimidine complexes (Davidson *et al.*, 1975) (Tousimis Research Corp., Rockville, Maryland) stain nucleic acid-containing structures (Aggarwal,

1976). *cis*-Dichloro-diammine platinum(II) (Polysciences, Inc.) likewise reacts mainly with nucleic acids, but it can also bind to proteins (Heinen, 1977). DNA shows a higher contrast than RNA.

Thin sections of aldehyde-fixed specimens are mounted on grids and then stained by immersion with the *cis*-Pt solution (200–1000 μm/ml in 0.01 M NaClO$_4$) for 1–5 days in darkness at 4, 37, or 60°C.

Selective staining of DNA-containing structures can be obtained by exposing aldehyde-fixed tissues to 3 N HCl hydrolysis for 1 hr at room temperature before embedding them in a resin. Thin sections are exposed to Schiff's reagent for 30 min and then stained with 1% platinum–pyrimidine in 0.05 M phosphate buffer or in 50% ethanol. This solution is freshly prepared and filtered through a 0.22-μm Millipore filter before use.

Potassium Permanganate

Osmium tetroxide-fixed tissues treated with KMnO$_4$ show intense staining of tonofibrils, myelin sheath, basement membranes, terminal bars, desmosomes, and cytoplasmic membranes in general. Monolayer tissue cultures of blood leukocytes, feather keratins, and α-keratins have also been stained with KMnO$_4$. This stain is useful for showing the true surface of viruses forming in infected cells, the lignin component of cell walls, and fungal walls. After KMnO$_4$ staining, a distinct difference in contrast between collagen and elastin can be obtained; elastin is more intensely stained (Wiedmer-Bridel *et al.,* 1978). For certain studies, Ba(KMnO$_4$)$_2$ may be preferable.

Staining is accomplished by placing a drop of 1% aqueous KMnO$_4$ on the sections mounted on a grid. Since surface oxidation occurs, the staining liquid is withdrawn from below the surface with a disposable pipette. Noncopper grids or acid (dilute formic acid) washing of copper grids before use may minimize the contamination. Staining is completed in a few minutes followed by a thorough rinsing in distilled water. Certain tissues (e.g., some leaf tissue) are better stained with KMnO$_4$ followed by lead citrate than with uranyl acetate and lead citrate.

If contamination is present, it can be removed by exposing the grid to 0.05% citric acid solution for 30 sec; too long a treatment will result in destaining. Filtration of the KMnO$_4$ solution is undesirable. Permanganates should not be used for staining sections of Epon containing the hardener NMA. Other resins present no problem.

Potassium Permanganate and Lead Citrate (Bray and Wagenaar, 1978)

This method is useful for staining plant tissues embedded in Spurr medium. One g of KMnO$_4$ is dissolved in 100 ml of distilled water. For staining, the liquid is withdrawn from below the surface with a Pasteur pipette. A drop of the stain is placed onto the section side of the grid for 2 min. After rinsing with distilled water, the grid is counterstained with lead citrate.

Potassium Pyroantimonate–Osmium Tetroxide

This technique has been claimed to precipitate and localize Ca^{2+}, Na^+, Mg^{2+}, K^+, Ba^{2+}, H^+, Zn^{2+}, and Fe^+ as well as organic compounds such as amino acids, histones, and glycogen. However, the retention of these materials is incomplete. Calcium seems to be the most frequent cation to be precipitated, either alone or in combination with other cations. This technique is widely used to study Ca^{2+} distribution in muscles.

A 5% solution of K-pyroantimonate ($K_2Sb_2O_7 \cdot 4H_2O$) is prepared by boiling 5 g of the reagent in 100 ml of distilled deionized water. The solution is allowed to cool to room temperature, diluted to 100 ml with distilled deionized water, and filtered through Whatman no. 5 filter paper to remove pyroantimonate precipitate.

When 25 ml of 2% aqueous OsO_4 solution is added to 25 ml of 5% K-pyroantimonate solution, the final concentrations are 1 and 2.5%, respectively. The pH of this mixture is 9–9.5, which is lowered to 7.6–7.8 by adding up to 5 ml of 0.1 N acetic acid. The fixative has about 0.04 N acetic acid. Before use, the fixative is passed under pressure through a Millipore (0.22 μm) filter to remove any fine particles or precipitated pyroantimonate. At all stages the fixative is maintained at 4°C. Tissue specimens are fixed for 1 hr at 4°C and then rinsed twice (for a total of 5–10 min) with 0.05 M potassium acetate buffer (pH 7.8 at 4°C). Thin sections are poststained with uranyl acetate and lead citrate.

Alternatively, a mixture of glutaraldehyde and K-pyroantimonate can be used to improve ultrastructural preservation. The fixative is prepared by breaking a 2-ml vial of purified 70% glutaraldehyde into 3 ml of distilled water, yielding a 28% glutaraldehyde solution (Weakley, 1979). Sufficient 2% aqueous K-pyroantimonate is added to 2.9 ml of this solution to yield a total volume of 20 ml. The pH is adjusted to 8.5–10.0 with 0.2 ml of 1% acetic acid. The final concentrations of glutaraldehyde, K-pyroantimonate, and acetic acid are 4%, 1.7%, and 0.01 N, respectively. The fixation time is 4 hr at 4°C.

Ruthenium Red

Ruthenium red stains mucopolysaccharides (surface coat of cells), junctional sarcoplasmic reticulum in skeletal muscle, and plant cell walls. It is known to bind to isolated DNA, and it interferes with the bonding and transport of Ca^{2+} in mitochondria.

Solution A

Glutaraldehyde (4% in water)	5 ml
Cacodylate buffer (0.2 M, pH 7.3)	5 ml
Ruthenium red stock solution	5 ml
(1500 ppm in water)	

Solution B

Osmium tetroxide (5% in water)	5 ml
Cacodylate buffer (0.2 M, pH 7.3)	5 ml
Ruthenium red stock solution	5 ml
(1500 ppm in water)	

Tissues are fixed and stained with solution A for 1 hr at room temperature (Luft, 1971). They are then rinsed with the buffer, and postfixed and stained with solution B (which is prepared immediately before use) for 3 hr at room temperature. Poststaining of sections is not necessary.

Silicotungstic Acid

Tungstic and molybdic heteropolyanions (stable only at a low pH) are strong precipitating agents for positively charged molecules such as proteins in acid solutions. Thus silicotungstic acid (STA) is useful for staining proteins in both plant and animal specimens. It also reacts with other positively charged groups.

Simultaneous treatment with STA and OsO_4 stains both presynaptic and synaptic membranes. When STA is employed directly on fresh tissues and then fixed with OsO_4, small black precipitates are observed in the synaptic vesicles, but none in other synaptic structures. The tissue is treated first with 5% aqueous STA for 30 min and then with a mixture of 2% OsO_4 and 5% STA in water for 1 hr. In another method, treatment of neuromuscular junctions with STA and phosphomolybdic acid permits the visualization of electron-dense precipitates in the synaptic vesicles of the cholinergic motor nerve terminals (Tsuji *et al.*, 1983). Dwarte and Vesk (1982) have used 5% STA in 6.25% aqueous sodium sulfate for 2–3 hr at room temperature for localizing biliproteins (which constitute a protein complex of accessory photosynthetic pigments that are associated with the thylakoids in algae) in the aldehyde-fixed tissues. Osmication is not used.

Silver

Generally, silver is deposited in regions containing mucopolysaccharides and diffuse proteins, including glycoproteins.

Silver Methenamine

This solution interacts with cystine-containing proteins (Swift, 1968).

Solution A

Silver nitrate (5%)	5 ml
Methenamine (3%)	100 ml

Solution B
Boric acid (1.44%) 10 ml
Borax (1.9%) 100 ml
Final staining solution
Solution A 25 ml
Solution B 5 ml
Distilled water 25 ml

The staining solution should have a pH of 9.2. It can be stored in the dark at 4°C for up to 1 week. Thin sections are stained by immersing the grid in the staining solution in a covered Petri dish placed in a light-free oven at 45°C for 30 sec to 2 hr.

Periodic Acid–Silver Methenamine

This method stains carbohydrates (structures that give a positive PAS reaction) (Thiéry, 1967; Rambourg, 1967). Glycogen fails to stain, but some nonspecific staining occurs. Fixation with formaldehyde is preferred, because glutaraldehyde may introduce aldehyde groups. Postfixation with OsO_4 cannot be recommended. The staining solution is prepared as follows:

Silver nitrate (5%) 5 ml
Methanamine 100 ml
Sodium tetraborate (5%) 12 ml
Double distilled water 100 ml

Immediately before the solution is to be used, the silver nitrate and methanamine are agitated together, and then sodium tetraborate and water are added.

Thin sections mounted on inert grids are treated with 1% aqueous periodic acid for 20–25 min at room temperature. (The periodic acid solution is stable for a month at 4°C.) The grid is thoroughly rinsed for 20 min in several changes of distilled water and then transferred, in a diffuse light, to the surface of the staining solution in a small, clean glass container and covered. The container is placed in an oven at 60°C for 40–60 min (sections will appear pale brown) and then transferred to a refrigerator. When cooled, the grid is placed on the surface of distilled water, rinsed with 5% sodium thiosulfate for 5 min, and then rinsed with distilled water (two changes).

Periodic Acid–Chromic Acid–Silver Methenamine

This method differs from the preceding one in that nonspecific staining of ribosomes and nuclei is minimized and glycogen is stained more intensely (Hernandez et al., 1968). Thin sections on an inert grid are exposed to 1% periodic acid for 20 min. After rinsing in distilled water for 30 min, the grid is treated with 10% aqueous chromic acid solution for 5 min. Following a brief rinse in distilled water, the grid is exposed to 1% aqueous sodium bisulfite solution for 1 min. The grid is then rinsed again in distilled water for 30 min, stained with silver methenamine solution for 25–40 min at 60°C, and rinsed for 20 min in distilled water.

Golgi Impregnation Method

This method is used for analysis of the structural organization of nervous system (neurons) with the light and the electron microscope. The following procedure has been used to study insect neurons (Ribi, 1976). Tissues are fixed with a mixture of 2% formaldehyde and 2.5% glutaraldehyde in phosphate buffer (pH 7.2) for 4 hr at 4°C. After being rinsed in the buffer, specimens are postfixed with a mixture of 2% potassium dichromate, 1% OsO_4, and 2% glucose (pH is adjusted to 7.2 with KOH) for 12 hr. Specimens are transferred to 4% potassium dichromate solution overnight in the dark at room temperature (23°C). After being repeatedly rinsed with 0.75% $AgNO_3$ until the wash is clear instead of orange in color, they are left in this solution for 1–2 days in the dark at room temperature. Thin sections are poststained with uranyl acetate and lead citrate. The required area for thin sectioning can be selected in a thick section under the light microscope and then reembedded in a flat silicone mold.

A modified procedure (Braak and Braak, 1982) is useful for all parts of the brain. This procedure facilitates removal of a large proportion of the silver chromate precipitations that normally fill impregnated nerve cells. Thus only a fine scattering of electron opaque particles remains, allowing recognition of even small profiles of previously identified nerve cells with the light microscope as well as evaluation of cytological details and synaptic relations.

Brain tissue is fixed with aldehydes by vascular perfusion or immersion. Blocks of about 4 mm^3 of the fixed tissue are placed in a freshly prepared mordant (30 g $K_2Cr_2O_7$, 125 g sucrose, 12.5 ml 40% formaldehyde, and 1000 ml distilled water) for 8–10 days at 27°C in the dark. After being rinsed several times in 0.75% aqueous solution of silver nitrate, the tissue blocks are allowed to remain in this solution for 2–6 days at 27°C in the dark. They are then dehydrated through a graded series of glycerol in water (20, 40, 60, and 80%; 15 min each step), and stored in 100% glycerol at 4°C. Thick sections (100–200 μm) can be cut with a razor blade and viewed with the light microscope so that cells of interest can be sketched and photographed.

These sections are rehydrated through a graded series of glycerol in water (80, 60, 40, and 20%), rinsed in distilled water, and placed in a freshly prepared lixiviation fluid [1 ml of 25% liquid ammonia, sp. wt. 0.91 (Merck), in 10,000 ml of distilled water] for 12–24 hr. A fine brownish precipitation in the perikaryon remains, allowing for recognition of impregnated cells at light and electron microscope levels. Lixiviation time is shorter if the sections are processed within a few days after impregnation. After a rinse in distilled water, the sections are postfixed with 2% aqueous OsO_4 for 30–60 min.

Golgi Impregnation Method with Gold

Tissue specimens that have been fixed with an aldehyde are rinsed in 0.15 M cacodylate buffer (pH 7.2) for 15 min. They are then immersed in a mixture of

OsO_4 and potassium dichromate for 4 days (Fairén *et al.*, 1977). This mixture is prepared by dissolving 12 g of potassium dichromate and 1 g of OsO_4 in 500 ml of distilled water. At least 12 ml of this mixture is used for each tissue block. Specimens are rinsed in 0.75% silver nitrate and stored in a fresh solution of silver nitrate for 2 days. They are gradually transferred to pure glycerol and can be stored at 4°C.

Specimens are embedded in 7% agar and cut at 150–200 μm with a tissue chopper. During thick sectioning, specimens must be kept wet with glycerol. Portions of these sections containing impregnated neurons are transferred to small glass vials containing glycerol. Sections are gradually transferred with gentle agitation to distilled water.

Gold toning is accomplished by placing the sections in 0.05% solution of yellow gold chloride [hydrogen tetrachloroaurate(III), $HAuCl_4 \cdot 4H_2O$] for 10–15 min in an ice bath in a refrigerator. Effective agitation is carried out every 2 min. The time for gold chloride treatment is critical for achieving optimal gold deposition. Sections are rinsed in three changes of cold distilled water to remove excess gold chloride. While the vials are still in the ice bath, 0.05% oxalic acid is poured into them to reduce gold chloride to metallic gold; this reduction takes 2 min. Sections are rinsed and vials are gradually brought to room temperature. Deimpregnation of silver chromate is accomplished by treating the sections (with periodic agitation) with freshly prepared 1% sodium thiosulfate for 60–90 min at 20°C.

To remove sodium thiosulfate, sections are rinsed in distilled water. They are poststained with 2% OsO_4 in 0.1 *M* cacodylate buffer (pH 7.2) for 30 min at room temperature. After being rinsed in distilled water, thick sections are stained *en bloc* for 2 hr with 1% uranyl acetate in 70% ethanol during dehydration. These sections are embedded in flat trays.

Silver Lactate–Osmium Tetroxide

This method is purported to localize intracellular anions (Cl^-) (Komnick, 1962). After this treatment, chloride ions are precipitated as AgCl. The fixation and staining mixture is prepared by mixing 0.5–1.5% silver lactate ($AgC_3H_5O_3 \cdot H_2O$) with 1–2% OsO_4 in equal parts in a suitable buffer (pH 7.2). Tissues are fixed for 2 hr in this mixture at 4°C, using a red safelight. During dehydration, tissues are treated with nitric acid (50% acetone containing 0.1 *N* HNO_3) to dissolve unspecific precipitates.

Sodium Tungstate

The reagent sodium tungstate is a general stain for nucleic acid-containing structures such as chromatin, nucleoli, and ribosomes. Moreover, it has been suggested that acidic solutions of sodium tungstate increase the contrast of polysaccharides and glycoproteins, as does acidified PTA. For nucleic acid staining, thin sections of glutaraldehyde-fixed specimens are floated on 10% aqueous sodium tungstate (pH 5.5, adjusted with 1 N HCl) for 1–2 hr at room temperature. Poststaining is not required (Stockert, 1977).

Tannic Acid–Ferric Chloride

This poststaining procedure is useful for the demonstration of carbohydrates in the epiphyseal cartilage (Takagi et al., 1983). Matrix granules (proteoglycan monomers) and chondrocyte secretory granules are stained intensely, but collagen fibrils and glycogen are not stained. The staining reactions are pH-dependent.

Cartilage tissue is fixed with 2.7% glutaraldehyde in 0.1 M cacodylate buffer (pH 6.8) containing 2% tannic acid for 2 hr at 22°C. After being rinsed several times in the buffer containing 7% sucrose, the specimens are embedded. Thin sections on stainless-steel grids are treated for 10 min with a filtered 5% aqueous tannic acid solution (pH 2.6–2.8). The grids are rinsed three times in distilled water and then treated for 1 min with a freshly prepared solution of ferric chloride (5 ml of 40% ferric chloride added to 95 ml of distilled water) (pH 1.4–1.6). The grids are thoroughly rinsed in distilled water.

Tannic Acid–Uranyl Acetate

This poststaining method is useful for strongly staining matrix granules, intracellular glycogen and chondrocyte secretory granules, and moderately staining collagen fibrils in the cartilage matrix (Takagi et al., 1983). Tissue specimens are fixed and thin sections are treated with tannic acid as described in the preceding paragraph. The grids are rinsed in distilled water and then treated for 5 min with freshly prepared 1% uranyl acetate solution (pH 4.1–4.3).

Tetraphenylporphine Sulfonate

When tetraphenylporphine sulfonate (TPPS) is complexed with a heavy metal, it contributes significantly and specifically to the electron density of elastica (Albert and Fleischer, 1970). The preparation of TPPS is given by Albert and Fleischer (1970). Thin sections of tissues fixed with glutaraldehyde followed by OsO_4 are poststained with uranyl acetate and lead citrate. After a rinse in distilled water, sections are stained with 5% gold TPPS for 5 min, and washed with water. Alternatively, sections are stained with 10% silver TPPS for 45 min. The staining solution should be centrifuged before use, and gold or stainless-steel grids should be used.

Thallium Ethylate

This method is used for selective staining of DNA-containing structures, including prokaryotic DNA and DNA viruses (Moyne, 1973). Tissue blocks are fixed with 4% formaldehyde in 0.1 M phosphate buffer containing sucrose. After the tissues have been rinsed in the buffer and dehydrated in acetone, pyridine is added gradually to bring them to absolute pyridine over a period of 15 min. Specimens are rinsed thoroughly in pyridine at 4°C and then acetylated overnight at room temperature with a freshly prepared mixture of 6 volumes of pyridine and 4 volumes of acetic anhydride with sodium acetate. Following three rinses (10 min each) with acetone at room temperature, specimens are rinsed with 1 : 1 mixture of acetone and pyridine and infiltrated with Epon.

Thin sections on gold or titanium grids are hydrolyzed on the surface of 5 N HCl for 20 min to 1 hr at 20°C and then rinsed with distilled water. Grids are floated for 30 min at room temperature on a decolorized pararosaniline solution in a solid watch glass with a lid. This solution is prepared by dissolving 0.75 g of pararosaniline hydrochloride in 100 ml of distilled water and adding 0.75 g of potassium metabisulfite. When the pararosaniline has been dissolved, 1.5 ml of HCl is added and allowed to remain at room temperature for 4–6 hr. The solution is further decolorized by shaking it with 0.25 g of finely powdered activated charcoal for 1 min before filtering it through a Millipore filter (1.2 μm pore size). Grids are rinsed in distilled water and dried on filter paper.

Thallium ethylate solution is prepared by placing 2 g of thallium metal (cut into 1-mm cubes) in the bottom of a clean, dry 25-ml bottle with a stopper. After being completely filled with 100% ethanol, the bottle is tightly stoppered and kept thus for 10–20 days until a fine precipitate of thallium hydroxide appears. The grid is immersed in a solid watch glass containing a mixture of 1 ml of 100% ethanol, 1 drop of thallium stock solution, and 1 drop of distilled water. The watch glass is covered with a lid; staining time is 8–15 min at room temperature. Grids are rinsed with 100% ethanol for 5 sec, dried, and examined within 12–24 hr.

Thiosemicarbazide and Thiocarbohydrazide

The following methods are analogous to the PAS reaction in common use for the localization of vicinal hydroxyl groups of mucosubstances at the light microscopic level.

Periodic Acid–Thiocarbohydrazide or Thiosemicarbazide–Silver Proteinate (Thiéry, 1967)

Thin sections of fixed tissues are transferred by means of a plastic loop onto gold grids that are floated on 1% periodic acid in double-distilled water for 5–20 min at room temperature. Sections are treated with 0.2% thiocarbohydrazide in 20% acetic acid for 1–2 hr at room temperature. They are rinsed first in three changes (10 min each) of 10% acetic acid and then in three changes (10 min each) of distilled water. Alternatively, thin sections are floated on 1% thiosemicarbazide in 10% acetic acid for 1 hr at room temperature. Sections are rinsed in three changes (5 min each) of 10% acetic acid and then in three changes (5 min each) of double-distilled water.

Sections are floated on a 1% silver proteinate solution (Protargol) in double distilled water for 20 min in the dark at room temperature. They are transferred under a red safelight to double-distilled water and thoroughly rinsed.

Uranyl Acetate

Uranyl acetate stains specifically nucleic acid-containing structures in fixed specimens. It also acts as a postfixative for nucleic acids, phospholipids, and plant cell walls. It can be used either for *en bloc* staining or for section staining. *En bloc* staining is preceded by a thorough rinse of the fixed specimens to remove phosphate or cacodylate buffer, which are precipitated by uranyl salts. *En bloc* staining may adversely affect glycogen staining.

Preparation of Staining Solutions

Saturated aqueous solutions of uranyl acetate are widely used for moderate staining. Dissolution of this stain in water is very slow. If accelerated dissolution and penetration of the stain and a heavier staining are needed, the staining solution can be prepared in an alcohol. For example, an excess of uranyl acetate is placed in a vial (5–15 ml) containing 50% ethanol or acetone and allowed to remain overnight at room temperature. Intermittent or continuous agitation will accelerate the dissolution of the stain and saturation of the solution. Uranyl acetate is more soluble in methanol than in water or ethanol. A 7% solution of this stain in absolute methanol is recommended (e.g., for elastin in animal and human tissues); the duration of staining varies from 5 to 15 min. Grids stained in methanolic uranyl acetate are rinsed with three changes of methanol prior to staining with lead citrate. Prolonged treatments with methanolic solutions may cause extraction of cellular materials.

En Bloc Staining

Fixed specimen blocks are immersed in 0.5–2.0% aqueous solution of uranyl acetate (pH 3.9) for 10–50 min before dehydration. Alternatively, staining can be carried out with uranyl acetate in 10–50% ethanol or acetone during dehydration.

Section Staining

The grid is floated (section side down) on a drop of the stain for 5–20 min. The stain should be removed from the vial for use with a capillary pipette without disturbing the undissolved crystals at the bottom. After the staining, the grid is thoroughly rinsed in two changes of distilled water. This staining is usually followed by staining with lead citrate.

Specific Staining of Neuroendocrine Granules

Uranyl acetate, under specific conditions, specifically stains neuroendocrine granules (Payne *et al.*, 1984) and adenine nucleotides in organelles storing biogenic amines (Richards and Da Prada, 1977). Tissues fixed with glutaraldehyde and OsO_4 are thoroughly rinsed in a buffer and then in 0.9% NaCl at 4°C for 72 hr with several changes of fresh NaCl. This rinsing is necessary to wash out the buffer salts used during fixation and storage steps; otherwise uranyl acetate may form precipitates with these salts. The specimens are treated with 4% aqueous uranyl acetate (pH 3.9) for 48 hr at 4°C. They are rinsed in three changes (15 min each) of 0.9% NaCl. Poststaining is not necessary.

Semiselective Staining of RNA (Bernhard, 1969)

Thin sections of an aldehyde-fixed specimen are mounted on copper grids and stained with 5% aqueous uranyl acetate for 1 min at room temperature. The grid is floated on 0.2 M EDTA (pH 7.0) for 20–60 min before being stained with lead citrate for 1 min. This method is useful for distinguishing between RNA and DNA.

Vanadium

Vanadium sulfate ($VOSO_4 \cdot 2H_2O$) or ammonium molybdate ($(NH_4)_6Mo_7O_{24}$ is a general-purpose stain (Callahan and Horner, 1964), but imparts less intense contrast than uranyl and lead salts do. Thin sections are stained for 15–30 min at room temperature. For general purposes, a 1% aqueous solution of vanadyl sulfate (pH 3.6) is adequate. This solution is stable for 2 weeks. A more stable solution can be prepared by mixing 20 ml of 1% vanadyl sulfate and 80 ml of 1% ammonium molybdate. The dark purple solution first formed undergoes further oxidation to form a clear yellow solution, which is stable for up to 1 year.

LIPID PRESERVATION AND STAINING

Homogeneously dense lipid droplets indicate satisfactory preservation, whereas the presence of droplets with dense rim and less dense or clear center is an

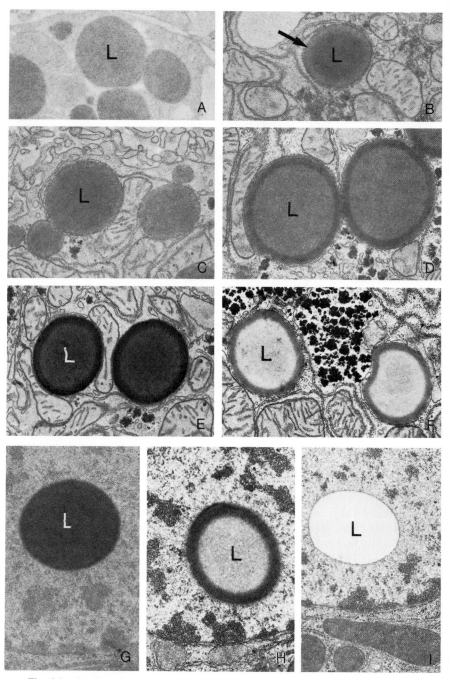

Fig. 4.1. Rat liver fixed by vascular perfusion, showing the quality of lipid (L) preservation after different staining techniques. (A) Homogenously dense lipid droplets; postfixation for 30 min in

indication of lipid extraction. During fixation, lipid extraction begins usually in the center of lipid droplets and then toward the periphery. In other words, the sequence of extraction is centrifugal. This is expected since lipid reaction with OsO_4 is minimal in the center of the lipid droplet. On the other hand, the sequence of lipid extraction during dehydration and embedding is usually centripetal in intact lipid droplets. After complete extraction, lipid droplets appear as empty vacuoles (Fig. 4.1). Unlike other vacuoles, these vacuoles are without a membrane, although a dense rim may look like a membrane.

Several methods are available for preserving and staining lipid droplets. One of the two reliable methods is the application of OsO_4 buffered with imidazole (p. 197). When this method is used poststaining of ultrathin sections with lead salts should be avoided. These salts tend to extract lipids at a high pH (Fig. 4.1) (Neiss, 1983). Treatment of sections with distilled water at a high pH, obtained by adding a few drops of 10 N NaOH, also removes the electron density of lipid droplets as well as membranes (Nickerson, 1983). Apparently, the removal of lipid opacity is attributable to exposure to highly basic solutions of lead. Poststaining of sections with uranyl acetate prior to staining with lead salts does not prevent the extraction of lipids. Poststaining of sections with uranyl acetate alone does not extract lipids. Prolonged staining with salts of lead and/or uranium is known to cause general destaining.

The other reliable method to preserve and stain lipids (both unsaturated and saturated) is en bloc treatment with p-phenylenediamine after fixation with glutaraldehyde and OsO_4 (p. 197) (Fig. 4.2). This procedure considerably minimizes extraction of intracellular lipids during dehydration, and allows poststaining of ultrathin sections with lead and uranium salts without adverse effects (Ledingham and Simpson, 1972; Boshier *et al.*, 1984). Figure 4.2 shows con-

---

1% OsO_4, rinsing for 30 min in cacodylate buffer, and a second postfixation for 30 min in 1% OsO_4; all treatments at pH 7.4; no section staining. (B) Lipid droplets with a dense center and pale halo (arrow); postfixation for 30 min in 1% OsO_4 containing 18 mM $K_4Fe(CN)_6$ (pH 9.8); no section staining. (C) Homogenously dense lipid droplets; postfixation for 30 min in 1% OsO_4 containing 36 mM $K_4Fe(CN)_6$ (pH 10.4); no section staining. (D) Droplets with dense rim and more translucent center; postfixation for 30 min in 1% OsO_4 containing 36 mM $K_4Fe(CN)_6$ (pH 10.4); section staining for 10 min in saturated uranyl acetate in 50% ethanol. (E) Droplets with dense rim; postfixation for 30 min in 1% OsO_4 containing 36 mM $K_4Fe(CN)_6$ (pH 10.4); section staining for 10 min in 0.2% lead citrate (pH 11.8). (F) Droplets with dense rim and very translucent center; section staining for 10 min in uranyl acetate followed by 5 min in 0.2% lead citrate. (G) Homogenously dense intranuclear lipid droplet; postfixation for 30 min in 1% OsO_4 containing 2 mM $K_4Fe(CN)_6$ and 5 mM $MgCl_2$ (pH 7.4); no section staining. (H) Intranuclear droplet with dense rim and clear center; postfixation for 30 min in 4% OsO_4 containing 18 mM $K_4Fe(CN)_6$ (pH 9.9); section staining for 10 min in uranyl acetate followed by 5 min in 0.2% lead citrate. (I) Completely extracted intranuclear lipid droplet; postfixation for 30 min in 1% OsO_4 containing 2 mM $K_4Fe(CN)_6$ and 5 mM $CaCl_2$ (pH 7.5); section staining for 10 min in uranyl acetate followed by 5 min in 0.2% lead citrate. All figures ×20,000. (From Neiss, 1983.)

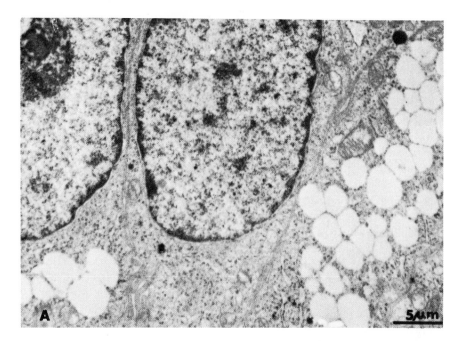

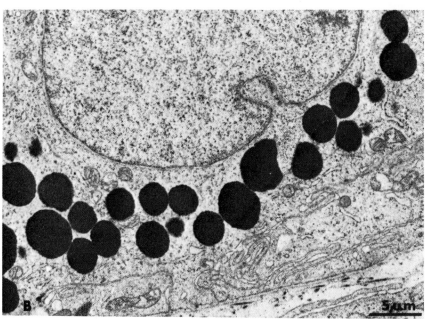

Fig. 4.2. Rat uterine epithelial tissues fixed by vascular perfusion with glutaraldehyde (2.5%) in 0.07 M sodium cacodylate buffer (pH 7.5) for 5 min, and then fixed by immersion for 6 hr at 4°C. The specimens are rinsed overnight in the buffer at 4°C. (A) Poststained with uranyl acetate and lead citrate; empty-looking lipid droplets are present. (B) After *en bloc* staining with alcoholic *p*-phenylenediamine, well-defined electron dense lipid droplets are present. (From Boshier *et al.*, 1984.)

vincingly that after treatment with *p*-phenylenediamine rat uterine epithelial tissue retains significant amounts of lipids, whereas these lipids are extracted during conventional fixation and staining.

Other approaches used to minimize lipid extraction are: (1) OsO_4 in combination with potassium ferrocyanide (p. 197); (2) tannic acid in an aldehyde fixative (p. 19); (3) malachite green in glutaraldehyde (p. 18); (4) poststaining with PTA.

Many lipids are bound to proteins in a way that the former are inaccessible to OsO_4 and other lipid stains. Treatment with lipid-soluble terpenoids such as farnesol and myrcene seems to enhance lipid binding to OsO_4 (see p. 17) (Wigglesworth, 1981). This method is based on the assumption that lipids accumulate these terpenoids by partition and thereby show an increase in osmium uptake. Thymol can be added to this procedure to unmask lipids bound to proteins and make them accessible to the terpenoids, without visible damage to the ultrastructure. This approach also allows distinction between the osmium bound to lipids and that bound to proteins.

MULTIPLE STAINING

Ultrathin sections of specimens fixed with aldehydes and OsO_4 are usually poststained with uranyl acetate and lead citrate. This protocol is more accurately described as "triple staining" because osmium also contributes to the increase in contrast. In many studies, uranyl acetate is also used as an *en bloc* stain before dehydration. Such multiple staining either enhances the contrast already imparted by another stain or increases the contrast of a very-low-density structure that does not react with another stain, or both. In other words, multiple staining enhances the contrast of all cell components more than does any single stain alone. Specificity of staining is apparently not the primary purpose in this approach. Another advantage of multiple staining is that many of these heavy metals act as fixatives to various degrees. Moreover, interaction between heavy metals plays an important role in better preservation and enhanced contrast of cellular structures. The following procedure is recommended for general morphological staining.

Specimens are fixed with glutaraldehyde or a mixture of glutaraldehyde and formaldehyde followed by OsO_4; they are then immersed in a 2% solution of uranyl acetate in 10% acetone or ethanol for about 10 min. After dehydration and embedding, ultrathin sections are poststained by floating the grid (section side down) on a drop of 1% aqueous uranyl acetate for 5–20 min, followed by lead acetate for 3–20 min (see p. 214). Additional contrast can be obtained by poststaining with 2% aqueous tannic acid for 10 min between the uranyl acetate and lead citrate poststainings. The grid should not be allowed to dry completely between the different staining treatments. Caution is warranted in the use of *en bloc* staining with uranyl acetate, for it may extract glycogen. Poststaining with uranyl and lead salts can be expedited by treatment of thin sections with 10% H_2O_2 for 1 min (Pfeiffer, 1982). This treatment, however, may cause extraction of cellular materials.

If conventional staining with uranyl acetate followed by lead citrate is less than satisfactory, an approach called the double lead citrate method can be used. Sections are stained first with lead citrate for 1–5 min, rinsed in double-distilled water, and dried. They are then stained with a saturated solution of uranyl acetate for 40 min and finally again with lead citrate, this time for 20 min (Daddow, 1983).

DOUBLE STAINING OF THIN SECTIONS WITH URANYL ACETATE AND LEAD CITRATE

The following method is used to double stain thin sections. A small quantity of fresh NaOH pellets is placed at one side of an absolutely clean plastic Petri dish (Fig. 4.3) to produce a CO_2-free atmosphere. Since drops of suitable size and shape are not easily obtained on a glass surface, a plastic dish is preferred. Alternatively, a piece of a clean sheet of dental wax is placed in a glass Petri dish; the wax is the staining surface (Fig. 4.3). (Plastic and wax surfaces are equally useful.) The dish is covered. The grid with mounted sections is floated section side down, on a drop of 0.5–1.0% aqueous uranyl acetate solution for about 5 min in a second dish. By the time the staining with uranyl acetate is completed, the atmosphere in the first dish will be CO_2-free. The grid is rinsed in distilled water and kept for subsequent staining with lead.

With the aid of a clean Pasteur pipette, a small quantity of Reynold's lead citrate stain is drawn up from below the surface of the staining solution (freshly prepared solution is preferred), and 2 drops, each slightly larger than the grid, are placed in the first dish. The first drop of stain from the pipette is not used. Immediately, the grid wet by immersion in CO_2-free distilled water is placed, section side down, on the drop of staining solution. (Alternatively, the grid can be stained by immersion.) One drop of staining solution is used for each grid. No time should be wasted between placing the drops of the stain in the dish and placing the grids on them. Each drop should be small enough to allow the grid to float on the top of the dome of the drop instead of sliding down to the sides. The dish is covered immediately after the grid is placed in it. The duration of staining ranges from 3 to 15 min. Coated grids and thicker sections are stained longer.

After the grid has been stained, it is quickly and thoroughly rinsed consecutively in CO_2-free 0.02 N NaOH (0.08%) solution and CO_2-free distilled water. While the grid is being rinsed, the lid should be replaced because the other grids are still being stained. In fact, the Petri dish should be kept covered as

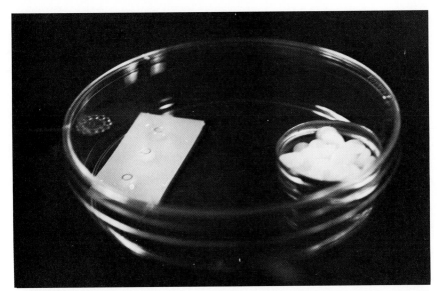

Fig. 4.3. A Petri dish used as a chamber for staining grids with Reynold's lead citrate. The dish contains pellets of NaOH and a small sheet of dental wax. Three drops of the staining solution are placed on the sheet, each showing a grid floating on it.

much as possible. The rinsing can be accomplished either by holding the grid at its edge with a forceps and dipping it rapidly in the fluid by agitation in a small beaker or by dousing the grid under a jet from a plastic wash bottle. Care should be taken that the grid is quickly immersed in the CO_2-free washing fluid and that all the stain is washed off the grid and the forceps.

The washed grid is blotted dry; the section side must not touch the filter paper. If the forceps carry excess water when the grid is being placed on a filter paper, the grid is drawn up between the tines of the forceps. This problem can be avoided by placing a piece of filter paper between the tines and advancing it toward the grid. It is desirable not to stain more than three grids simultaneously with this method. Breathing very close to the Petri dish should be avoided. If lead precipitates are found, they can be removed by exposing the sections (mounted on the grid) to 10% acetic acid for 1 min (see also p. 225).

To prevent the grid from sliding down to the sides of the staining solution drop and dragging of the drop across the wax surface when the grid is being placed onto or being removed from it, round depressions of 5 mm in diameter and 1 mm deep can be made in the dental wax by pressing a metal rod against the wax. One grid is placed in each depression, which contains one drop of the staining solution.

The contamination of lead stains during their storage can be significantly reduced by a simple procedure introduced by Richardson *et al.* (1983). The staining solution can be stored for at least 5 months without contamination. The stain from the container can be used repeatedly without contaminating the remaining stock solution. About 30 ml of lead citrate solution is filtered with a syringe-operated Millex GV-filter (Millipore Corp.) into a 60-ml serum bottle. The bottle is stoppered with a flange rubber, and a 25-gauge disposable Luer-lok needle is inserted into the rubber stopper (Fig. 4.4). A plastic tube attached to a can of compressed inert gas (Freon) is inserted into the hub of the needle. A few short bursts of gas are applied to purge air from the serum bottle through the same needle. The plastic tube inserted in the hub does not completely seal it. While the needle is being removed, gas is slowly released to replace most of the atmospheric air. This step reduces the amount of CO_2 that can form lead carbonate. Immediately before use, the staining solution is removed from the bottle with a disposable Luer-lok needle (20–25 gauge, 1.5 in.) and a 3-ml syringe (Fig. 4.4).

1. Prepare lead stain (4)

2. Filtering stain

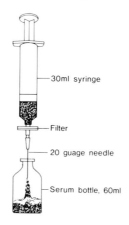

30ml syringe

Filter

20 guage needle

Serum bottle, 60ml

3. Insert flange stopper

Fig. 4.4. Schematic diagram of the storage and removal of lead staining solution without contamination. (From Richardson *et al.*, 1983.)

MULTIPLE-GRID STAINING

Several methods have been devised for holding and staining groups of grids; most of these methods have been reviewed by Hayat (1981a); more recently introduced methods are presented here. The Grid-All method (Giammara, 1981) (Polysciences, Inc.) allows grids to be collected serially in the same module from the ultramicrotome through the processes of transport, staining, rinsing, and storage. Holders of different sizes can be made to support various numbers of grids (Fig. 4.5A). Each holder is made of a flexible RTV Silastic material that allows insertion of grids (including tabbed or slotted ones) when it is flexed (Fig. 4.5B). When relaxed, the material grips the grids tightly in its slots. The holder has a channel that permits longitudinal flow of solutions over the grids. The dimensions of the holder permit its insertion into a modified flexible staining pipette (Fig. 4.5C).

4. Insert 25 gauge needle

5. Freon is released in space above liquid

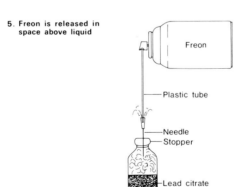

6. Removing stain for use

Fig. 4.4. *Continued.*

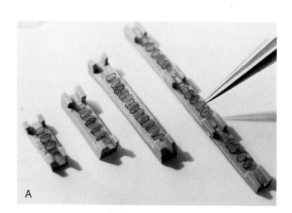

A

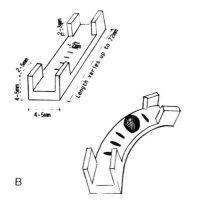

2-3mm

4-5mm · 2-5mm

Length varies up to 7cm

4-5mm

B

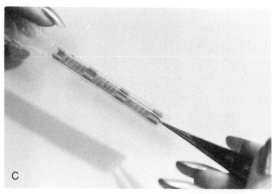

C

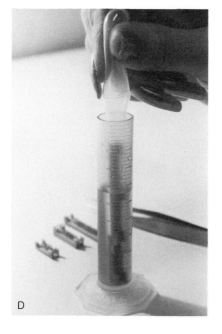

D

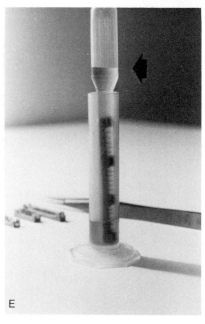

E

Once the required number of grids have been inserted into the slits, the holder is dipped into distilled water and then placed into its flexible staining pipette. The air is expelled by squeezing the bulb, and the pipette is introduced into a special stain container that contains 4 ml of stain (Fig. 4.5D). When the flexible bulb is gently released, the stain is drawn up into the pipette (Fig. 4.5E). Any entrapped bubbles can be dislodged by tapping the holder. The solution can be expelled by squeezing the flexible bulb. Flushing is done through several rinsing solutions. After the final rinse, the holder is removed from the staining pipette with forceps and placed on an absorbent towel for drying.

An inexpensive and simple multiple-grid staining device can be constructed from Tygon tubing rings and a vial (Fig. 4.6A) (Cardamone, 1982). The specifics of the tubing and rings are as follows:

Quantity	Length	Width (outer diameter)
1	50 mm	6 mm
2	8 mm	10 mm
2	8 mm	12.2 mm

First the 10-mm rings are placed on each end of the 50-mm tubing, and then the 12.2-mm rings are fitted over the 10-mm rings. These rings function as spacers that keep the main body of the tubing away from the wall of the vial. A razor blade is used to cut a series of 10–12 slits in the tubing, just deep enough to reach the hollow center of the tubing. These slits, when squeezed laterally, open to accomodate the edge of a grid. Conversely, when the lateral pressure is relaxed, the tubing firmly grips each grid (Fig. 4.6B). To keep track of grid placement, a notch is cut at one end of tubing, thus denoting position 1. Once the grids are in place, the entire holder can be immersed in the staining solution in a vial (Fisher no. 03-340-2D) that is stoppered.

Another simple, disposable, and inexpensive device was introduced by Kobayasi (1983). A piece of plastic or synthetic rubber tube is cut. The length of the tube is about one-fourth the length of the syringe, and the diameter is a little smaller than that of the syringe. After a quarter of the tube is removed with scissors, a scalpel is used to make slits 2–3 mm long in the tube parallel to its long axis. To avoid buildup of air bubbles between the grids during staining, the slits should not be cut too close together. The grids are inserted by their edges in the slits and the clip is gently removed. The tube is carefully placed in the syringe so that the axes of the tube and the syringe are parallel to each other. The syringe plunger is replaced, and the staining solution is sucked up and flushed out for

---

Fig. 4.5.   The Grid-All method. (A) Holders of various sizes are shown with grids. (B) The holder is flexed to allow insertion of grids into slots. (C) The holder is inserted into the staining pipette. (D) Less than 4 ml of staining solution is taken into the flexible staining pipette. (E) Staining of sections occurs within the closed system. (From Giammara, 1981.)

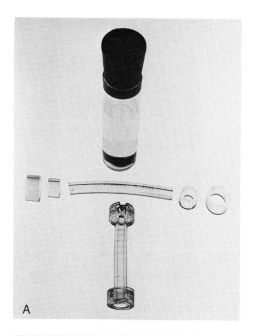

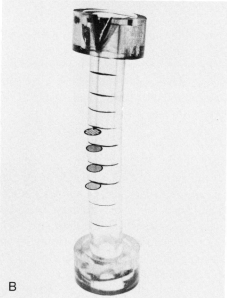

Fig. 4.6. Multiple-grid staining device. (A) Tygon tubing, rings, and assembled device with slits and a stoppered shell vial. (B) Close-up of the device, showing grids firmly held in slits. (From Cardamone, 1982.)

staining. After the staining and rinsing are completed, the syringe plunger is removed and the tube is taken out of the syringe. The tube is allowed to stand dry. It is then carefully unrolled and held flat by a paper clip, allowing the grids to be removed. If necessary, a Millipore filter and a suction apparatus can be connected to the syringe.

Simultaneous staining and rinsing of a large number of grids can be accomplished by using a device introduced by Brown (1983b) (Fig. 4.7). This is a closed staining system having a multigrid holder with a flow-through tube. Filtration through an external filter ensures that clean staining solutions are drawn into the staining chamber. The staining chamber consists of a Pyrex glass tube (14 × 0.35 cm) fused onto the stopcock outlet of a 25-ml separating funnel. This funnel functions as the reservoir for the rinsing fluid. Just below the stopcock a side arm (curved glass tube) is formed for the attachment of a vacuum pump. A 5-ml syringe is used to provide the necessary vacuum.

The multiple-grid holder is made from a polyethylene tube (12 cm long, 0.32 cm outside diameter) in which slots are cut to hold 17 grids (3.05 mm in

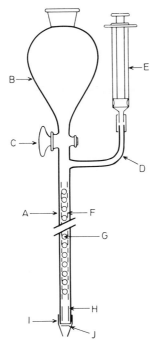

Fig. 4.7. Cross-sectional diagram of staining apparatus: A, staining chamber; B, reservoir; C, stopcock; D, side arm; E, syringe; F, grid holder; G, specimen grids; H, retaining collar; I, filter holder; J, filter disk. (From Brown, 1983b.)

diameter) in linear order along its length. A collar cut from the tube retains the grid holder within the specimen chamber. The filter holder is formed from a 6-mm polyethylene embedding capsule (Taab) by removing the cap and cutting a hole in the cone. The filter (which sits inside, covering the opening in the cone) consists of a disk (6 mm in diameter) of 0.45-μm Millipore filter supported by a similar-sized disk of prefilter. If the holder is a tight push fit on the staining chamber, the filter will remain clamped in position. If the filter material is sandwiched between thin plastic sheets, a three-hole paper punch can be used to cut out the filter disks.

Grids bearing the sections are placed in the grid holder, which is then inserted in the staining chamber, followed by the retaining collar. The filter holder, fitted with new filter disks, is firmly attached to the apparatus. With the stopcock in the closed position and with the syringe operating as a vacuum pump, the staining solution is drawn through the filter into the staining chamber, completely covering the grids. After removal of the filter, the outside of the staining chamber is washed with distilled water. The apparatus should be mounted in a vertical position in a test tube stand and the reservoir should be filled with distilled water.

After the staining, the grids are washed free of stain by slowly opening the stopcock and allowing the distilled water in the reservoir to flow gently through the staining chamber. By making a hole of an appropriate size in the embedding capsule, one can control the rate of flow from the reservoir. Before Formvar-coated grids are stained, the apparatus can be purged with a slow stream of nitrogen prior to affixing the filter. This procedure permits staining in an air-free atmosphere and avoids the "etching" and "precipitation" effects typical of Formvar film. Multiple-grid staining devices are available from E. F. Fullam, Inc., Polysciences, Inc., and Ted Pella, Inc.

REMOVAL OF BOUND OSMIUM FROM THIN SECTIONS

To remove bound osmium from thin sections, the unmounted sections are floated on 10% aqueous solution of periodic acid for 20–30 min at room temperature. Alternatively, sections are exposed at room temperature either to a saturated solution of potassium periodate for 30–60 min or to a 15% solution of hydrogen peroxide for 10–15 min. A thorough washing in distilled water follows. Either treatment may extract cellular components.

STAINING FOR HIGH-VOLTAGE ELECTRON MICROSCOPY

Thick sections can be used for transmission electron microscopy, provided very high accelerating voltages are used. For example, the unit membrane structure can be resolved in Epon sections of 0.1, 1.0, 1.5, and 2.0 μm thickness at 100, 500, 800, and 1000 kV, respectively (Hama and Kosaka, 1981). If specimens are stained, very thick resin-embedded sections can be observed. Stained sections thicker than 5 μm have been examined at 1000 kV. Longer durations and elevated temperatures are required to stain thick sections. To facilitate thorough penetration into the sections, some staining solution (e.g., uranyl acetate and PTA) can be prepared in alcohols instead of in water. Both sides of the section should be exposed to the stain by immersing the section in the staining solution. *En bloc* staining is also useful. Adding a detergent (Tween 80, 0.3%) or dimethyl sulfoxide (DMSO) to the staining solution facilitates the penetration. If precipitates cling to the surface of the section as a result of prolonged staining, they can be dissolved by exposing the section for 1 min to 0.05% citric acid solution.

The highest overall contrast is achieved by *en bloc* staining with uranyl acetate before dehydration, followed by prolonged poststaining with uranyl acetate and lead citrate; the damaging effect of uranyl acetate when used *en bloc* on glycogen should be kept in mind. For relatively selective staining, either uranyl acetate or lead citrate may be appropriate: uranyl acetate emphasizes the membranous components and nucleic acid-containing structures, especially chromatin, whereas lead citrate alone results in the staining of ribosomes and cytoplasmic ground substance. Selective staining can also be achieved with molecular tracers such as lanthanum, ruthenium red, and thorium dioxide.

Staining Methods

1. Uranyl acetate (1–2%) in 95% ethanol is used to stain thick sections for 20 min to 24 hr at 60°C, depending on the tissue type, embedding medium used, and objective of the study. If desired, uranyl acetate staining can be followed by staining with standard lead citrate for 20–90 min. The grids should be stained by immersion.

2. Thick sections of aldehyde-fixed specimens can be stained with 2% PTA in 95% ethanol for 30 min at 60°C. Hot alcoholic PTA solution yields excellent contrast to depths of about 15 μm, and thick sections thus stained can be observed without additional staining. However, staining can be intensified by poststaining the sections with 2% alcoholic uranyl acetate for 12 hr at 60°C.

3. *En bloc* staining can be carried out by immersing tissue blocks in 2% alcoholic uranyl acetate for 12 hr at 60°C and then staining them in 2% PTA in 95% ethanol at 60°C for 30 min.

4. Selective staining of the Golgi apparatus and of endoplasmic reticulum can be obtained by impregnating the tissue with an unbuffered aqueous solution of OsO_4 (1%) for periods as long as 4 days. The staining solution should be renewed after 24 hr. The sections are examined without additional staining.

5. Specific staining of neurofibrils in nerve cells can be achieved by simultaneously fixing and impregnating tissue slices (2–4 mm) with silver by immer-

sion in an aqueous solution of silver nitrate at 37°C for periods as long as 14 days. After reduction by either pyrogallol or hydroquinone, the specimens are dehydrated and embedded. Thick sections are examined without further staining.

6. The T system in skeletal muscle can be stained by fixing the tissue first with glutaraldehyde and then with OsO_4 according to standard procedures. The specimens are infiltrated by lanthanum. Sodium hydroxide is added during fixation to form lanthanum hydroxide.

7. The rapid Golgi method can be used for the CNS. It consists of fixation with glutaraldehyde followed first by immersion in a mixture of OsO_4 (0.19%) and potassium dichromate (2.33%) and then by immersion in an aqueous solution of silver nitrate (0.75%).

8. Cells growing on coated grids attached to coverslips can be fixed with glutaraldehyde and then with OsO_4 according to standard procedures. After a wash, they are stained in 0.2% uranyl acetate in a mixture containing acetone (15%) and DMSO (5%) for about 15 min. After rapid dehydration, the cells are dried by the critical-point drying method (Hayat, 1978) and coated with carbon.

9. Staining *en bloc* with a mixture of lead and copper is effective in visualizing cell walls, plastid envelopes, and Golgi vesicles in plant cells (Hawes and Horne, 1983). Aldehyde-fixed tissue specimens are treated *en bloc* with 5% aqueous uranyl acetate (pH 3.5) for 1 hr at 40°C, rinsed in two changes of distilled water, and then stained *en bloc* with a lead and copper citrate solution for 1 hr at 40°C. This solution is prepared by adding 0.5 ml of 10% copper sulfate solution to 2 ml of 1 M lead nitrate (Thiéry and Rambourg, 1976). The white precipitate formed is mixed with 19 ml of 4.6% sodium citrate solution. The precipitate is dissolved by slowly stirring in 4 ml of 1 M NaOH. The specimens are rinsed in distilled water and postfixed with 1% aqueous OsO_4 for 2 hr. The staining is weaker toward the center of the tissue block.

PRECAUTIONS TO MINIMIZE ARTIFACTUAL STAINING PRECIPITATES

A thorough rinsing between aldehyde and OsO_4 fixation is necessary. The entire staining procedure should be carried out extremely carefully. Artifactual staining precipitates can be minimized by following these guidelines:

1. Make sure all containers used during staining are absolutely clean and free of dust and lint.

2. Use only clean, boiled distilled water for preparing lead and rinsing solutions.

3. Avoid staining longer than necessary.

4. Wet the grids before transferring them to any staining solution.

5. Do not allow the grids to dry between various staining steps.

6. Wash grids thoroughly with *warm* distilled water after each staining step.

7. If desirable, submerge sections into the staining solution instead of floating them.

8. Add one drop of glacial acetic acid to 10 ml of uranyl acetate solution in water or 50% ethanol. Acetic acid prevents photodegradation of uranyl acetate solution, and such solutions are less likely to produce artifactual fine granules on the sections.

9. If possible, use freshly prepared staining solutions; allow undissolved crystals of uranyl acetate to settle down before using the solution.

10. If desirable, filter uranyl acetate solution before use.

11. Use CO_2-free NaOH pellets from a freshly opened container to prepare 1 N or 0.02 N NaOH in a bottle with a rubber cap. Use a syringe to remove the desired aliquots through the rubber cap. If the container of NaOH pellets is opened frequently, it gets contaminated with carbonate.

12. Prepare and store lead solutions in plastic rather than glass containers.

13. Maintain a CO_2-free environment during lead staining. Avoid breathing on sections during lead staining.

14. If possible, eliminate the air–stain interface during lead staining.

15. If a white precipitate ($PbCO_3$) is visible in the bottle containing the lead solution, discard the solution instead of filtering it. However, this precipitate can be sedimented by centrifugation.

REMOVAL OF ARTIFACTUAL STAINING PRECIPITATES FROM THIN SECTIONS

If lead precipitates are present, they can be removed by treating the grid with 2% aqueous uranyl acetate for 2–8 min at room temperature. Alternatively, lead precipitates can be eliminated by treatment with 10% aqueous acetic acid for 1–5 min. Precipitates resulting from poststaining with both uranyl and lead salts can be removed by treating the grid with 0.5% oxalic acid for 12 sec, 10% acetic acid for 1 min, or 2% aqueous (or ethanol) solution of uranyl acetate for 1 min. Aqueous uranyl is preferred over acetic acid. Uranyl acetate precipitation can be eliminated by rinsing the grid in *warm* distilled water. Acetic acid (10%) is also effective in general removal of stains. All the aforementioned treatments may damage the fine structure. These precipitates are difficult to remove after prolonged exposure of the grid to a strong electron beam.

REMOVAL OF EPOXY RESINS FROM SEMITHIN SECTIONS BEFORE STAINING

Semithin sections can be stained either intact or after the embedding resin has been removed with a suitable solvent. A mixture of sodium, benzene, and methyl alcohol is effective in removing the resin without detectable cytologic distortions. A total of 2.5 g of metallic sodium is slowly added in small pieces to 25 ml of methyl alcohol at 55°C under a fume hood. The level of the solution during preparation should be maintained at 25 ml by adding methyl alcohol. After the sodium is completely dissolved, 25 ml of benzene is added. Alternatively, 205 g of commercially available sodium methoxide is dissolved in 25 ml of methyl alcohol and then 25 ml of benzene is added. The clear mixture can be stored in a dark bottle for several months.

The mixture is used either at full strength or suitably diluted with equal parts of methyl alcohol and benzene, depending on the section thickness. Floating sections are transferred to a drop of water in the center of a glass slide, using a fine brush, a wire loop, or a sharpened wooden applicator stick. The slide is heated on a hot plate at 58°C and allowed to dry at this temperature. An exposure of the slide for 1–3 min to the mixture just described is adequate to remove resin from the sections. The slides with sections are rinsed first in a 1 : 1 mixture of benzene and methyl alcohol and then in two changes of acetone and water. At this stage, the slides are ready to be stained.

STAINING METHODS FOR SEMITHIN SECTIONS

Solutions for rapid staining of epoxy sections can be obtained by mixing 1% azure II in 1% borax, 1% toluidine blue and 1% azure II in 1% borax, or 1% basic fuchsin in 50% acetone, or by using 1% aqueous paraphenylenediamine (specifically for specimens fixed or postfixed with OsO_4). A few selected methods of polychromatic staining are given next. The duration of staining is dependent on section thickness.

Azure B for Plant Tissues

Semithin sections are transferred to a drop of 10% acetone on a glass slide, which is then briefly heated to expand the sections and evaporate the acetone. The sections are flooded with 0.2% azure B solution in 1% sodium bicarbonate at pH 9.0, and the slide is placed on a hot plate at 50°C for 2–5 min. After being rapidly rinsed in water, the slide is air-dried and the coverslip is mounted in Epon. Nucleoli and primary walls are blue, secondary walls are light blue, cytoplasm is gray, nuclei are blue-gray, and chloroplasts are blue-green.

Basic Fuchsin and Methylene Blue

Staining solution
Sodium phosphate (monobasic)   0.5 g
Basic fuchsin   0.25 g

Methylene blue	0.2 g
Boric acid (0.5%)	15 ml
Distilled water	70 ml
NaOH (0.72%, pH 6.8)	10 ml

Semithin sections are heat-fixed on a glass slide. About 1 ml of the staining solution is placed on the slide with the attached sections and heated for 4–5 sec at 45–50°C. The slide is rinsed in running tap water and air-dried. The sections are covered with a drop of immersion oil and a coverslip and then sealed with Epon. Mitochondria, myelin, and lipid droplets are red; erythrocytes in blood vessels, glomeruli and tubules in kidney, smooth muscle cells, axoplasm, and chondroblasts are pink; collagen is brilliant pink; elastic lamina and zymogen granules are reddish purple; nuclei are bluish purple; and collagen and connective tissue are blue.

Methylene Blue–Azure II–Basic Fuchsin

Solution A

Methylene blue	0.13 g
Azure II	0.02 g
Glycerol	10 ml
Methanol	10 ml
Phosphate buffer (pH 7.0)	30 ml
Distilled water	50 ml

Solution B

Basic fuchsin	0.1 g
Ethanol (50%)	10 ml
Distilled water	90 ml

Slides with attached sections are immersed first in solution A for 20–30 min at 65°C and rinsed in water. They are then immersed in solution B for 1–3 min and again rinsed in water. Nuclei and goblet cells are dark blue; striated muscle and endothelial cells are blue; microvilli, smooth muscle, cytoplasm, and lipid droplets are light blue; red blood cells and fat are blue-green; glycogen, collagen, and mucus granules are red; connective tissue, fibers, collagen, and cartilage are pink; and mucus granules, elastin, and leukocytes are violet.

Methylene Blue–Azure II (Schroeder *et al.*, 1980)

The following method is useful for staining PAS-positive structures in semithin sections of tissues fixed with an aldehyde and OsO_4.

Solution A: Freshly prepared 5% periodic acid in distilled water.

Solution B: Schiff's reagent stored at 4°C. It is prepared by pouring 100 ml of 0.15 *N* HCl on 0.5 g each of basic fuchsin and potassium metabisulfite. The mixture is shaken at intervals of 2–3 hr or until dye is converted to fuchsin–

sulfurous acid. About 300 mg of fresh decolorizing charcoal is added and then the mixture is shaken for at least 5 min and filtered through a Millipore filter (pore size 1.2 μm) or a hard filter paper. The filtrate should be clear and colorless; if it is not, the last two steps are repeated. This reagent can be stored in the refrigerator.

Solution C: 0.5% methylene blue, 0.5% azure II, and 0.5% borax in distilled water. This solution can be stored at room temperature for several months.

Semithin sections are floated on solution A in a small glass dish placed on a hot plate at 50°C for 30 min. After being washed with distilled water, the sections are immersed in solution B for 30 min at 50°C. After a thorough wash with distilled water, the sections are transferred to solution C, where they remain for 20 min at room temperature. Following another wash with distilled water, each section is transferred onto a drop of water on a glass slide and dried on a hot plate at 50°C. Each section is mounted in a synthetic resin (Permount). PAS-positive structures stain red, while all other tissue components appear differentiated in shades of blue.

Giemsa (Cañete and Stockert, 1981)

Semithin sections of animal or plant tissues are transferred to a glass slide and dried on a hot plate. The slide is immersed in a Coplin jar containing the Giemsa solution (Giemsa, Merck, 2 drops/ml of distilled water, pH 5.5–6.0) for 1–2 hr. The staining solution has been previously heated to a temperature of 60°C. After cooling, the slide is washed in distilled water, air-dried, and mounted.

Starch, connective tissue, collagen, smooth muscle, parietal cells, pellucide zone, and salivary glands stain pale blue; chromatin, internal elastic membrane, spermatids, acrosome, and keratohyalin granules are dark blue; eosinophils, filiform papillae, and tips (tongue) are light blue; erythrocytes, striated muscle, zymogen granules, chief cells (stomach), juxtaglomerular granules (kidney), secretion content of glands (uterus), myelin, chromosome bands, and cell walls are blue; nucleoli, basophilic cytoplasms, fibroblast, plasmocytes, endothelial and epithelial cells, lymphocytes, stroma cells (spleen), megakaryocytes, lymphoblasts, microvilli, goblet cells, mucin granules, oocytes, neurons, neuroglial cells, and vitellin plates are violet; mast cells and granules (connective tissue) are red–purple; and Leydig cells, neutrophils, and lipid droplets (testes) are unstained.

Hematoxylin–Phloxine B (Shires *et al.*, 1969)

Semithin sections are transferred to a glass slide and dried on a hot plate at 50°C. The slide is immersed in xylene in a Coplin jar for 1 hr and then in a mixture (1 : 1) of xylene and 100% ethanol for 2 min; the latter step is repeated. The slide is next immersed in 95% ethanol for 2 min and then rinsed in three changes of distilled water for a total of 10 min (alcohol should be completely removed). If OsO_4 has been used, the slide is exposed to 0.1% $KMnO_4$ for 2 min and then bleached with 1% aqueous oxalic acid until brownish residues are removed; the slide is washed in three changes of distilled water for a total of 7 min.

The slide is incubated in hematoxylin (Harris's or Ehrlich's) at 60°C for 12–24 hr. After being thoroughly rinsed in six changes of distilled water (pH 9.0) for a total of 15 min, the slide is counterstained with 0.5% aqueous phloxine B for 10 min. Excess stain is removed with a filter paper and the sections are allowed to air-dry. The slide is washed in two changes of 100% ethanol until unbound stain is removed. Then the slide is cleared with a mixture (1 : 1) of xylene and 100% ethanol for 2 min followed by three changes of xylene for 1 min each. A coverslip is applied with synthetic resin.

Chromatin, nucleoli, karyolymph, basophilic cytoplasm, mitochondria, plasma and nuclear membranes, anisotropic myofibrils, mast cell granules, and elastic membranes of blood vessels are pale blue to deep blue or black; collagen, endoplasmic reticulum, goblet-cell mucin, hyaline cartilage matrix, stereocilia of epididymal epithelial cells, cytoplasm of some cell types, and erythrocytes are pale pink to dark red; fat droplets and perichondrocyte matrix are of varying shades of green; and Golgi apparatus is unstained.

Hematoxylin–Malachite Green–Basic Fuchsin (Berkowitz et al., 1968)

Iron chloride hematoxylin

Distilled water	75 ml
HCl (concentrated)	0.5 ml
Ferric chloride crystals ($FeCl_3 \cdot 6H_2O$)	0.62 g
Ferrous sulfate crystals ($FeSO_4 \cdot 7H_2O$)	1.12 g

After the preceding salts have been dissolved, 25 ml of freshly prepared 1% alcoholic solution of hematoxylin is added.

Malachite green

Ethanol (30%)	100 ml
Azure B	0.4 g
Malachite green	1 g
Aniline	1 ml
Phenol crystal	1 g

The stain does not dissolve completely. The solution is refrigerated for 3 days. The supernatant is the staining solution, which should not be filtered.

Basic fuchsin

Ethanol (50%)	5 ml
Basic fuchsin	2 g
Distilled water	45 ml

Semithin sections are heat-fixed to a glass slide. The slide is immersed in 2% NaOH in 100% ethanol for 10 min to dissolve the resin. After being washed in four changes of 100% ethanol for a total of 16 min, the slide is placed in phosphate buffer (pH 7.0) for 5 min. The slide is rinsed in three changes of distilled water and then placed in buffer (pH 4.0) for 5 min. A wash for 5 min in tap water follows, and then the slide is stained in iron hematoxylin for 20 min; staining should be controlled microscopically.

The slide is rinsed rapidly in tap water and examined. Chromatin and mitochondria must be light gray and the cytoplasm clear; if the slide is overstained, it is destained with 4% ferric alum. The slide is stained in malachite green for 2 min and then rapidly rinsed in tap water. Finally, the slide is immersed in basic fuchsin on a hot plate at 37°C for 20–120 sec and rinsed rapidly in tap water. After air-drying, the slide is mounted in a resin (the coverslip is pressed to obtain the thinnest possible layer of resin) that is polymerized at room temperature. Myelin, lipid droplets, nuclei, and oligodendrocytes are bright blue-green, and nuclei and astrocytes are purplish pink.

Sudan III (Angold, 1980)

In this method, which produces an effective stain for lipids, 500 mg of Sudan III is dissolved in 100 ml of 70% ethanol. Thick sections dried on a glass slide are stained for 10 min with this staining solution and then treated with two changes of 50% ethanol for 2 min each. The sections are air-dried and mounted in glycerol jelly. The stain will be extracted if the sections are mounted in xylene-based mountant.

Chromotrope 2R–Methylene Blue (Dougherty and King, 1984)

The following procedure is useful for staining within 6 min semithin sections of glycol methacrylate. This method does not require the use of ethanol or acetone before or after staining.

Solution A: 1% aqueous chromotrope 2R, adjusted to pH 3.0 with glacial acetic acid.

Solution B: 0.1% aqueous methylene blue.

A semithin section of aldehyde-fixed and glycol methacrylate-embedded tissue is mounted on a glass slide and dried overnight on a warming plate. The section is immersed in solution A for 2 min, rinsed in running water to remove excess stain, and then *immediately* dried with bursts of compressed air or gas (Freon). The section is counterstained with solution B for 3 min, again rinsed in running water, and immediately dried with compressed air.

In the striated squamous epithelium of esophageal mucosa, plasma membrane and desmosomes of the epithelium are deep red. Chromatin is blue with a faint reddish background in the nucleoplasm. The cytoplasm of cells in the basal layer of the epithelium is bluish and becomes progressively more red in cells toward the luminal surface. The striations of the skeletal muscle fibers are alternately red and purple. Nuclei of smooth muscle cells are blue, the cytoplasm is faint purple or lilac, while the sarcolemmal borders are deep purple. Endothelial cell nuclei of vessels are blue. The cytoplasm of proximal convoluted tubule cells is red, brush border is faintly purple, and nuclei are blue. Glomeruli are bluish due to the staining of nuclei of capillary endothelial cells and podocyte epithelial cells. The glomeruli basement membrane remains unstained. Mucus of goblet cells is bright red. In the simple columnar epithelium of the appendix, nuclei of epithelial cells and cells in the lamina propria are blue.

5

Negative Staining

INTRODUCTION

Negative staining is a simple and rapid method to study the morphology and structure of particulate specimens such as viruses, cell components (e.g., ribosomes), cell fragments (e.g., membranes), and isolated macromolecules (e.g., protein). Negative staining allows determination of shapes at the molecular level using high-resolution electron microscopy. Negatively stained specimens often show well-preserved order. The reason is that in addition to providing contrast, negative stains preserve macromolecular structure as well as minimize irradiation damage to the specimen.

The specimen is embedded in a negative stain, which is a metal such as phosphotungstic acid (PTA). On drying, the electron-dense metal atoms envelop the specimen. The difference between the specimen and the surrounding heavy metal atoms with respect to their density produces the necessary contrast. The specimen appears light surrounded by a dark background of dried stain. The electron beam passes through the low electron density of the specimen, but not through the metallic background. The specimen substructure is revealed by the penetration by the stain into its holes and crevices. In other words, the structure is inferred from the distribution which the specimen imposes upon the stain. The clarity of specimen detail depends on the degree to which the stain remains amorphous as it dries, as well as on the thickness of the dried negative stain envelope.

The negative staining method is based on the principle that there is no reaction between the stain and the specimen. This is accomplished by using the stain at a pH at which the attraction between proteins and stain is negligible. In positive staining, on the other hand, the atoms of heavy metals are attached to specific sites on the stained molecule. Some of the commonly used metal salts (e.g., uranyl acetate) are applied both for negative and for positive staining. Images of exclusively negatively stained specimens are difficult to achieve because positive staining is a small part of these images. The details of contrast mechanisms responsible for negative and positive staining are presented elsewhere (Hayat, 1986).

Many factors affect the appearance of negatively stained specimens. The shape and size of specimens are influenced by both the mode of negative stain application and the stain itself. Prefixation with an aldehyde allows the revelation of internal components. The pH and concentration of the stain, concentration of the specimen, temperature, duration of staining, nature of the support film, and length of time it takes for the specimen to dry on the support film influence the final image of the specimen.

Extreme precautions should be taken in handling infectious materials. All potentially infectious specimens must be inactivated. Ultraviolet (UV) irradiation kills most viruses. It is desirable to keep a container of a disinfectant (e.g., hypochlorite solution containing an anionic detergent) on the workbench to dispose of contaminated materials such as filter paper.

NEGATIVE STAINS

A large number of negative stains are available and their preparations are presented here; their properties, advantages, and limitations are discussed elsewhere (Hayat, 1986). The choice of a stain is determined by the objective of the study, the type of specimen under study, and previous experience. Uranyl acetate and potassium (or sodium) phosphotungstate are the most widely used negative stains.

Ammonium molybdate is used in concentrations of 0.5–5.0% at a pH ranging from 5.0 to 8.0. The pH can be adjusted with 0.1 N NH_4OH.

Methylamine tungstate is used at a concentration of 2% in double-distilled water at a pH range of 6.5–8.0.

Potassium (or sodium) phosphotungstate (PTA) is prepared by titrating phosphotungstic acid to a pH near neutrality with 0.1 M KOH. It can be applied at concentrations ranging from 0.5% to 3.0%.

Uranyl acetate is used in concentrations of 0.2–4.0% at a pH range of 4.0–5.5. This stain takes about 30 min to dissolve in distilled water. Owing to its radioactivity, uranyl acetate should be handled with utmost care.

Uranyl formate is used in concentrations of 0.5–1.0% at a pH range of 3.5–5.0.

WETTABILITY OF SUPPORT FILMS

A serious problem in the negative staining method is the unequal spreading of the specimens on support films. This problem is inevitable because support films are normally hydrophobic and specimens are usually in aqueous suspensions. Although particle suspensions of sufficiently low surface tension can be spread adequately, most biologic preparations do not wet the support films uniformly. Wetting of support films is a very complex phenomenon and is difficult to assess quantitatively. Even traces of contamination and aging may drastically alter the properties of a given support film. Freshly evaporated carbon films are hydrophobic, but aging makes them hydrophilic. On the other hand, aluminum oxide films behave oppositely.

Hydrophobic films usually result from the coating process. If films are carbon-coated before use or are made of carbon alone, their hydrophobic state is influenced by the operating conditions and cleanliness of the coating unit. Although the vacuum conditions during evaporation are known to affect the degree of hydrophobicity, no information is available about exactly what occurs during evaporation. An extremely clean unit with a baffled oil-diffusion pump produces better, stronger, and less hydrophobic films than a less clean unit.

Some biologic specimens contain soluble protein or surface-active impurities, both of which act as a surfactant. In the absence of such an agent, two main approaches are available. One approach is to modify the surface of the support film by irradiation (ion bombardment, UV irradiation). This modification can be accomplished by exposing the film to a high-voltage glow discharge passed between metal electrodes in a continuously evacuated vessel. This treatment renders the support film readily wettable even by preparations having a high surface tension. Moreover, the treatment reduces the ability of the film to "hold" virus particles as a result of the interaction between the virus particles and the surface charge of the film.

Glow discharge treatment in air makes films hydrophilic, negatively charged, and uniformly thinner. On the other hand, films treated with glow discharge in organic vapors (e.g., alkylamine) become hydrophobic, positively charged, and uniformly thicker. The latter approach involves essentially the deposition of basic radicals on carbon films in a reducing atmosphere produced with a glow discharge in organic vapors. Negatively charged films are ideally suited for adsorbing positively charged specimens such as ferritin or cytochrome c. Conversely, positively charged films strongly adsorb negatively charged specimens such as nucleic acids or glutaraldehyde-fixed proteins. A positively charged hydrophilic film has not been obtained by glow discharge. This difficulty can be circumvented by neutralizing the negative charges of the hydrophilic surface by treating it with a solution of magnesium acetate before adding the specimen solution to be adsorbed (Portmann and Koller, 1976). A brief description of the glow discharge procedure follows.

Most coating units are equippped for routine glow discharge. The best glow discharge is achieved when an alternating voltage is applied to a continuous stream of vapor. An electric field of about 100 V/cm is required. Grids carrying support films are placed in a bell jar, in which the pressure is reduced to either 10^{-1} torr (for glow discharge in air) or 10^{-2} torr (for glow discharge in organic vapors). The efficiency of the glow discharge for making surfaces more wettable is higher at lower pressures. To produce a glow discharge in organic vapors without stopping the action of the pump, the organic liquid is let in at a very slow rate until the pressure is stable at a reading of about 10^{-1} torr on the Pirani gauge.

The glow discharge is obtained by applying a tension of about 500 V to the electrodes for 10 sec. During glow discharge, the space between the electrodes should be filled with a uniform glow. [An excessive discharge will burn away the film, thus making it so fragile that it can withstand neither the use of macrodrops deposited with a pipette or transferred with a wire loop nor the virus-stain mixture sprayed with a glass Vaponefrin nebulizer (Vaponefrin Co., Edison, New Jersey.)] At this stage the high-tension supply is turned off, the valve connecting the bell jar to the pump is closed, and air is immediately admitted to the system. After glow discharge, the grids should be used as soon as possible to prevent atmospheric contamination of their surfaces; for example, for DNA adsorption they should be used within 1 hr. Special care must be taken to exclude the possibility of injury by electricity or implosion of the bell jar. A detailed discussion of mounting macromolecules on support films and the treatment of these films by glow discharge is presented by Dubochet et al. (1982).

Alternatively, carbon-coated films can be made hydrophilic by exposure to strong ultraviolet radiation from a xenon arc for 40 min. Some type of ionization process probably produces the change, since the irradiation is accompanied by a strong smell of ozone.

The second approach to modify the surface of the support film is the addition of a surface-active compound to the specimen suspension, thus lowering its surface tension and facilitating spreading. Such compounds include sucrose, glycerol, and bovine serum albumin (BSA). Although the use of BSA (0.005–0.05%) is preferred by some, it may increase the background granularity because of the large size of its molecule. The polypeptide antibiotic bacitracin (PB) is superior to BSA as a wetting agent. It is also more effective than BSA for lowering the surface tension, and its molecule is too small to be visualized by negative staining at a moderate magnification. Furthermore, PB allows specimen deposition by a pipette, loop, or spray on the support film. Solutions of PB are stable for 2–3 weeks at 4°C (Gregory and Pirie, 1973). As a guide, 1% PTA containing 50 μg/ml of PB is satisfactory. PB allows some extraction of virus particles from tissue when such tissue and stain (with PB) are ground up simultaneously. At higher magnifications, PTA– or uranyl acetate–PB may give background noise.

Another approach is the attachment of cells to a positively charged surface via electrostatic interactions. Glass or plastic coverslips, freshly cleaved mica, or carbon films can be coated with polylysine (Mazia *et al.,* 1974), ruthenium red, or alcian blue (Sommer, 1977). Carbon-coated plastic films or plain carbon films on grids are floated on a drop of 0.25% polylysine (mol. wt. 4000) or 1% aqueous alcian blue for about 5 min. The grids are transferred onto a large drop of distilled water and then placed on clean filter paper (coated face down). Cell suspension is applied to the grids by quickly spreading a drop over the surface (Nermut, 1982b). Care should be taken to avoid the formation of air bubbles.

Choice of Support Film

The choice of the support film depends on the type of specimen and the objective of the study. In routine work, carbon-coated plastic films (Parlodion, Formvar, Butvar, and pioloform) mounted on 400-mesh copper grids can be used. For high-resolution work, plain carbon films (5–10 nm thick) are desirable. Carbon films are hydrophobic and negatively charged. As stated previously, they can be made hydrophilic by means of glow discharge in air (Reissig and Orrell, 1970). This is the most widely used method. Glow discharge in vapors of organic compounds (amylamine), on the other hand, produces positively charged hydrophobic carbon films, which are ideal for the adsorption of nucleic acids (Dubochet *et al.,* 1982).

If glow discharge is not available, carbon films can be made hydrophilic by exposing them to UV light for 10–40 min. Hydrophilicity can also be achieved by floating carbon-coated grids on 1% alcian blue, which has been filtered through a low-porosity membrane filter, for 5 min, and then briefly rinsing in filtered distilled water and air-drying (Nermut, 1982a). Hydrophilic surface can also be obtained by floating carbon films from mica directly onto a virus suspension (Valentine *et al.,* 1968).

GENERAL METHODS

Basic Considerations

Because negative staining in conjunction with electron microscopy is used for determining the basic structure as well as for identifying specimens such as viruses, the following remarks apply primarily to such specimens. Most virus particles range in diameter from 20 to 500 nm and can be easily studied by use of negative staining. For the majority of viruses, a concentration of 10^6 virions/ml of starting viral material is required if the specimens are to be easily visualized with negative staining.

Both viable and nonviable virus particles in a sample are seen in the electron microscope. Since concentration of almost all components present in a sample occurs on the grid during the drying phase, any low-molecular-weight proteins and organic salts present in the sample should be minimized or eliminated (Almeida, 1980). These extraneous materials tend to obscure the specimen image. If specimens are taken from a liquid culture, the grid should be briefly rinsed with distilled water after specimen adsorption and before staining in order to remove most of the extraneous materials. For dilution of a concentrated sample, distilled water should be used. It lyses cellular structures but usually leaves viruses undamaged. If needed, concentration of the virus by centrifugation should be carried out at the lowest speed (15,000 g for 1 hr). After the final pellet has been obtained, the last drops of fluid in the tube should be removed; otherwise when the pellet is resuspended in distilled water, the low-molecular-weight material present will contaminate the specimen.

For general purposes, the pieces of equipment and supplies needed are as follows: fine forceps, Formvar–carbon-coated 400-mesh copper grids, glass slides, Pasteur pipettes, filter paper, a washing bottle containing distilled water, 4% solution of PTA (pH 6.0) in a drop bottle, and a Petri dish. For acid-labile specimens the pH of the stain is raised to 8.0 with 1 N KOH.

After the specimen grid is placed into the electron microscope, it should be left there for some minutes before turning on the electron beam because the grid needs to be vacuum-dried before being irradiated; otherwise wet stain will boil under electron bombardment and impair the structural details. It is desirable to first stabilize the grid by irradiation from the over-focused condenser system for some minutes before it is focused (Lickfeld, 1976). This practice prevents sagging (due to heating during focusing) of the square of the grid that is directly under the electron beam. Focusing should be performed at the lowest possible beam current giving adequate screen brightness. An image intensifier is very helpful.

One-Step (Simultaneous) Method

The one-step method is the simplest and most rapid means of negative staining, and it is especially suitable for specimens that are unstable at a low ionic strength. The choice of this method depends on the stability of the specimen in the stain environment. Certain unfixed specimens when mixed with the stain tend to be damaged.

A 1 : 1 mixtuxe of specimen and stain is used. As a guide, 1 volume of virus suspension (10^6–10^{10}/ml) is mixed with 1–2 volumes of 2% PTA (pH 7.5). If needed, an appropriate volume of a wetting agent (e.g., bacitracin in a final concentration of 50 μg/ml) can be added to this mixture. A carbon-coated (6–10 nm) 400–mesh copper grid is picked up by the edge with self-clamping forceps and placed on a flat surface with the support film side up. With the use of a fine Pasteur pipette (or a 5-μl Eppendorf pipette), a small drop of the mixture is placed onto the grid to form a "bead" extending nearly to the edge of the grid. The pipette is held vertically. After about 30–90 sec, the droplet is touched from one side with the torn edge of a filter paper to remove excess liquid, leaving a thin monolayer that is allowed to dry at room temperature; alternatively, the grid can be vacuum-dried. The amount of stain left on the grid is empirically determined for each type of specimen. The residual film rapidly dries and is ready for examination in the electron microscope. The grid should be examined as soon as possible. When uranyl acetate is used, the grid can be stored without adverse effects.

The negative stain is deposited very rapidly over both the exposed support film and the surfaces of the specimen, and the stain penetrates any existing cavities or crevices so that they can be observed. If the specimen is allowed to remain in contact with the staining solution for longer periods, binding of the stain with the specimen can occur, resulting in positive staining. A lapse of as little as 90 sec can cause positive staining. Positive staining is usually complete after 2 min.

The "drop" method described has a limitation in that small particles (e.g., capsomeres) and large particles (e.g., viruses) adsorb to the film at different rates. Small particles adsorb rapidly because they show faster Brownian motion. Erroneous conclusions may thus be made regarding the relative proportions of particles of different sizes in a sample. This difficulty can be circumvented by using a nebulizer to deposit the mixture, an approach termed the spray method.

Spray Method

Equal volumes of a virus suspension and 2% PTA are mixed. This mixture is placed in a glass nebulizer (Horne, 1965) and sprayed as fine droplets (5–20 μm in diameter) onto a carbon-coated, 400-mesh grid (Fig. 5.1). Alternatively, a high-pressure spray gun (Höglund, 1968) can be used. This approach produces a droplet pattern in which some drops are completely contained within the area of a single grid square, allowing quantitative determinations of particle ratio. The droplets dry in a fraction of a second, thus preventing time-related structural changes. Another advantage of this approach is that it prevents the preferential adherence to the grid of one type of particle over another type in a mixture. Either a premixed volume of specimen and stain is used or the stain is drawn into the capillary first and is followed by an air space and finally the specimen. The latter approach assures minimal contact of a stain with the specimen.

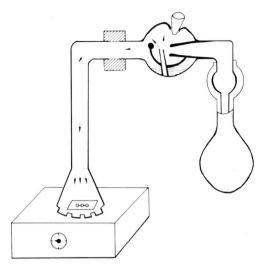

Fig. 5.1.   Schematic diagram of the spraying apparatus, positioned over the carbon-coated grids to be sprayed with a mixture of virus suspension and the stain.

Negative Stain–Carbon Method (Horne and Pasquali-Ronchetti, 1974)

The negative stain–carbon method allows the preparation of viruses and other small particulate specimens from very highly concentrated or crystalline suspensions. It provides ordered paracrystalline monomolecular arrays of protein molecules of viral or nonviral origin. Thus image analysis by optical diffractometry, filtering, and image reconstruction is facilitated, and a greater quantitative understanding of the molecular structure is achieved. The method provides a means for obtaining a very thin specimen thickness (1.5–1.0 nm) for high-resolution viewing.

Suspensions containing virus material (e.g., brome mosaic virus or cowpea chlorotic mosaic virus) are added to an equal volume of 3% ammonium molybdate (pH 5.2) (or 4% methylene tungstate) and then mixed mechanically in a Whirmixer for 10 sec. (Use of 3% ammonium molybdate might etch the virus particles at high pH.) The final concentration of the virus in the negative stain ranges from 0.4 to 8.0 mg/ml, depending on the concentration of the virus in the original suspension. A small volume of the negative stain and virus mixture (0.2 ml) is spread carefully on the surface of a precut and freshly cleaved mica sheet having one end pointed and the other flat (Figs. 5.2 and 5.3). The pointed end assists in the initial release of the final specimens at the liquid/air interface (as explained later). The excess liquid is carefully removed with a pointed filter

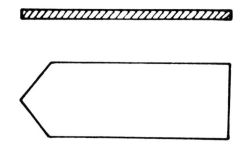

Fig. 5.2.   Diagram showing the shape of a cut mica piece before cleaving with a razor blade. The diagram on the top is a side view. (From Horne and Pasquali-Ronchetti, 1974.)

paper, leaving a thin liquid film at the mica surface, and then allowed to dry at room temperature. After drying, the mica sheet is placed in the evaporator and coated with a very thin layer of carbon (Figs. 5.4 and 5.5). The carbon film is marked by placing four black spots (Fig. 5.4) with a soft felt-tipped pen at suitable positions in order to locate the film at a later stage, because the film deposited at the correct thickness is normally invisible to the naked eye. If the film is easily visible after being released from the mica, it is too thick for high-resolution work.

The carbon–virus film is separated from the mica surface according to the standard procedure used for carbon films, except that the film is floated on a second negative stain (e.g., 1% uranyl acetate at pH 4.0), instead of distilled water, in a Petri dish (Fig. 5.6). Grids covered with holey films are placed

Fig. 5.3.   The mixture of the first negative stain and virus suspension is spread with a fine pipette on the freshly cleaved mica surface. (From Horne and Pasquali-Ronchetti, 1974.)

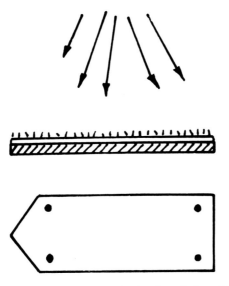

Fig. 5.4. Carbon is evaporated onto the air-dried stain and virus. After removal from the evaporator, the carbon surface is carefully marked with four black spots. (From Horne and Pasquali-Ronchetti, 1974.)

underneath the floating specimen film and then raised with a pair of forceps to collect the virus and carbon film.

The success of this method depends on the purity and stability of the viruses in suspension. When virus suspensions stored at 4°C for 24 hr or longer are processed by this method, a considerable reduction is seen in the areas formed by the regular arrays (Fig. 5.7B) (Wells *et al.*, 1981). The addition of polyethylene glycol (PEG) (mol. wt. 6000) to stored virus suspensions results in the formation of extensive continuous areas of crystalline arrays (Fig. 5.7C). About 4 μl of 10% PEG in distilled water is added to 4 μl of virus suspension, and then 100 μl of 3% ammonium molybdate adjusted to pH 5.6 is added. The addition of PEG facilitates the preservation of virus structure.

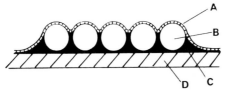

Fig. 5.5. Schematic drawing of the mica piece (D), negatively stained virus (B and C), and thin carbon layer (A) before being floated onto the surface of the second negative stain. (From Horne and Pasquali-Ronchetti, 1974.)

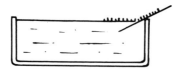

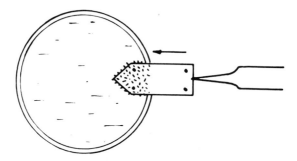

Fig. 5.6. The mica piece is slowly immersed into the Petri dish containing the second stain, and the specimen film areas are identified by the four black spots. (From Horne and Pasquali-Ronchetti, 1974.)

As is true of every method, this method has certain limitations. Air-drying during the first negative staining may cause virus particles to collapse. Rehydration during second negative staining (uranyl acetate) may be damaging to specimens such as large viruses (adenoviruses). Furthermore, dry specimens are penetrated by the stain, which may obscure the surface details. This method has been most successful for isometric and filamentous plant viruses; it has not yet produced satisfactory images of enveloped animal viruses.

Two-Step (Sequential) Method

The two-step method involves the deposition of specimens onto a carbon-coated grid followed by negative staining. This approach allows a longer duration of adsorption (1–30 min), which is required if either the particle concentration is very low or diffusion (Brownian motion) is reduced because of the

Fig. 5.7. Broad bean mottle virus negatively stained with 3% ammonium molybdate (pH 5.6). (A) Freshly prepared sample mixed with the stain, showing noncrystalline aggregates at the mica surface. (B) The sample was stored for 4 days at 4°C before staining. The virus particles are randomly dispersed with an increase in the amorphous background material, which is attributed to virus capsid components from dissociated particles. (C) A two-dimensional crystalline array of the virus is formed from the same stored suspension as in (B), but with PEG added to the mixture of virus and negative stain. (From Wells *et al.*, 1981.)

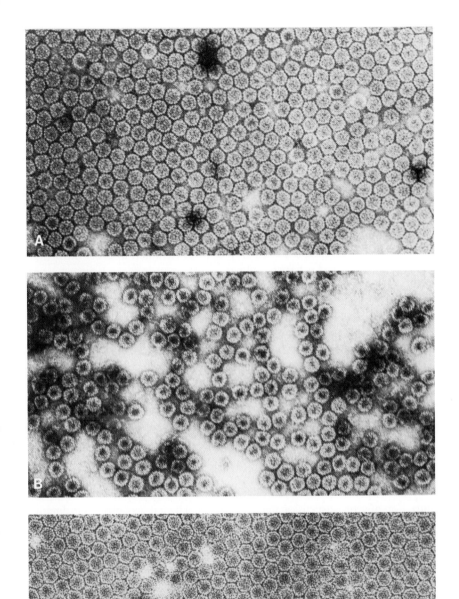

presence of gradient components such as glycerol and sucrose (Nermut, 1982a). Another advantage of this approach is that the specimens can be suspended in buffers or gradient components and washed after adsorption to the support film. After adsorption, specimens can be treated on the grids with various reagents such as enzymes, detergents, or fixative solutions before negative staining.

In this method, a carbon-coated, 400-mesh grid is floated, with the support film side facing down, on a few drops of the specimen suspension on a watch glass. Alternatively, a drop of the specimen suspension can be put on a carbon-coated grid that has been placed, with the support film side facing up, on a flat surface. A needle-gridholder can be used to handle several grids simultaneously. After enough time has elapsed to permit specimen particles to adsorb to the support film, the grid is washed with the aid of a Pasteur pipette. Another method is to transfer the grid onto a large drop of distilled water; to avoid breaking thin carbon films, the grid is transferred with a wire loop. Adequate rinsing is necessary to prevent possible precipitation caused by a reaction between sucrose or phosphate (present in the sample) and a negative stain such as uranyl acetate. Water is replaced by an appropriate negative stain that is drained off after about 10 sec. Longer staining (3–5 min) is needed if penetration of the stain into the specimen is required. At least two grids should be treated with one type of stain and two grids with another type of stain.

Very dilute specimens require special methods. The necessary concentrations of very dilute specimens can be obtained on Formvar–carbon-coated grids either by allowing prolonged adsorption time (about 20 min) or by using the agar- or paper-filtration technique. For agar filtration, a drop of the virus suspension is placed on an agar block and allowed to soak until a thin film remains. The grid is placed on top of the film for a few seconds; it is then removed, rinsed, and stained.

According to one modification, a carbon-coated grid is deposited for 10 min on a drop of virus suspension that is placed on a dental wax plate; this duration is determined by the virus concentration. Washing with distilled water and negative staining are carried out underneath the grid so that the virus suspension drop is drained off into a capillary pipette, and distilled water is simultaneously added with another pipette. In this approach the liquid is replaced underneath the grid without exposing it to surface tension forces during transfer from one drop to another. Subsequently, the distilled water is replaced by the negative stain (e.g., 1% uranyl acetate, pH 4.4). The grid is picked up with forceps and is dried by touching it with a piece of filter paper for about 2 sec. If the film has a strong charge, the specimen particles will cling to the surface and withstand considerable washing. However, all washings should be done carefully.

Staining after Fixation

A carbon-coated grid is placed on a drop of virus suspension. After the necessary adsorption time, the grid is transferred onto a drop of 1% glutaraldehyde in phosphate-buffered saline (pH 7.2) for 5–10 min. The grid is then placed on 1–2 drops of distilled water and finally onto a drop of negative stain such as 3% ammonium molybdate (pH 6.5). In every step the grid is placed with the support film side facing down. For quick fixation before negative staining of certain specimens such as bacteria, the grid carrying the adsorbed specimens is held for several seconds in the neck of a bottle containing OsO_4 solution.

The following simple device used in my laboratory allows manifold treatments of several tabbed grids simultaneously. This method is especially useful when each of the treatments is of a very short duration. Moreover, capillary problems encountered when the grid is floated on a drop of the fluid are avoided. A piece of an appropriate length of double adhesive tape is attached onto a microscope glass slide, so that the longitudinal edge of the tape is exactly over the edge of the slide (Fig. 5.8). The carbon-coated tabbed grids (coated side up) are carefully attached to the edge of the tape-slide, so that the tab rests on the tape while the grid is hanging free in the air; as many as 13 grids can be attached simultaneously. With the use of a Pasteur pipette, a drop of the specimen suspension is placed on the grid and allowed to stand long enough (about 2 min) to be adsorbed on the grid surface. Excess fluid is removed by touching the rim of the grid with a piece of filter paper, so that a thin film of the fluid remains on the grid. A drop of the staining solution is placed on the grid with another Pasteur pipette. After a very short period of time (20–40 sec), the excess stain is removed by touching the rim of the grid with a filter paper. The removal of the stain should be complete. If needed, the grid with the adsorbed specimens can be rinsed with a drop of distilled water before negative staining.

Single- or Double-Carbon-Layer Method

Another variation of the two-step method was developed by Valentine *et al.* (1968) to visualize antibodies. This is referred to as the single-carbon-layer

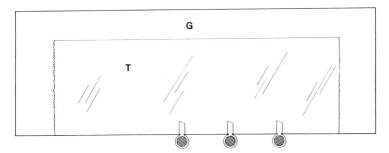

Fig. 5.8.   Diagram of the device that is easy to prepare and facilitates negative staining of as many as 13 grids at a time. The rim of the grid should be attached to the adhesive tape very carefully. The slide–tape–grid setup should be placed on a support about 26 cm above the workbench prior to its use. G, glass slide; T, double adhesive tape.

technique. Because thin, hydrophilic carbon films are used, this technique is suitable for high-resolution electron microscopy of smaller specimens such as ribosomes and parts of purified virus (capsomeres). It involves partial flotation of a small carbon film from a piece of freshly cleaved mica onto the specimen suspension. The carbon film is withdrawn by lifting the mica from the suspension, and the attached layer of carbon traps some of the specimen particles between it and the mica. This mica is inserted into a solution of the negative stain so that the carbon film is completely floated off the mica substrate.

In the double-carbon-layer technique (Lake, 1979), the first step is performed as in the single-layer technique. In the second step, however, the carbon layer with adsorbed specimen particles is rapidly floated off the mica and allowed to touch the wall of the uranyl acetate container so that the particles become sandwiched in a shell of negative stain between two carbon films (Fig. 5.9C).

In both variations a 400-mesh grid, coated with an adhesive (its preparation is described in the next paragraph), is placed on top of the carbon film(s), which is lifted from the staining solution by placing a piece of absorbent paper (for example, a newsprint with very dark printing) on top of the grid. The grid is lifted from the staining solution after the solution has begun to wet the paper (Fig. 5.9D).

Adhesive-covered grids are prepared by dissolving the adhesive tape from a 3-mm strip of Scotch brand transparent tape (not Magic mending tape) in 5 ml of chloroform (Lake, 1979). This solution is poured over grids arranged on the bottom of a Petri dish, and the chloroform is allowed to evaporate. After com-

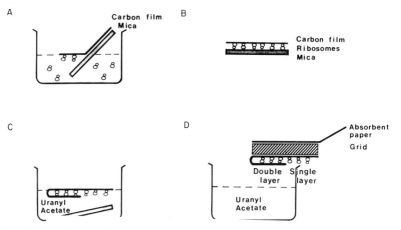

Fig. 5.9. A diagram of the technique for preparing double-carbon-layer negatively stained specimen grids. (A) Adsorbing the ribosomes to the carbon film. (B) The carbon film after withdrawing the mica. (C) Floating the carbon film off the mica and onto the surface of the negative stain. (D) Lifting the grid from the negative stain, showing a double-carbon-layer grid. (From Lake, 1979.)

pletely drying, the grids are picked up with finely tipped forceps. In this technique the specimens are fully embedded in stain, giving a two-sided image. This technique is particularly useful for studying ribosomes, but it has not been used systematically in virus research.

One-Side Negative Staining Method

On drying, most specimens are surrounded by the negative stain from below as well as from above. Important structural information about certain specimens can be obtained by staining either from above or from below only. For example, by staining only the lower half of icosahedral viruses, very useful information on their geometry can be obtained. Negative staining from below can be accomplished by drying a film of 2% PTA on carbon-coated Formvar films under an infrared lamp. A dense suspension of specimen is transferred to these dry grids and then drained off immediately. Alternatively, virus particles can be adsorbed to Formvar-coated grids, which are then floated on a negative stain (3–5%) for 30–60 sec (Nermut, 1982a); an increase in staining time would lead to staining also from above. After drying, the grids are coated with carbon to strengthen the Formvar film.

Negative staining from above can be obtained by first adsorbing the specimen on carbon-coated grids, and then staining with, for example, PTA (pH 6.0) for 10 sec (Müller and Peters, 1963). This approach causes the specimens to collapse on support films during air-drying so that the stain has very little access to the near-side face of the specimen. Alternatively, specimens such as Orf virus can be fixed on carbon-coated grids with formaldehyde vapors and then sprayed with PTA (Nagington et al,, 1964). Duration of spraying is a critical factor in achieving negative staining only from above. The pitfall is the presence of artifacts caused by air-drying from water; specimens such as enveloped viruses are especially prone to drying damage.

Paper-Filtration Method (Nermut, 1972, 1982a)

In the paper-filtration method, a collodion (0.2%) film floating on water is picked up with a clean Whatman no. 1 filter paper and semidried at room temperature. Squares of about 3 × 3 cm are cut off and floated on water to see whether the film is intact. The square is placed on the lower part of a Millipore filter holder connected to a water pump. A drop of virus suspension is placed in the middle of the square (and spread over a larger area if required), and the water pump is turned on. After the drop has disappeared and a wet film is left, the filtration is stopped and the square is removed from the filter holder. The area with the material is cut off (usually as two to four small pieces) and immediately floated on negative stain such as 2% PTA (pH 6.8). After about 40 sec, each piece of film is picked up on a grid and dried on filter paper. A thin layer of carbon is then evaporated onto the grids. The staining occurs from below, so there is a chance that a one-sided image can be obtained.

Pseudoreplica Method

The pseudoreplica method is probably the best method among other negative staining methods used for the detection of viruses such as herpesviruses, arboviruses, hemorrhagic fever virus, and rotaviruses. This method has a high sensitivity for virus detection and is more sensitive than the ultracentrifugation method and the enzyme-linked immunoabsorbent assay. The pseudoreplica method has the following advantages over some other rapid methods (McCombs *et al.*, 1980). As the fluid is absorbed on the agarose or agar surface, any virus present is concentrated on that surface. The agarose diffusion reduces the salt content of the specimen, preventing salt from crystallizing and confusing interpretation. This method requires very small quantities of specimen (0.025 ml) and takes less than an hour to complete. Ultracentrifugation is not needed because virus concentration is obtained by fluid dialysis into and evaporation from the agarose. Also, partial purification of viruses occurs through diffusion of particles up to 15 nm in diameter, macromolecules, and interfering salts (Doane and Anderson, 1977), resulting in clearer background during examination in the electron microscope.

One drop of virus suspension containing more than 10 particles/ml is placed on an agarose block (1.5 × 1.5 × 0.6 cm, solidified from 1.5% agarose solution), spread with a glass rod, and allowed to semidry (Kellenberger and Arber, 1957; Palmer *et al.*, 1975). (Agarose is preferred over agar since some viruses tend to stick more strongly to the latter, resulting in diminished virus recovery on Formvar film.) The preparation is covered with 2 drops of 0.1% Formvar or collodion and placed on edge to drain and dry the film. All four edges are trimmed with a razor blade, and the Formvar film is floated onto the surface of the stain (e.g., ammonium molybdate, pH 6.5). The grids are then placed on the film and removed with a low-absorption filter paper. After excess stain has been drained off, the grids are air-dried and given a light carbon coating. A modification of this method is explained in Fig. 5.10.

Agar Filtration Method (Kellenberger and Bitterli, 1976)

The conventional method for preparing negatively stained particles suffers from selectivity; i.e., different species of particles in a mixture do not adsorb with equal efficiencies to the support film. This selectivity might alter the proportion of observed particles by factors of up to 10^3, as compared to the initial proportion in the mixture (Dubochet and Kellenberger, 1972). Although the

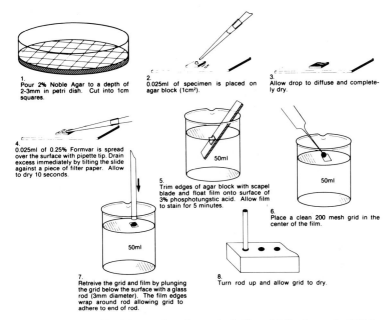

Fig. 5.10.   Details of the pseudoreplica method. (From McCombs *et al.*, 1980.)

spray droplet method does not show any selectivity, particles in the size range of viruses have to be in a concentration of about $2-4 \times 10^{11}$ particles/ml, and they must be suspended in negative stain dissolved either in distilled water or buffers of low concentrations of volatile salts. The agar filtration method requires a concentration of about $2-4 \times 10^{10}$ particles/ml in any physiologic medium. This method consists of filtering a particle-suspension through a collodion film into an underlying, slightly dehydrated 1.0–1.5% agar in physiologic medium. The film acts as a filter, which is then used as the support film.

Agar plates are prepared by adding 15 g of agar to 1 l of a 1% solution of tryptone in distilled water. After being sterilized in an autoclave, the agar is poured into glass Petri dishes (10 cm in diameter) while its temperature is still about 50°C. Several L-shaped stainless-steel carriers are positioned into the dishes before pouring. The agar layer should be 5–8 mm thick above the carriers. Utmost care should be taken to avoid contamination from various sources such as fingerprints. The solidified and cooled agar plates are dehydrated 15–20% in a ventilated oven at 37–45°C. These plates are immediately cleaned with distilled petrol ether to remove contaminants from the agar surface and the sides of the Petri dish. After a few minutes have been allowed for the ether to evaporate, 0.5–0.8% collodion in amyl acetate is poured over the surface of the agar. The plate is immediately turned upside down and deposited in an inclined position

(Fig. 5.11) on a table covered with wet filter paper. Within 3–4 hr, all the amyl acetate has evaporated. The plates can be stored at room temperature or in the refrigerator after being sealed in polyethylene bags or aluminum foil.

The collodion–agar plates are prepared for filtration by cutting agar blocks over the L-carrier with a razor blade (Fig. 5.12A). The remaining agar is removed except for these blocks. A drop (10 μl) of the particle suspension is deposited on each of the three blocks (Fig. 5.12B) and spread with a glass spreader made out of a Pasteur pipette (Fig. 5.12C). To achieve uniform spreading, the plate is moved back and forth on a flat table while the spreader is held stationary. The spreader should not touch the collodion film. The Petri dish is closed to avoid drying by evaporation.

Within 10–30 min, filtration is completed on 80–100% of the surface which is visible in reflected light. Wet parts on the block will not be used later. The block is exposed to formaldehyde vapors for 10 min. Alternatively, the block can be exposed to OsO_4 vapors for 10 min for the observation of nucleoids of gram-negative bacteria such as *Escherichia coli*. These types of bacteria are relatively small. The L-carrier is seized with a forceps and obliquely introduced into distilled water. The film floats off on the surface of the water (Fig. 5.12D), and is picked up onto the grids from below with a "fishing device" (Fig. 5.12E). This device is placed on a filter paper, so that the film becomes stretched over the grids. While still soaked underneath, each grid is transferred to fresh filter paper to remove the liquid below the film (Fig. 5.12F). The grid now is ready for observation. Floating the film on water is not recommended for counting particles. For this purpose, the film is floated onto 8–10% solution of PTA. The procedure for particle counting is described on p. 256.

Freeze-Dry Negative Staining

Freeze-dry negative staining combines good preservation and high-resolution surface images of enveloped viruses (Nermut, 1977a, 1982a). Freeze-drying

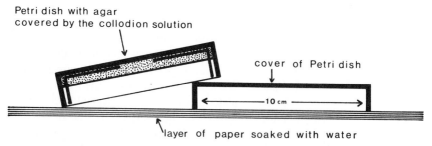

Petri dish with agar
covered by the collodion solution

cover of Petri dish

10 cm

layer of paper soaked with water

Fig. 5.11. Disposition of the Petri dish over a layer of very wet paper during evaporation of the amyl acetate. (From Kellenberger and Bitterli, 1976.)

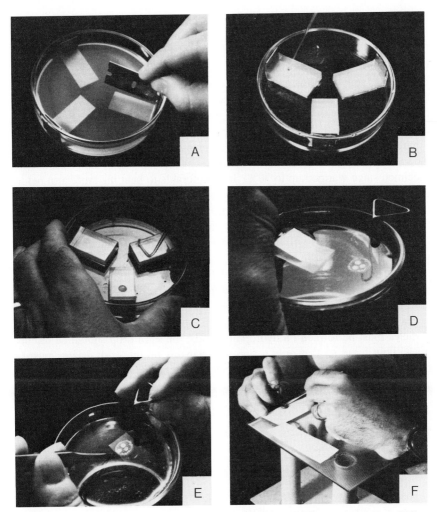

Fig. 5.12. The main steps of the agar filtration method as described in the text. (From Kellenberger and Bitterli, 1976.)

avoids formation of drying artifacts due to surface tension forces, and the spreading technique prevents orientation artifacts. About 0.05 ml of a virus suspension of high concentration is mixed with 0.01 ml of 0.1% cytochrome c in 0.1 M ammonium acetate (pH 7.0). This mixture is spread on the surface of distilled water in a plastic Petri dish (4 cm in diameter) and, if necessary, slightly compressed. The film is touched with a glow-discharged, carbon-coated grid. The grid with an adsorbed monolayer of virus particles is washed on 3 drops of

0.1 M ammonium acetate or distilled water for 2 min or longer. It is transferred onto a drop of negative stain (e.g., 3% ammonium molybdate, pH 6.5) for 10–30 sec. The excess stain is drained off over a Dewar bottle containing liquid nitrogen, and the grid is immediately immersed in the liquid nitrogen. With hydrophilic support films and a high concentration of particles on the grid, the drain time is 2–3 sec. After 10 sec the grid is quickly transferred onto a pre-cooled ($-150°C$ with Freon) specimen stage of a freeze-etch unit (Balzer). The chamber is evacuated and the specimen stage temperature brought to $-80°C$. Simultaneously, the knife arm is cooled down to $-150°C$ and then moved over the specimen stage for 20–30 min. The knife arm cooling is switched off and the specimen stage warmed up to 30°C. It is preferable to heat the knife arm before opening the chamber. The dried negative stain, in the form of a white powder on the grid, is removed by gentle blowing from a Pasteur pipette, so that only a fine layer of it remains attached to the virus particles. Too much or too little powder remaining on the virus will result in poor images. The grid should be observed immediately or stored in a vacuum.

The described method is more difficult than air-dry negative staining, and only about 50% of the grids processed yield good results. The percentage of useful grids increases if the following conditions are strictly observed (Nermut, 1982a). When the spreading technique is used, a reasonably dense monolayer of virus particles must be formed on hydrophilic grids; the stain drop should be drained off without haste so that only a fine film remains on the grid; drying must be completed in 20–30 min; and the vacuum chamber is opened only when the specimen stage has been warmed up to room temperature.

VIRUSES

General Methods for Animal Viruses

The following protocols are recommended for preparing viruses from a wide variety of sources (Almeida, 1980). Cloudy or contaminated *urine* (as large a volume as possible) can be clarified by centrifugation at 3000 g for 15 min in a bench centrifuge. The pellet is discarded, and the supernatant is centrifuged at 15,000 g for 1 hr. The pellet is used for negative staining. A 10% suspension of *stool* is made up in distilled water, which is clarified in a bench centrifuge. The pellet is discarded, and either the supernatant is directly stained or about 2 ml of it is centrifuged at 15,000 g for 1 hr, and the resulting pellet is stained. *Serum* (0.2–1.0 ml) is diluted with an equal volume of distilled water and centrifuged at 15,000 g for 1 hr. The supernatant is discarded, and the pellet, which may not be visible, is brought to its original volume in distilled water. This is centrifuged again, the supernatant is discarded, and the pellet is stained. *Vesicle fluid* is obtained by puncturing unbroken lesions with a sterile Pasteur pipette or tuberculin syringe that contains a small amount of distilled water. This water can be used to irrigate the lesion and obtain a larger volume of fluid than the vesicle itself could yield. This suspension can be applied directly to the grids and stained. The sample should be handled with caution to avoid laboratory infections.

Cells of a *cell culture* are disrupted either by several cycles of freezing and thawing or by sonication. This suspension is centrifuged at 15,000 *g* for 1 hr. Preparations of *allantoic fluid* are treated in a similar way. For *soft tissues* (e.g., brain), a 10% suspension is made in distilled water with a glass Teflon homogenizer. A few strokes will suffice for brain tissue, while *liver* tissue will need more effort. This homogenate is clarified by centrifugation for 10 min in a bench centrifuge. The pellet is discarded and the supernatant is centrifuged at 15,000 *g* for 1 hr. *Skin scabs* can be handled in the same way.

Hard tissue is cut into small fragments and placed in a glass or porcelain mortar with a little silver sand. After a small amount of distilled water has been added, the specimen is ground with the pestle until the tissue is further fragmented. More water is added with continued grinding until a relatively smooth suspension is obtained. The homogenate is processed in the same way as in the case of soft tissues. *Tissue scrapings* (e.g., conjunctive cells) are diluted with sufficient distilled water to disrupt any cells present and then are stained. *Cerebrospinal fluid* is diluted with an equal part of distilled water and stained on the grid. *Sputum* is diluted 1 : 4 with PBS, and a homogenous suspension is obtained in a homogenizer. The suspension is clarified by centrifugation for 10 min in a bench centrifuge. The pellet is discarded and the supernatant is centrifuged at 15,000 *g* for 1 hr to obtain a pellet for staining.

Viruses from Skin Lesions (Narang and Codd, 1979)

This method is useful for observing virus particles in very small samples containing low virus concentration and much tissue debris. Skin lesions suspect of virus infection are uncapped with a scalpel blade, and the vesicle fluid (2–5 µl) is aspirated with a finely drawn Pasteur pipette. Flat-bottomed, 6-mm-diameter polyethylene tubes are cut to a length of 15 mm. To each of the two tubes, 8 drops (200–250 µl) of distilled water are added. Carbon-coated grids are lowered into the water at the side of the tubes and tilted so that the grids are dropped to the bottom with the Formvar film facing upward. The vesicle fluid is expelled into one tube by gently rinsing in distilled water, and a single drop is transferred to the second tube, thereby giving a final dilution of 1 : 9 between the tubes. When the lesions have dried, the material on the tip of the scalpel blade is washed in 8 drops of distilled water, which is transferred into an empty plastic tube. A grid is then emplaced as previously described.

These tubes are placed into 5 × 1-cm glass tubes and centrifuged horizontally in a bench centrifuge for 30 min at 3700 rpm (2100 g). The plastic tubes are lifted out of the glass tubes, and the grids are picked out with fine forceps after tilting the tubes on the side. The grids are dried on filter paper for a few minutes and then floated on a drop of PTA (pH 6.6) for 30–60 sec. Excess stain is blotted off with a filter paper. The grids are then left under ultraviolet light for 15 min to inactivate the viruses before examination. Care should be taken not to deposit too much of the material on any of the grids; otherwise the grid will appear black.

Viral Detection in Fecal Specimens (Rice and Phillips, 1980; Gbewonyo, 1982)

Direct Staining Method

The fecal material is suspended in sterile distilled water to obtain a concentration of 10–20% (v/v) and is centrifuged at 3000 rpm for 30 min at 4°C in a minicentrifuge. The supernatant (5 ml) is collected in a small viral bottle. A drop of this supernatant is placed on a Formvar-coated 200-mesh copper grid. After a few minutes the edge of the grid is touched with a piece of filter paper to remove excess fluid. The specimen is stained for 45 sec with 2% PTA (pH 7.2), and excess stain is removed. Although this method is rapid, a considerable amount of time is needed to actually examine the specimen, and the method is less than sensitive. Alternatively, the fecal material can be mixed with PBS to give 5 ml of a 20% suspension. This suspension is centrifuged for 10 min at 1500 rpm, corresponding to a calculated gravitational force of 490 g. A drop of each clarified preparation is negatively stained. Further centrifugation is unnecessary.

Microsolute Concentration Method

About 1.5 ml of the supernatant (obtained as in the preceding method) is pipetted into the cells of a concentrator (Minicon-B15 Disposable Multiple Microconcentrator, Amicon B. V. Oosterhout, Holland). The sample concentrates onto a Formvar film on a 200-mesh copper grid. After a few minutes, excess fluid is drained off by touching the grid with a filter paper, and the sample is negatively stained with 2% PTA (pH 7.2) for 45 sec; excess stain is removed with a filter paper. The whole process is completed in 1 hr. This method is very reliable.

Pseudoreplica Method

A 2% solution of purified agar is poured into a Petri dish to a depth of 4 mm and allowed to harden. A 10-mm-square block is cut from the agar and placed on the edge of a glass slide. One drop of the supernatant (obtained as in the direct staining method) is then placed on the agar block surface. After 50 min the drop diffuses into the agar. About 0.2% Formvar is pipetted onto the agar block, and the excess is drained onto filter paper. Within 40 sec the Formvar dries, and the edges of the agar block are trimmed with a razor blade. The Formvar film is floated off the agar over 3% PTA (pH 6.0) for 4 min. Copper grids (200 mesh) are placed face down on the Formvar film, and the grids are retrieved from the stain with flat dental wax plates (15 mm square). The specimens are dried and examined. The whole process is completed in 90 min.

Ultracentrifugation Method

About 3 ml of the supernatant (obtained as in the preceding methods) is pipetted into 5 ml of viral centrifuge tubes, and 2 ml of saline is added to each tube. The samples are centrifuged at 45,000 rpm for 2 hr in a 50.1 swinging-bucket rotor on the Beckman 65B ultracentrifuge. Pellets are resuspended in 3 drops of sterile distilled water. A drop of the suspension is placed on Formvar-coated 200-mesh copper grids, and after a few minutes the excess fluid is removed with filter paper. The specimen is negatively stained. The whole process is completed in 4–6 hr. This method is less sensitive than the microsolute method.

Other Methods

The following procedure is useful for purifying rotaviruses and noncultivable enteric adenovirus from large volumes of fecal extracts (Beards, 1982). Feces samples, for example from diapers containing feces, are collected from virus-infected children. The diapers are soaked overnight at 4°C in 0.1 M Tris buffer (pH 7.2) containing the antibiotic crystamycin (320 mg/ml) and fungizone (10 μg/ml) and 1.5 mM $CaCl_2$; rotaviruses are relatively stable in this solution. Sufficient buffer is added to each diaper (500 ml/diaper) to give about 20% (w/v) suspension of feces. The fecal suspension is extracted from the diaper by hand while wearing two pairs of rubber latex gloves inside a class 1 safety cabinet. The crude fecal extract is clarified by centrifugation at 10,000 g for 30 min.

To 500 ml of fecal suspension, enough polyethylene glycol (mol. wt. 6000) and NaCl is added to obtain final concentrations of 8% and 0.5 M, respectively. The suspension is kept overnight at 4°C while being continuously stirred. The precipitate is collected by centrifugation at 10,000 g for 30 min, then redissolved in 20 ml of 0.1 M Tris buffer containing 1.5 mM $CaCl_2$.

With the use of a Beckman SW-50 rotor at 35,000 rpm for 90 min, 3 ml of concentrated virus extract is centrifuged through 1 ml of 45% sucrose in 0.002 M Tris buffer, forming a cesium cushion consisting of two layers of 0.5 ml each. For rotaviruses, the density of the lower layer is 1.4 g/ml and that of the upper layer is 1.34 g/ml. For adenoviruses, the density of the upper layer is 1.3 g/ml. If only one band of virus is visible in the cushion after centrifugation, this band is removed with a needle and a 1-ml syringe by piercing the side of the tube. This fraction is dialyzed against a suitable buffer (e.g., 0.001 M Tris) for 2 hr at room temperature or overnight at 4°C. One drop of 1% bacitracin and 1 drop of 4%

ammonium molybdate (pH 7.0) are added to 1 drop of purified virus. The bacitracin is required to facilitate uniform spreading of the negative stain on the grid, which otherwise may not occur when very pure virus preparations are being mounted on carbon–Formver support films.

Direct Detection of Viruses with the Beckman Airfuge

Unpurified virus preparations or clinical specimens containing viruses can be examined by the following method (Hammond *et al.*, 1981). Crude cell cultures containing viruses or a 5% suspension of clinical material such as fecal specimen in distilled water are clarified in a benchtop centrifuge (Eppendorf S42 microcentrifuge or Beckman Microfuge 12 centrifuge) at about 12,000 *g* for 5 min. Filter paper strips, 5 mm wide, are used to retrieve 400-mesh copper grids spaced about 5 mm apart on a Formvar film in a water bath. The strips are allowed to dry and are then cut into 5 × 5-cm squares. The Formvar-coated grids backed with filter paper are placed at the bottom of the EM-90 rotor cell base. Alternatively, Formvar-coated grids can be placed in the new 3-mm wide EM-90 rotor wells without filter paper backing. The clarified virus supernatant or purified virus preparation is placed in the rotor cell.

To avoid leakage during centrifugation, the well is filled 2–3 μl below its capacity, and thick rotor gaskets are used. Ultracentrifugation is carried out at 30 lb/in^2 air pressure for 20 min. The grid is removed and stained. Excess stain is removed by filter paper absorption, and the grid is allowed to dry. Occasionally the clarification step from direct clinical specimens contains too much debris and may require a more dilute initial suspension. Care should be taken to avoid retrieval of the sedimented pellet after the completion of the clarifying centrifugation. It is advisable to clean the rotor wells and sterilize them with a cotton swab after immersion in 2.5% glutaraldehyde.

Virus Particle Counting

With the use of a specially designed rotor (EM90 Electron Microscopy Particle Counting Rotor), virus particles can be sedimented for counting in the electron microscope either by a thin-section method (Miller *et al.*, 1973; Miller, 1979) or by a direct grid method (Miller and Rdzok, 1981). The rotor (Fig. 5.13A) operates in the Beckman Airfuge tabletop ultracentrifuge and was specifically designed (Miller, 1979) for uniformly sedimenting particles to be counted in the electron microscope. The direct grid method is desirably fast to perform and works best for counting relatively pure suspensions of virus particles. Although the thin-section method takes somewhat longer, it yields accurate counts of partially purified or crude virus suspensions because viral internal structure aids in recognition and counting of particles among cellular debris.

A particle count is performed by (1) placing either a plain 5-mm^2 Millipore filter support (thin-section method) or filter support to which a 500-mesh copper

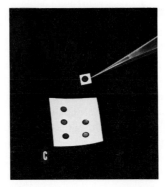

Fig. 5.13. (A) Disassembled EM90 electron microscopy rotor for virus particle counting. Note sector-shaped cells in the rotor core that holds mounted grids. (B) A 5-mm² filter support being inserted into one of the sector cells. (C) Grids attached to the filter, which holds them in place in the rotor. Virus particles are sedimented onto the support, which is then processed for thin-section electron microscopy and subsequent virus counting. (From Miller and Rdzok, 1981.)

grid has been attached (direct grid method) at the back of sector cells in the rotor core (Fig. 5.13B), (2) filling cells with 0.1 ml of appropriately diluted virus suspension, (3) sedimenting particles onto supports, (4) counting in the electron microscope the number of particles deposited over a measured area of the support, and (5) calculating the number of particles per unit volume of the original suspension. The sedimentation time in the EM90 rotor for all known viruses is less than 15 min. The 5-mm^2 filter supports are cut from larger type VSWPO25 Millipore filters, using the template provided with the rotor.

In the direct grid method, 500-mesh copper grids are sandwiched between Parlodion support films and 5-mm^2 filter supports (Fig. 5.13C), which hold grids in a vertical position at the base of sector-shaped cells in the rotor core (Fig. 5.13A). The first step is to strip a Parlodion film (0.5% Parlodion in amyl acetate) from a glass microscope slide onto the surface of triple-distilled water. Six grids are arranged on the floating film so that they are about 10 mm apart, and precut 5-mm^2 filter supports are placed on top of each grid. The Parlodion film with adherent grids and filter squares is quickly picked up from the top with Whatman filter paper and allowed to dry thoroughly. The filter squares with attached grids are then placed (grid down) for 30 sec on droplets of polylysine (1 μg/ml, mol. wt. 100,000), after which they are washed with water and air-dried. This step promotes both the adherence of particles to support films and the uniform negative staining of specimens. Following sedimentation of the virus, the grids are picked off the surface of the filter supports and negatively stained by floating on droplets of 2% PTA. Sedimented virus is then photographed at the lowest magnification permitting clear visualization and counting of particles.

For the thin-section method, the filter supports with sedimented virus are fixed by immersion in 3% buffered glutaraldehyde, postfixed in 1% OsO$_4$, dehydrated in a graded series of isopropyl alcohols, cleared in toluene, and flat-embedded in silicon rubber molds. Epoxy blocks in which filter squares are embedded are oriented in the microtome so that thin sections are cut perpendicular to the plane of the filter. Resultant cross sections of filter supports with adherent virus are then picked up from beneath on 300-mesh parallel wire grids. Sections are double stained with uranyl acetate and lead citrate, and virus particles deposited along the surface of the filter support are photographed at the lowest magnification at which particles can be clearly distinguished and counted among cellular debris.

Virus particles are counted in measured areas on the surface of the grid (direct grid method) or along measured segments of the surface of the filter support (thin-section method) by superimposing a grid of fine lines over negatives and viewing particles with the aid of a dissecting microscope. The virus concentration is calculated by relating the number of particles in a microscope field of known surface area to the surface area of the entire filter. The number of virus particles per milliliter (N) is calculated from the equation

$$N = \frac{250 \, V_f D}{A}$$

where V_f is the average number of virus particles per field, A is the surface area of the field in millimeters, and D is the dilution factor of the virus suspension. For the direct grid method, A is calculated from direct measurements of field length and width. For the thin-section method, A is calculated by multiplying the measured field length by the estimated effective section thickness. The latter serves to compensate for particles that lie only partially within a given thin section. Effective section thickness of 9×10^{-5} and 16×10^{-5} mm may be used for particles having diameters of about 70 and 140 nm, respectively.

Virions in Bacteriophage Plaques

The following method is useful for rapid surveys of virion morphology in new isolates or in individual plaques. Its success depends on the concentration of virions within a plaque; a concentration of about 10^{10} virions/ml is satisfactory (Bell and Roscoe, 1982). Carbon-coated Formvar grids (200-mesh) are placed (carbon side down) onto the bacteriophage plaque and pressed gently. Each grid is then removed and negatively stained by rinsing it with about 20 drops of 2% PTA (pH 7.0); the final drop is held on the grid for 15 sec before blotting it with filter paper.

General Methods for Plant Viruses

Rapid Procedures (M. J. W. Webb, 1982, personal communication)

The following rapid procedures are recommended for preparing plant viruses.

1. A 2-mm^2 piece of leaf or any other plant tissue is ground up in 0.1 ml of PTA on a clean glass slide, using a glass rod with a blunt tip. One small drop of 0.05% bacitracin may be added to aid virus release from the tissue. A grid is floated, with the support film side down, on a droplet of this suspension for 30 sec. Alternatively, the suspension droplet can be picked up in a capillary tube and transferred onto a carbon-coated grid clamped in forceps. After 30 sec most of the droplet is drawn off by using a piece of filter paper between the forceps blades until the grid appears dry.

2. Plant tissue (2 mm^3) is placed in 2 drops of methylamine tungstate and 1 drop of wetter on a glass slide. The tissue is squashed in the stain and the droplet of suspension is drained to the edge of the slide, leaving the coarse debris behind. This droplet is transferred to a carbon-coated 400-mesh grid.

3. Plant tissue is ground, and a grid (film side facing down) is floated on a drop of this suspension for 30 sec to 15 min. The grid is rinsed with 10–20 droplets of a negative stain. This treatment removes debris and usually leaves distinct virus particles on the grid. To avoid the production of precipitates, it is recommended that uranyl salts not be ground up with the tissue.

4. A fragment of cigarette tobacco is ground up in 0.01 ml of PTA on a glass slide. A droplet of this suspension is transferred to a coated grid by using one of the preceding or other means. This simple procedure gives a good preparation of tobacco mosaic virus.

Viruses in Crude Extract

A small drop of 2% PTA (pH 6.5) is placed on a carbon-coated 400-mesh grid (Hitchborn and Hills, 1965). A piece of epidermis, peeled from the undersurface of a virus-infected leaf, is laid, with torn surface downward, onto the stain for a few seconds, is gently drawn over the surface of the stain, and is then removed. Excess stain is removed from the grid by momentarily touching its edge with filter paper. The grid is examined immediately. The cell contents deposited on the grids are mainly those from broken spongy mesophyll cells, most of the epidermal cells remaining intact. The pH of the stain can be varied from 5.5 to 6.9 without an apparent effect.

An alternative method involves the use of sap, extruded from macerated infected leaf tissue. The sap is clarified by centrifugation at a low speed either directly or after being frozen for 2 hr at $-10°C$ and then thawed. The supernatant is diluted 10 times with distilled water; undiluted sap is too dense to be useful. Specimens are prepared as before. The tissue-strip method is preferred over the sap method.

The following method is useful for extracting plant viruses from small amounts of infected tissue (Duncan and Roberts, 1981). A watch glass serves as a mortar and a matched glass rod as the pestle for 100–500 mg of tissue. For smaller amounts of tissue (0.1–70 mg), a 0.5-ml microanalysis tube acts as the mortar and a tapered-to-match glass rod as the pestle. About 300 mg of tissue and 25 mg of 600-mesh washed Carborundum powder (acting as an abrasive) are placed in the watch glass. After the tissue has been ground for 1–2 min, cells and organelles are disrupted and the virus particles are released. Only enough buffer or fixative solution is added initially to produce a fine paste. The extract (about 1.5 ml) is transferred to a tube for centrifugation for 5–15 min at 8000 g in a micro-angle centrifuge. The supernatant is discarded; the pellet can be further fixed and/or washed with a buffer. A drop of virus suspension is stained as usual.

SPECIFIC METHODS

Apoferritin

Single droplets of apoferritin suspension (0.01–0.2 mg/ml) are placed on a clean Parafilm surface and mixed with an equal volume of 2% ammonium molybdate (pH 7.0–9.9, adjusted with NaOH) (Harris, 1982). A small amount of this mixture is applied to the surface of freshly cleaved mica and allowed to drain onto the edge of a filter paper. After drying at room temperature, the mica pieces are coated with a thin layer of carbon. The carbon layer plus adhering protein is floated off on the surface of 2% uranyl acetate (pH 4.5), and portions are picked up on specimen grids coated with perforated carbon films.

Although the formation of paracrystalline monolayers is not easy, they can be produced by varying the protein concentration, the specimen pH and concentration, the type and pH of the negative stain used during the mica spreading stage, the drying conditions, the type and pH of the second negative stain used at the floating-off stage, and the length of time the carbon layer–protein is left in contact with the second negative stain. By trial and error, one can determine the appropriate conditions for the formation of monolayers of proteins such as apoferritin, human erythrocyte cylindrin, and *E. coli* glutamine synthetase (Harris, 1982).

For this method to be successful, it is imperative that the protein used be pure and stable with respect to dissociation of its subunits during processing. Charge interactions between the mica surface, the stain, and the protein macromolecules are of considerable importance. The mica surface carries bound K^+, which may influence the ordering of negatively charged protein macromolecules as the aqueous solution of stain and protein dries on the mica. Since some properties of proteins depend on the arrangement of hydrophilic and hydrophobic residues in their constitution, a negative stain composed of hydrophilic and hydrophobic components may prove superior for visualizing details of protein structures (Fabergé and Oliver, 1974). Such stains may promote wetting of the support film on which the specimen is prepared.

Bacteria

Upon removal of a sample from a culture, 0.1 volume of 2.5% glutaraldehyde in phosphate-buffered saline (PBS) (0.2 M K_2HPO_4 in 0.85% NaCl, pH 7.2) is added to the cells, which are fixed for 24 hr at 4°C (Chan *et al.*, 1974). To free the cells of as many surface adherents as possible, the cells are washed three times with aqueous 0.004% sucrose solution at 4°C in a Sorvall refrigerated centrifuge. Sucrose is used for its wetting properties. Washed cells are diluted with 0.004% sucrose to a light suspension, giving a just visible turbidity in the capillary part of a Pasteur pipette. One drop of the suspension is placed on a 3-mm Formvar-coated 400-mesh grid and allowed to dry. The grid is stained with 2% PTA (pH 6.5).

Fibrinogen

By using very dilute fibrinogen solutions and allowing a long adsorption time for them, one can visualize the morphology and dimensions of this protein (Estis and Haschemeyer, 1980). Fibrinogen is diluted to a concentration of 1–2 µl/ml in either the 0.02 M sodium phosphate buffer (pH 7.0) or 0.01 M acetic acid. A copper grid bearing a carbon film (10–15 nm), which has been subjected to a low-vacuum (10^{-1} torr) glow discharge at 5000 V for 2 min, is floated on the surface of the fibrinogen solution for 10–15 min. The grid is transferred to the surface of either 2% PTA or 0.5% uranyl acetate and remains there for 1–5 min; excess stain is then removed by touching the edge with a piece of filter paper. Areas near the edge of a smooth wavelike depression (channel) in the carbon film should be photographed. Areas near the periphery of the depression provide excellent contrast.

Microsomes

The following method preserves ribosomal association with microsome membranes in both sections and whole microsomes following negative staining (Mrena, 1980). Rat liver homogenate is prepared in 0.25 M sucrose buffered to pH 6.5 by 50 mM Tris-maleate buffer, using a glass–Teflon Potter-Elvehjem homogenizer operating at low speed (100 rpm). The homogenate is centrifuged at 2000 g for 10 min, and the sediment is discarded. The resulting supernatant is centrifuged at 5000 g for 15 min, and the pellet containing the mitochondria is discarded. This second supernatant is centrifuged at 100,000 g for 1 hr. The resultant pellet corresponds to the microsomal fraction.

Preparation of Negative Stain

Two grams of uranyl acetate are dissolved in 50 ml of twice-distilled water, the uranyl ion being precipitated by slowly adding 11 ml of 1 N NaOH while stirring. The uranyl precipitate is then separated by centrifugation at 200 g for 5 min, after which it is given three washings with 0.1 N NaOH to eliminate the acetate. The washed precipitate is redissolved by citric acid in the proportion of 0.5 g of $C_6H_8O_7 \cdot H_2O$ in 2 ml of distilled water. The resulting solution is diluted to 100 ml with distilled water to a final concentration of 2%. The pH of the solution is 5.0, which is adjusted to 6.5 by adding 1 N NaOH. This solution is used without further dilution.

Negative Staining

A drop of the diluted microsomal fraction is placed for 1–3 min on a Formvar-coated grid; the excess liquid is removed by blotting it on filter paper. The grid is floated immediately, with the specimen side down, for 3 min on the staining solution. The grid is then withdrawn, blotted on filter paper, and air-dried.

Thin Sectioning

One volume of the first supernatant is mixed with 1 volume of negative staining solution for 15 min and immediately centrifuged. The resulting pellet is fixed with 1% OsO_4 in 0.15 M phosphate buffer (pH 7.2) for 2 hr at 4°C. The pellet is dehydrated and embedded according to standard procedures.

Protein Macromolecules

The following method is useful for determining the shape of macromolecules in the range of 10^5-10^6 daltons (e.g., ribosomes, catalase, fibrinogen, ferritin, IgM, C-reactive protein, and glyceraldehyde-3-phosphate dehydrogenase) (Malech and Albert, 1979). The method is less useful for large vesicles, membrane fragments, or specimens larger than 100 nm. The main advantages of this method are simplicity, speed, and avoidance of artifact formation due to the support films and of stain distribution related to poor wetting.

Copper grids (400–500 mesh) are cleaned by immersion in 10% formic acid for 10 min and then rinsed several times in distilled water. Subsequently, they are rinsed three times first in 100% acetone and then three times in chloroform and finally are placed on filter paper in a Petri dish to dry. The negative staining solution consists of 2% PTA (pH 7.3) containing 20 μg/ml of bacitracin. An aliquot of the specimen is mixed with an equal or greater volume of the staining solution. A 20- to 40-μl final volume is sufficient to stain two or three grids. One side of the clean grid is touched to the surface of a drop of stain–specimen mixture. The excess mixture is drawn off with a filter paper, and the grid is allowed to air-dry. This mixture forms an extremely thin film over portions of the grid holes. However, some workers indicate that stain on grid holes is thicker than on Formvar surfaces, and therefore more prone to shrinkage. Nevertheless, contrast is much better than on Formvar surfaces. A protein concentration of 20 μg/ml in the stain–protein mixture results in good film production. This film stabilizes to some extent after exposure to the electron beam. The grids are initially scanned at a low-intensity illumination. The concentration of NaCl, sucrose, or other solutes in the stain–specimen mixture should not exceed 20 mM; otherwise crystals may develop, obscuring specimen details.

As stated previously, this method is unsuitable for the examination of bulky specimens, which tend to be swept to the edges. Moreover, the type of film just described will not accommodate high-salt conditions, detergents, or common buffers. These difficulties can be circumvented by using specially prepared carbon films, although such an approach results in a slight loss of resolution and demands more time. A freshly cleaved sheet of mica (1 × 3 cm) can be glow-discharged for several minutes in a vacuum evaporator and immediately carbon-coated with a flash evaporator. The specimen is diluted in the bacitracin–PTA stain mixture (10 vol stain + 1 vol specimen + 100 μg protein/ml) and placed as a drop on a wax surface. A small piece of the carbon-coated mica (5 × 5 mm) is cut, and the carbon film is floated onto the drop by immersing the mica into the drop. A good wetting of the carbon surface is achieved because this surface is not exposed to air. The carbon film is picked up on a 200-mesh grid that has been cleaned with formic acid. After the excess liquid has been drawn off with filter paper, the grid is allowed to dry. The carbon-coated mica can be kept in a desiccator for quite a long time without a change in the properties of the carbon film. Pieces of the coated mica can be cut off when needed.

6

Support Films

INTRODUCTION

Support films should be used only when absolutely necessary. Ultrathin sections of epoxy and polyester resins are sufficiently strong without any support film to withstand electron bombardment. Therefore, to gain contrast and resolution and avoid contamination, uncoated grids (300–400 mesh/in.) should be used for sections of these resins. Sections adhere tightly to the unsupported grids when the sections are mounted on the matte (dull) surface of a grid rinsed in acetone before use.

However, in certain cases the use of support film is necessary. Sections thinner than 60 nm should be mounted on filmed grids for high-resolution work; a resolution better than 2.5 nm is not expected with sections thicker than 60 nm. Unless a section is free of holes and covers the entire opening of a bare grid, it may drift and break up under the electron beam. Grids having larger openings than 200 mesh usually require a support film irrespective of the type of embedding resin used. Sections of water-miscible resins (e.g., glycol methacrylate) require support films on the grids. Particulate specimens (bacteria, viruses, cell fragments, and macromolecules) must also be mounted on filmed grids. Serial sections require grids having one large hole or slit (Fig. 3.6), and these must have a support film. Support films can make a relatively large area available for observation, and thus a rarely occurring object can be located much more easily.

Relatively thick films are needed for grids having large openings in order to avoid splitting, whereas very thin films can be mounted on grids containing very small openings. In the case of organic materials (collodion or Formvar), films of 10–20 nm thickness can be obtained. These plastic films are useful for moderate-resolution electron microscopy. Collodion films show little hydrophobicity, while Formvar films are slightly hydrophobic, causing some aggregation of particulate specimens on the surface. The latter are more radiation-resistant than the former.

Very thin, electron transparent, and stable films on the order of 1–2 nm can be prepared from carbon. At present, evaporated carbon films are the most commonly used supports. However, even very thin carbon films show focus-dependent characteristic structure. This becomes a problem in high-resolution electron microscopy; for example, faint contrast of single atoms is obscured by the pronounced phase-contrast structure of carbon films. Another disadvantage of carbon films is their hydrophobic behavior. Various approaches used to produce hydrophilic carbon films are discussed on p. 234. Although other materials (e.g., mica, graphite, and MgO) have been employed to prepare supports, their use has remained limited. Detailed properties of support films have been presented by Baumeister and Hahn (1978) and Hayat (1981a).

PLASTIC SUPPORT FILMS

Formvar Films Cast on Glass

A 0.3–5.0% solution of Formvar is prepared at room temperature in ethylene dichloride (solvent) and transferred to a Coplin jar; the jar is filled 7–10 cm deep (Fig. 6.1). It is essential that the solvent should be dry. New standard glass microscope slides are thoroughly cleaned with soap and water to remove the surfactant. They can be then bathed overnight in *aqua regia,* rinsed with water followed by weak solution of ammonia (to neutralize any residual acid), and finally with distilled water.

After drying the slide, it is dipped about two-thirds of the way into the Formvar solution, and after 2–3 sec it is removed from the jar. (The jar is kept covered when not in use. Water in the polymer solution will result in holes in the film.) The slide is drained onto a filter paper (Fig. 6.2) and air-dried vertically for 1–10 min in dust-free air. This procedure is carried out by propping the slide up at an angle against an object, so that only the uncoated end of the slide touches the object. The drying should be accomplished under cover. Alternatively, the slide can be dried in a desiccator.

A large trough is filled to the brim with distilled water and, immediately before use, the water surface is cleaned by sweeping it with a sheet of lens paper. The separation of the film from the slide occurs when water penetrates the space between the film and the glass surface by capillary force. This penetration is facilitated by scraping the four edges of the slide with a razor blade (Fig. 6.3); a

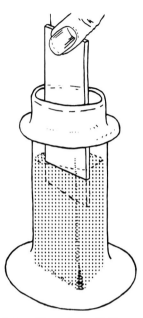

Fig. 6.1. A Coplin jar filled about 10 cm deep with Formvar solution and a glass slide in the process of being dipped into it.

few lines scribed across the slide affords additional routes for capillary permeation of the water. During scraping, the contamination of the film surface with glass and plastic particles should be avoided.

The slide is held at its uncoated end and lowered *slowly* into the water at an angle of about 45° from the horizontal (Fig. 6.4). To release the film from the

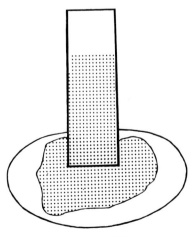

Fig. 6.2. The glass slide is drained onto a piece of filter paper for air-drying in a dust-free place.

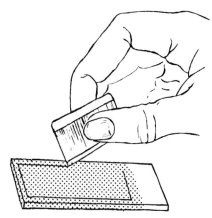

Fig. 6.3. The four edges of the film deposited and dried on the glass slide are scraped with a razor blade.

slide, the water surface tension is teased for a few seconds with the slide until the film begins to be released. If the film is incompletely released, it should be teased away from the slide corner with a needle. Under an appropriate reflective lighting and dark background, film release can be observed. The film surface appears shiny white, while the water surface appears nonreflective. Some workers prefer to breathe onto the cooled slide before lowering it into the water; this treatment may be sufficient for the water to penetrate. The slide is gradually submerged, leaving the film from its upper side floating on the water. If stripping of the film from the glass slide is a problem, a piece of LKB glass strip used for the preparation of glass knives is recommended. Easy separation of the film from the glass slide can also be accomplished by keeping the scraped slide overnight in

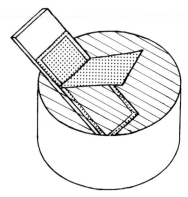

Fig. 6.4. The glass slide with the scraped film is lowered at an angle of 30–40° into the water surface in a suitable container.

a moist chamber prior to lowering it into water. A simple method is to support the slide on glass rods and place wet filter paper under the slide in a Petri dish, which is then covered. Under reflected light, the thinnest films appear gray, while gold films are too thick. The concentration of Formvar can be varied to obtain the desired thickness.

With the use of forceps, 200-mesh grids (previously washed in acetone) are quickly but carefully placed matte side down on good areas of the floating film (Fig. 6.5); wrinkled and dusty portions of the film should be avoided. Exerting light pressure on each grid by *gently* tapping it with the forceps ensures good adherence to the film. A piece of lens paper or Parafilm is placed over the floating film-grids and lifted off the water surface. The lens paper is placed, with grids upward, in a covered Petri dish or a desiccator, where they can remain until needed. Each grid can be picked up with the forceps as the film tears around the grid, leaving the film on the grid intact. If the film tears on the grid, the film around the grid should be perforated with a needle before picking up the grid. If needed, these grids can be coated with a thin layer of carbon to increase the stability of the plastic film.

If difficulty is encountered in coaxing the film off the glass slide in this method, the slide can be pretreated with Victawet, a commercially available wetting agent (Edwards *et al.*, 1984) (E. F. Fullam, Inc.). Slides are dipped in 100% ethanol and then wiped dry. A tungsten wire basket is secured in a vacuum evaporator, and with fine forceps, a small piece of Victawet (1 mm^3) is placed in the basket and then gently forced toward the bottom, or conical, end. Three slides are placed directly under and 5 cm from the tungsten wire basket. The bell jar is evacuated and the current is *slowly* increased across the tungsten wire. When heated properly, the Victawet melts and evaporates over the slides. About 30 sec is needed to completely evaporate the Victawet. These slides are now ready for use.

The collection of grids without damaging the film after placing them on the delicate floating Formvar film is not easy. Methods currently applied for collecting these grids include the use of wire loops, glass slides, and sheets of tissue, lens, or filter paper. The success of these methods depends largely on the skill and patience of the worker. The following method is recommended for collecting large numbers of coated grids; this approach is especially useful for collecting slot and large-mesh hexagonal grids for serial sectioning studies (Trett and

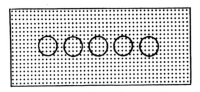

Fig. 6.5.   Five grids, with appropriate distance between them, are placed on the film.

Crouch, 1984). Cleaned grids are placed on the film floating on the surface of water. A strip of Parafilm (80 × 25 mm) is placed over a glass slide, which is held (Parafilm side downward) above one end of the film at 45° to the surface of the water. The slide is gently pushed beneath the water at the same angle. The film will adhere to the Parafilm and, along with the grids, is drawn beneath the surface. Once the film and grids are completely submerged, the slide is rotated until the Parafilm, support film, and grids are uppermost, and then slowly withdrawn, allowing the surface tension to remove any surplus water.

It is not easy to obtain clean, streak-free, and hole-free Formvar (or collodion) films. One of the major factors that controls the production of hole-free films is the moisture content of the atmosphere in which the slides dry after being dipped into the Formvar solution. Although the area in which the Formvar-coated slides are dried can be made locally dry by using dehumidifier or a heat lamp, these procedures are not always successful to produce a controlled environment. Moreover, dust from the air tends to contaminate the films. The danger of inhaling noxious fumes of ethylene dichloride persists.

These problems can be circumvented by using a controlled-environment chamber (Keene, 1984). The materials needed are: (1) a "glove bag" (inflated size is 42 × 42 × 26 cm) (Cat. 17-17, Instruments for Research & Industry, Cheltenham, Pennsylvania); (2) a regulated source of dry, pure nitrogen gas; (3) a 0.35–0.5% (w/v) solution of Formvar in ethylene dichloride in a *clean* bottle; (4) new glass slides cleaned in 90% ethanol and wiped dry with lint-free paper; (5) two covered Petri dishes containing enough phosphorous pentoxide (P_2O_5) powder to cover their bottoms; (6) filter papers; and (7) a flexible hose (outer diameter 13 mm).

The bag is formed in such a way as to allow the user to place the necessary materials inside through a large sealable opening, and to have access to these materials via gloves protruding into the chamber. The bag is then purged with dry nitrogen gas via a flexible hose inserted into a smaller opening in the bag. The purge is carried out by inflating the bag with dry nitrogen gas and then collapsing it; this procedure is repeated several times until the humid air in the bag has been replaced with the gas. Residual moisture can be removed by uncovering the dishes containing the desiccant P_2O_5. This desiccant is a strong skin irritant. The slides are coated with the Formvar and then dried inside the chamber. Floating the films from the slides onto the water surface and coating the grids are carried out outside the chamber according to the standard procedures described earlier. The fumes can be exhausted by opening the bag in a fume hood.

Formvar Films Cast on Water

Formvar films are difficult to strip from clean glass slides. If the glass surface is not clean, the film might be contaminated. Since the glass surface itself is structurally uneven, the films are highly surface-structured. Ultrathin Formvar films can be made at an air–water interface by the drop method (Davison and Colquhoun, 1985). These films show improved flatness and stability, and their surface is smoother than that of Parlodion and glass-stripped Formvar films. The films prepared by the drop method are comparable in quality to pure carbon films made off cleaved mica; both of these films are especially useful for shadowing and negative staining preparations. The quality of these Formvar films does not seem to be affected by humidity variations or even water in the Formvar solution. Formvar concentrations of 0.25–0.4% in ethylene dichloride are recommended. Lower concentrations (~0.1%) produce nets consisting of many tiny holes, which, after carbon stabilization, can be used for high-resolution studies.

One drop (about 13 μl) of Formvar solution from a Pasteur pipette is placed from a height of about 2 mm onto the surface of scrupulously clean distilled water. The film formed is almost invisible, but occupies a circular area about 8 cm in diameter. Grids (shiny side down) are placed on the central region of the film, at which time the film will become visible due to the presence of wrinkles around the grid edges. A piece of Parafilm is touched to the surface and then lifted; the film adheres to the Parafilm, sandwiching the grids in between. After obtaining one film, the water in the container is discarded and replaced with fresh water for making the next film.

If relatively high concentrations of Formvar (2%) need to be used, films can be made by placing a very small drop of the Formvar solution onto the surface of a mixture containing 30% sucrose, 1% propylene glycol, and 0.1% acetic acid (Day, 1984). Owing to its surface tension, pure water is not used for floating the film in this method. The drop of the Formvar solution formed at the tip of a Pasteur pipette is too large. Therefore a fine tip must be drawn out with heat before this pipette can be used. Grids are placed on the floating film (avoiding the central area), and the grids along with the film are picked up on a piece of Parafilm. Grids are washed in distilled water to remove traces of the mixture.

Formvar Films Cast on Mica

The drop method described earlier cannot be used for making thicker Formvar films (20–30 nm). Mica instead of glass is an excellent substrate for making such films. The cleaved mica surface is exceedingly clean and smooth and the films strip easily from it. These films are very clean and uniform in thickness and are less susceptible to methanol degradation than are Formvar films made off glass (Davison and Colquhoun, 1985). The latter characteristic is useful for poststaining the sections with heavy metal salts dissolved in methanol. Only freshly cleaved surface of high-quality mica (no dark inslusion spots) should be used. To make the Formvar film, the cleaved mica surface is submerged into 0.25–0.5% Formvar solution in ethylene dichloride, and then allowed to drain in a solvent vapor atmosphere. The film strips easily without sticking.

Collodion Films Cast on Water

Like Formvar films, collodion films can be cast directly on a water surface. A 12-cm Buchner funnel is filled with distilled water and a 5-cm circle of filter paper (or a piece of wire mesh) is placed on its bottom. Several grids (200 mesh), matte side up, are positioned on the filter paper. Amyl acetate is distilled and allowed to stand over a 4A molecular sieve before use so as to remove water. Amyl acetate should be used with caution; it is volatile and may cause liver damage with prolonged exposure. Collodion is baked before use. Two drops (totaling 25 μl) of a 1.5% solution of collodion in amyl acetate are dropped in the center above the water level; the amyl acetate is allowed to evaporate. The film is removed with a needle to clean the water surface of any dust particles and discarded.

A second film is formed in the same way. The funnel stopcock is slowly opened and the water is allowed to drain slowly from the funnel. As a result, the film gently drops down, coating the grids evenly. Under reflected light, the film should appear gray or silver-gray. The desired film thickness can be obtained by varying the number of drops of collodion solution used and/or by changing the concentration of the solution.

Collodion Films Cast on Glass

This method usually produces stronger films than those obtained by the preceding method and is preferred if only a few grids are needed at a time. A stock solution of 0.5% collodion is prepared in amyl acetate in a fume hood. Films are prepared and mounted on grids as described for Formvar films cast on glass (see p. 266). If plastic films show holes, their solutions have become contaminated with water.

CARBON SUPPORT FILMS

Carbon can be evaporated *in vacuo* to form a uniform amorphous film. Usually carbon films are deposited on the grids before collecting the sections, although sections can be collected on uncoated grids, stained, and then stabilized by evaporating a carbon film, For general purposes, carbon films are made by evaporating the carbon as a thin layer from a carbon arc under vacuum. To accomplish this, the availability of a vacuum evaporator is imperative. The vacuum evaporator consists essentially of a glass bell jar evacuated by a rotary and diffusion pump. Inside it are two carbon rods (each 4–7 mm in diameter); one is fixed and the other is spring-loaded and held lightly against it with the tips in contact. Spring tension provides the pressure to accomplish and maintain contact. The degree of tension controls film thickness because it affects the rate of carbon evaporation. Either both tips of the two carbon rods are pointed or only one tip is pointed and the other is squared. The advantage of the latter is that although the contact area remains small, the amount of misalignment that can be tolerated is increased. After sharpening the approximate final diameter of one tip is 0.25 mm and that of the other tip is 1.5 mm.

The target (e.g., freshly cleaved mica) on which the film is to be deposited is placed 10–15 cm from the source. To detect the thickness of the film deposited, a clean piece of white porcelain with a drop of vacuum oil on it is placed near the target. The area covered by the oil remains white, while the surrounding area of the porcelain shows shades of gray to brown. It is estimated that a film thickness of 5 nm shows a light brown color on the porcelain, which is a satisfactory thickness for most conventional studies. Gray carbon films are more stable than brown ones. For more accurate thickness estimation methods, see De Boer and Brackenhoff (1974) and Baumeister and Hahn (1978).

After evacuation to about 10^{-4} torr, an alternating current of 30–50 amp at 15–20 V is passed through the electrodes. This current heats the electrodes to white hot, and the thermal energy produced forces carbon atoms off the electrodes. The carbon atoms travel in straight lines, and the target will be gradually covered with individual atoms. After 10–30 sec, a continuous thin film is deposited onto the target.

More stable films are obtained after degassing the carbon rods before evaporation by preheating them with a current of 20–25 amp. The quality of the films can also be enhanced by applying current in short pulses instead of applying it continuously. Moreover, indirect evaporation of carbon yields smoother and less grainy films than those produced by direct evaporation. In the indirect evaporation unit, a stop is positioned between the evaporation source and the target so that there is no direct sight from source to target. Thus carbon atoms are deflected before they strike the target, ensuring that single atoms rather than clusters contribute to film production (Johansen, 1974). Whether or not rapid evaporation of carbon has any advantage over a slow evaporation is not clear.

SUBSTRATES FOR CARBON EVAPORATION

Substrates used for preparing carbon films include glass, mica, plastic films, organic glass, and glycerol. The films prepared on glass and mica are floated off onto a liquid surface, while in the other cases the substrate is dissolved after film formation. The advantages and disadvantages of various substrates have been summarized by Baumeister and Hahn (1978). Methods for the deposition of carbon films on various substrates follow.

Carbon Films Deposited Directly on Grids

Very thin films of carbon can be deposited directly on sections that have been mounted on grids. The advantage of this method is its convenience; however, it is not recommended for quality work. To facilitate the adhesion of carbon film to copper grids, they are treated with chloroprene rubber before mounting the sections (Fukami and Adachi, 1964; Wyatt, 1970). A thoroughly cleaned sheet of polyethylene (188 μm thick; 2.5 cm wide by 12.7 cm long) is pulled reasonably tight on the glass slide by taping it under the slide. The grids are placed on the surface of the polyethylene and a 0.3%–0.5% solution of Neoprene in toluene is dropped over them with a pipette. The excess solution is removed by tilting the slide, which is then placed in a desiccator for drying. The coated grids are detached from the polyethylene sheet by stretching it gently. The stretching also breaks the Neoprene film around the grids, which can be picked up with forceps. Sections are mounted on these grids.

Another method for facilitating the adherence of the film to the grids is to rinse the grids in concentrated HCl for 2 min; this treatment results in the slight etching of the grids. Alternatively, grids can be dipped in a 1% solution of polybutene in xylene, in an extremely dilute solution of rubber cement in carbon tetrachloride, or in a solution made by dissolving cellulose adhesive tape in chloroform. The treated grids with mounted sections are arranged on a glass slide that is placed in the vacuum evaporator. A very thin carbon film is sufficient to give necessary support to the sections.

Carbon Films Prepared on Glass

The simplest method for preparing carbon films is to deposit the layer on a cleaned glass slide and then to float it on a clean water surface. The advantage of this method is the ready availability of glass slides. At times it is difficult to strip carbon films from the glass surface. If needed, glass slides can be cleaned with a detergent, although slides as received from the manufacturer are already coated with a detergent layer. This is the reason why films usually float away from glass slides used without additional cleaning.

The best method to facilitate the release of carbon films was developed by Münch (1964). After the film is deposited, it is scratched into small squares a little larger than the size of the grid. A mixture of 0.5–1.0% hydrofluoric acid and 10% acetone in water is prepared in a flat plastic trough of an appropriate size. After the carbon layer has been scraped from the edges of the slide with a razor blade, the slide (with the carbon film on top) is immersed at a shallow angle to the liquid level in the trough. As the acetone reduces the surface tension and the hydrofluoric acid solution creeps between the glass surface and the carbon layer, dissolving the top surface of the glass, the carbon layer floats off. The floating carbon layer is picked up on grids from below.

Procedure

1. Without a detergent, clean and polish a glass microscope slide.

2. Deposit a carbon film of 10 nm thickness on the slide in a vacuum evaporator, as described on p. 272.

3. Float the carbon film onto a clean water surface by immersing the whole slide at a shallow angle.

4. Submerge the individual grid by holding it in forceps, and then bring it up through the floating film and dry it on a filter paper.

Carbon Films Prepared on Mica

Ultrathin (<3 nm) carbon films required for high-resolution electron microscopy should be prepared on exceptionally smooth surfaces such as mica. The removal of carbon films from the mica surface is relatively easy. Mica sheets are commercially available. Mica can be cleaved by gently introducing a razor edge between two crystal planes and exerting slight pressure until the natural cleavage planes separate. Gentle continuous pressure will produce two very smooth crystal surfaces; the crystals should not be forced apart. For easy handling, mica crystals somewhat thicker than paper are preferred. Mica should be cleaved immediately before use. Some workers prefer to keep freshly cleaved mica sheets for several hours under dust-protective conditions in a humidity box before evaporation. The thin water layer formed on the hydrophilic mica surface facilitates the ingress of water between the substrate and carbon film.

Procedure 1

1. Evaporate graphite pencils in an evaporator at 2×10^{-6} torr onto a freshly cleaved mica sheet cut to 12×15 mm.

2. Determine the amount of carbon deposited by placing a piece of white paper in close proximity to the mica.

3. Remove the carbon film by slowly immersing the mica sheet, with carbon side up, into distilled water at a $45°$ angle.

4. Grip one grid at a time with a pincer forceps and lock the forceps.

5. Immerse the grid under the floating carbon film and then lift it.

6. Rotate the forceps $180°$, and use a piece of filter paper to remove the droplet of water now located over the grid.

Procedure 2. A grid coating trough (Smith, 1981) (Ladd Research Industries) (Fig. 6.6) can be used for coating the grids with carbon film evaporated on a mica or glass slide. This device circumvents the problem of maneuvering the film over the grids that have been placed below the water surface in a container. The trough is filled to the top with *clean,* distilled water. About 5 ml and 1 ml of this water are drawn into the large and small syringe, respectively. These amounts should be determined empirically as one becomes familiar with the trough. The water level can be adjusted with the syringe throughout the pro-

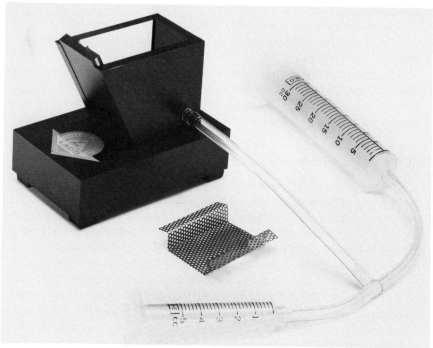

Fig. 6.6. Water trough used to facilitate the coating of grids with carbon and/or plastic. The device consists of a trough made of black opaque Plexiglas (with the exception of one side wall, which is made of clear Plexiglas to allow the progress of the slide to be seen inside the trough), stainless tray, large and small syringes, T-coupling, and tubing. (From Smith, 1981.) (Ladd Research Ind.)

cedure. The larger syringe is used for major adjustments in the water level and the smaller syringe for finer adjustments.

The stainless steel tray is positioned so that the hole is on the window side of the trough. The tray is covered with a piece of filter paper cut to roughly the same size as the tray. The water level is adjusted with syringes so that the surface bulges slightly above the top of the trough. Grids are sunk below the water surface and onto the filter paper. The position of the tray is adjusted so that its leading edge is about 2 mm from the rails on the front wall. The water surface can be cleaned with a Pasteur pipette attached to a vacuum line or rubber bulb. The water level can be lowered by drawing on the syringe until it is just below the grids. A mica or glass slide, upon which the substrate film has been deposited, is slid smoothly down the rails into the water until it is stopped by the retaining pin at the bottom of the rails. This procedure results in the thin film being partially floated off the slide to cover the grids beneath. It is still attached to the slide, however, and is therefore held in its correct position and does not float away.

The water level can now be lowered carefully by using the syringes so that the film comes to rest on the grids; lowering the level even further allows the filter paper to drain. The tray is removed from the trough, and the filter paper and grids are air-dried. Care should be taken to avoid contamination of the substrated grids while they are drying. Depending upon the condition of the water, it can either be pumped back into the trough from the syringes for reuse again or be discarded.

Carbon Films Using Plastic Substrates

A strong adhesion of carbon films to the copper grid can be accomplished by depositing them onto a plastic film, which subsequently is dissolved. The adhesion is due to plastic remnants on the grid bars. Collodion is preferred over other plastics. The disadvantages are that the carbon layer replicates the imperfect plastic surface and it is difficult to remove the plastic substrate completely.

Procedure

1. Coat the grid with a thin film of collodion as described on p. 272.

2. Arrange the grids on a clean glass slide (with coated side up) and then place the slide in a vacuum evaporator.

3. Allow the grids to be coated with a carbon film 3–10 nm thick.

4. Place the carbonized collodion-coated grids on a piece of wire gauze and then immerse the gauze in acetone at an angle of 45° or more for a few minutes to dissolve the plastic (for Formvar-coated grids, use chloroform instead of acetone). The duration of exposure to the solvent is crucial because the objective is to dissolve the plastic from the grid holes but leave the plastic between the grid bars and the carbon film.

Modified Method

A modified method, introduced by Stolinski and Gross (1969), has the following advantages: (1) about 79% of the total grid area constitutes carbon film and is available for observation; (2) the film is extremely thin (about 1.5 nm); (3) contamination is minimal since the carbon or grid does not come in contact with water; and (4) as many as nine grids can be prepared at one time. The film is useful for supporting thin sections as well as whole specimens such as bacteria, viruses, and cell fragments.

Procedure

1. Fill a glass trough (20 cm diameter, 6 cm deep) with distilled water to a depth of 4.5 cm and place it on a sheet of white paper.

2. Prepare the solution for the substrate film by adding 3 parts of Belco cellulose to 1 part of amyl acetate by volume.

3. Release several drops of this solution onto the water surface and allow them

to dry. The thick film thus produced will pick up the dust and other debris floating on the water surface.

4. Remove the film with a glass rod with a twisting motion.

5. Release 1 drop of cellulose solution onto the water surface from a height of 4 cm and allow it to dry. The film should have an area of 3–4 cm with an even silver-to-gold interference color. The colors can be observed by placing a tungsten bulb adjacent to the trough.

6. Bend Athene-type grids (3.05 mm diameter) having a hexagonal pattern by holding one part of the rim with fine forceps and pressing the opposite side onto a hard surface so that the shiny surface of the grid becomes concave. This procedure facilitates handling of the grids at a later stage.

7. Place as many as nine grids onto the floating film; the distance between the grids should not be less than 6 mm.

8. Push the film with grids to one side and position a lifting tool underneath the film so that the grids are allowed over the hole in the tool. The tool is made of stainless steel that has an outer diameter of 7 cm and a punched hole 2.5 cm in diameter. The top end with the hole has a rounded edge, and the handle is 12 cm long.

9. Raise the tool slowly to lift the film with grids out of the water. Before the tool is lifted away completely, gather the surplus film surrounding the tool by touching it with a glass rod.

10. Upturn the tool with the film and grids and place it on a clean glass slide.

11. After puncturing the film around the inner rim, lift the tool away so that the grids covered by the film are left on the slide.

12. Dry the film by placing the slide briefly on a warm plate.

13. Lift the grids individually and place them on another glass slide in a Petri dish.

14. Place the grids in a vacuum evaporator and coat them with carbon; a very faint gray color shown by a piece of white porcelain placed in the evaporator indicates a film thickness of about 1.5 nm.

15. Place the carbon-coated grids on a 50-mesh stainless-steel gauze (5 cm^2) and immerse them under pure amyl acetate in a Petri dish for 4–12 hr to dissolve the cellulose substrate.

16. Remove the gauze with the grids from the solvent and place both of them on a filter paper for drying in a dust-free area.

Carbonized Plastic Films

As stated earlier, plastic films can be made more stable by carbon coating. Carbonization of plastic films eliminates the problems of drift and shrinkage under the electron bombardment. The carbon layer protects both the plastic film and the section. Very thin composite films of plastic and carbon having the desirable features of both substrates can be prepared. It should be noted that carbon has a higher scattering power than collodion because the former has a higher density (about 2) than the latter (about 1.4). It is apparent, therefore, that carbon films prepared for stabilizing plastic films should be extremely thin. The ideal carbon film is shiny and shows no interference colors. One of the limitations of the combined film is that it is usually thicker than a pure carbon film and is thus unsuitable for high-resolution electron microscopy. For high-resolution work, pure carbon films or self-supporting specimens may have to be used. Plain collodion or Formvar films are not used anymore for quality work.

Procedure

1. Coat the grids with extremely thin collodion or Formvar film.

2. Arrange dry, plastic-coated grids on a clean glass slide (with coated side up) and then place the slide in a vacuum evaporator.

3. Allow the grids to be deposited with carbon film about 5 nm thick.

4. Store the grids in covered Petri dishes.

PERFORATED FILMS

For certain studies, such as particulate specimens, the mesh size (50–400) of commercial specimen grids or the hole diameters of drilled diaphragms (20–750 μm) are too large. Perforated films (microgrids) with holes of various diameters (0.05–100 μm) (Fig. 6.7) are helpful in such studies. Perforated films with large open areas are also called micronets. In perforated films, image contrast is higher in the regions of holes than in the areas with the film. Since all supporting films reduce image contrast, perforated films provide an increased image contrast as well as support for thin sections. Moreover, films with small circular holes are useful for testing the symmetry of the image and for correcting astigmatism in the lens.

Perforated collodion, Formvar, and carbonized plastic films are prepared by using three basic techniques: (1) the development and incorporation of local faults in the film, (2) the localized destruction of the film by physical or chemical treatment, and (3) the replication of perforated templates (e.g., etched eutectics or filters) (Baumeister and Seredynski, 1976). The numerous modifications of the three basic techniques reflect the difficulties in obtaining reproducible results. These difficulties include the presence of pseudoholes and unsuitable average hole diameter. Ideally, the technique should allow a control on the size and number of perforations as well as on the geometry of the hole.

It is desirable to select a film with a perforation size appropriate to the objective of the study. The perforation size is also determined by the magnification at

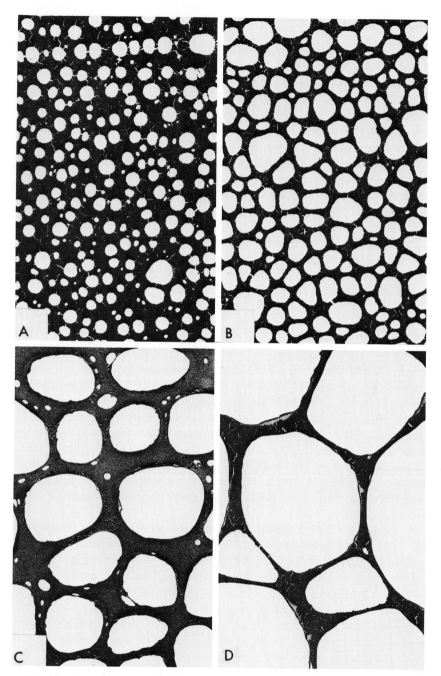

Fig. 6.7. Gold-coated perforated films prepared by the glycerol method ($\times 10,000$). The average hole size increases with the concentration of glycerol in the Formvar-chloroform/glycerol emulsion. (A) Glycerol content 0.1 ml/90 ml Formvar solution. (B) Glycerol content 0.3 ml. (C) Glycerol content 1 ml. (D) Glycerol content 3 ml. (From Baumeister and Hahn, 1978.)

which the observation is to be made. For example, perforations of 0.2–0.5 μm are ideal when the grids are to be examined at a direct magnification of 100,000 times, because the area of the viewing field is then comparable to the area of a single perforation.

Methods most commonly used to prepare plastic perforated films involve the addition of a substance that is not miscible with the solvent used to prepare the plastic solution. This treatment causes an interference in the uniform drying of the film. For example, to prepare perforated Formvar films, water is introduced into the Formvar–ethylene dichloride solution. Since water and ethylene dichloride do not mix, during the last stages of solvent and water evaporation a large number of air bubbles evolve, which eventually produce perforations in the Formvar film.

Perforations in the plastic film can also be obtained by blowing a stream of warm air saturated with water into a glass vessel containing the coated slide. Alternatively, minute water bubbles can be inserted into plastic films by gently breathing upon a freshly spread plastic film. Since the holes produced by such a treatment may be pseudoholes, they are converted into real holes by lightly heating the coated slide on a Bunsen burner. The perforations obtained have a mean diameter of 1 μm. The number and size of the holes depend on the duration and intensity of breathing. For an improved "breathing" method, see Drahos and Delong (1960).

Another method for obtaining films with perforations is based on cooling the hydrophobic glass surface to near freezing temperature and then forming, by condensation, minute water droplets on it. The size and number of perforations formed are determined primarily by the temperature of the glass surface, the surrounding environment's humidity and temperature, and the concentration of the plastic solution. By this method perforations ranging in size from 0.01 to 10 μm can be obtained. The total open area in each grid is estimated to be 40%, 60%, and 75%, with perforation sizes of 0.2–0.5 μm, 1–2 μm, and 3–6 μm, respectively.

Although the procedures just described produce excellent results, it is preferable to add water directly to the plastic solution to form an emulsion (simultaneous method). The disadvantage of the former procedures is the difficulty in controlling the rather critical timing for inserting water droplets into the drying film. Descriptions of the simultaneous method and other procedures for preparing perforated support films follow.

Perforated Formvar Films

Various methods are used to prepare perforated Formvar films. The following three methods (Harris, 1962; Elsner, 1971; Baumeister and Seredynski, 1976) are reliable, and perforations ranging from 0.05 to 25 μm in diameter can be obtained (Fig. 6.7).

Procedure 1

1. Prepare a 0.25% solution of Formvar in ethylene dichloride.
2. Add glycerol in the ratio of 1 : 32 and shake well to form an emulsion. This ratio should produce holes of about 7 μm in diameter.
3. Dip a clean microscope slide into the emulsion and then tilt it to allow it to drain and dry for about 10 min.
4. Expose the slide to a jet of steam for 1 min.
5. Float the film on the water surface and mount it on grids, as described earlier.

Note: If desired, these films can be coated with carbon. The proper size and number of holes in the film can be obtained by changing the proportion of glycerol (see Table 6.1).

Procedure 2. With the following method, small, round perforations with smooth edges can be produced consistently. This method employs solvent treatment for smoothing the edges and enlarging the perforations formed in high humidity. Perforation size and edge characteristic can be controlled easily, and useful perforations as small as 0.15 μm can be obtained.

1. Prepare a 0.2–0.4% solution of Formvar in ethylene dichloride.
2. Dip a clean glass slide into a Coplin jar containing the Formvar solution, and tilt it to allow it to drain and dry for 1 min. Room temperature should be 23°C with 60–70% relative humidity; lower humidity will result in fewer perforations.
3. Float the film on the water surface and mount it on grids, as described earlier.
4. Immerse the grids in chloroform–absolute methanol mixture (1 : 9) in the compartments of a depression plate. To reduce the solvent evaporation, place the plate on a blotted paper saturated with the solvent mixture in a covered Petri dish. An increased chloroform ratio causes the dissolving powder of the mixture to increase.

TABLE 6.1.

Relationship of Hole Size to Glycerol Content for Perforated Formvar Films

Formvar solution (parts)	Glycerol (parts)	Maximum hole diameter (μm)
8	1	24
16	1	14
32	1	7
120	1	4

5. Allow the grids to remain in the solvent mixture for 15–16 min at 23°C.

6. Wash the grids by dipping them in 95% ethanol, 50% ethanol, and distilled water (30 dips in each).

7. Place the grids on a filter paper for drying at room temperature.

Note: If desired, these films can be coated with carbon by placing the grids along the edge of double-stick tape on a glass slide. Perforated carbon films can be obtained by washing Formvar off the carbon-coated grids with chloroform.

Procedure 3

1. Prepare a solution by adding 0.18 g of Formvar to 90 ml of chloroform in an Erlenmeyer flask. The quantity of glycerol added ranges from 0.01 to 5.0 ml (the hole size increases with increasing concentrations of glycerol; Fig. 6.7). Stir the solution with a magnetic stirrer until the polymer is completely dissolved. Sonicate the solution by placing the flask in the water-filled tank of an ultrasonic generator for 30 min.

2. Clean a glass slide in a detergent solution, rinse it with distilled water, and rub it dry with a clean cloth (parallel to the long axis of the slide).

3. Immerse the slide into the emulsion for 5–10 sec and then slowly withdraw (13 mm/sec) and dry it under dust-free conditions.

4. Place the slide for about 5 sec in steam from a water bath and allow it to dry.

5. To aid complete perforation, immerse the slide into a beaker containing acetone for 10–20 sec.

6. After it has dried, float the film off on the water surface and transfer it to the grids.

7. To achieve good thermal and electrical conductivity and stability, coat the film with a thin layer (25 nm) of carbon.

Procedure 4. The method described here was developed by Reichelt *et al.* (1977) and produces microgrids having 93–95% of the useful grid area available for study. This method has advantages over most others. For example, the method developed by Hoelke (1975) is simple, but it produces microgrids whose holey area is only about 10–20% of the whole grid area.

A microscope slide is cleaned with a surface-acting agent to make it hydrophilic, thoroughly rinsed in tap water, and dried with a piece of cloth. The slide is immersed for 5–10 min in an emulsion consisting of 0.25% cellulose acetobutyrate in ethyl acetate and 1% glycerol. Upon removal, a thin film of cellulose acetobutyrate interspersed with glycerol droplets forms on the glass surface. The film is air-dried for about 10 min by allowing the slide to stand vertically. The glycerol droplets are dissolved by holding the slide in a stream of water vapor for about 2 min. The film is floated off on the surface of absolutely clean, double-distilled water, transferred to grids, and allowed to dry.

The diameter of the holes is enlarged by exposing the film on the grid to an atmosphere saturated with ethyl acetate. The size of the holes can be increased by prolonging the exposure time to the ethyl acetate. Their size can be controlled by observing the effect of this exposure under the phase-contrast light microscope. The maximum enlargement of the holes is obtained at a room temperature of 20°C and an exposure of 20–30 sec. These films can be strengthened by coating them with carbon in an evaporator and storing them until needed.

Support Films with Large Holes (Micronets) (Pease, 1975)

Procedure

1. Hydrophobize the glass slide by soaking it overnight in a saturated solution of ferric stearate in benzene. Prepare this solution by dissolving 1 part of ferric stearate in 100 parts of benzene and bring the mixture to a boil. Any ferric soap can be substituted.

2. Remove most of the original solution by dipping the slide in benzene and then vigorously rubbing it with thin paper wipes saturated with successive changes of benzene. Remove all visible traces of the film. The slide remains hydrophobic.

3. Dip the pretreated slide in 0.4–0.6% solution of collodion in amyl acetate, and dry it by holding it in a vertical position.

4. Condense the water on the surface by exposing the slide to furiously boiling water. Continue exposure to the steam until the milkiness disappears, indicative of complete evaporation of amyl acetate. Thin regions with large holes are found near the top of the film, while thicker regions with smaller holes are located near the bottom. The drainage pattern is responsible for the inhomogenous nature of the film.

5. Bake the grid with the dried film at 170–180°C to open up pseudoholes. The micronets can be stabilized by coating them with a thick layer of carbon.

7

Specific Preparation Methods

ACTIN FILAMENTS

Cells or tissues are fixed with diluted Karnovsky's aldehyde mixture in 0.1 M cacodylate buffer (pH 7.2) containing 4.5 mM $CaCl_2$ (which stabilizes microfilaments) at room temperature. After a rinse in the buffer, the specimens are postfixed with 2% OsO_4 in 0.1 M cacodylate buffer (pH 6.8) for 15 min. The specimens are thoroughly rinsed four times (3 min each) in 0.125 M (isomolar) cacodylate buffer and then in distilled water for 5 sec. Next, they are treated *en bloc* with a saturated solution of thiocarbohydrazide in distilled water for 5 min at room temperature. Following a thorough rinse in the same buffer, the specimens are exposed a second time to similarly buffered 2% OsO_4 for 15 min.

During fixation, F-actin *in vitro* can be protected against oxidative degradation by OsO_4 by using viroidin and viroisin peptides extracted from the mushroom. *Amanita virosa* (Gicquaud *et al.*, 1983). These peptides have a methyl sulfonyl group and are more effective than phalloidin in protecting the actin.

ALGAE (UNICELLULAR): GENERAL METHOD

Cells are concentrated by centrifugation and, after the removal of culture medium, are fixed with 3% glutaraldehyde in 0.2 M cacodylate buffer (pH 8.2) containing 0.01 M CaCl$_2$ and 0.25 M sucrose for 90 min at 4°C. The specimens are rinsed three times with the buffer; in each successive rinse the concentration of sucrose is gradually decreased. Postfixation is carried out with 2% OsO$_4$ in 0.2 M cacodylate buffer (pH 8.2) for 90 min at 4°C.

ALGAE (GREEN)

Method 1

Solution A

Glutaraldehyde (25%)	10 ml
Phosphate buffer (0.025 M, pH 7.2)	37 ml
Distilled water	100 ml

Solution B

OsO$_4$	1 g
Phosphate buffer (0.025 M, pH 7.2)	50 ml

Specimens are fixed in solution A for 1 hr at room temperature, rinsed with buffer for 5 min, and postfixed with solution B for 3 hr at 4°C.

Method 2 (Unicellular Marine Algae)

Solution A

Glutaraldehyde (25%)	24 ml
Distilled water to make	100 ml

Solution B

Solution A	50 ml
Seawater	50 ml

Solution C

Distilled water	50 ml
Seawater	50 ml

Solution D

OsO$_4$ (2%)	50 ml
Seawater	50 ml

After centrifugation, cells are fixed in solution B for 90 min at room temperature, rinsed three times in solution C, and postfixed in solution D for 1 hr at 4°C.

Method 3 (*Chlamydomonas* and *Volvox*)

The spheroids are collected on Miracloth and suspended in 3% glutaraldehyde in 10 mM HEPES buffer (pH 7.0) at 4°C for 18 hr. They are then collected on

Miracloth, rinsed for 1 hr in three changes of buffer, and resuspended in 1% OsO_4 in 4 mM potassium phosphate buffer (pH 7.0) for 24 hr at 4°C. The specimens are briefly rinsed with buffer and dehydrated.

ALGAE (RED)

Method 1

Solution A
Sodium cacodylate	2.14 g
Sucrose	8.558 g
Distilled water	100 ml

Solution B
Solution A	80 ml
Glutaraldehyde (25%)	20 ml

The buffer is added to adjust the osmotic pressure of the fixative to that of seawater. Cells are fixed in solution B for 3 hr at 4°C and then washed in a buffered series of sucrose concentrations (0.25, 0.15, and 0.05 M sucrose). Postfixation is accomplished in 2% OsO_4 in the same buffer without sucrose for 3 hr at room temperature.

Method 2

Cells are fixed for 1 hr at 4°C with 3% glutaraldehyde in 0.1 M Millonig's phosphate buffer (pH 6.6) containing 0.25 M sucrose and 50 mg/ml $CaCl_2$. After being rinsed twice in buffer containing $CaCl_2$ and in a graded series of buffer containing decreasing amounts of $CaCl_2$ until only buffer remains, the cells are postfixed with 1% OsO_4 for 1 hr at 4°C. Fresh specimens yield the best results.

AMOEBA

Cell suspension in growth medium is prefixed by dilution with an equal volume of 2.5% glutaraldehyde in 0.05 M cacodylate buffer (pH 6.8) containing 2 mM $CaCl_2$ and 0.2 M sucrose for 1 hr at room temperature. After centrifugation, the cells are fixed in the same fixative for 12 hr at 4°C. They are then rinsed twice for 30 min at 4°C in buffer and centrifuged. A dense suspension of cells is suspended in a drop of 3% agar. After cooling, the agar is cut into small cubes and postfixed with 1% OsO_4 in the same buffer for 1 hr at room temperature.

ANTHERS

Excised anthers are cut into 1-mm lengths and fixed with 3% glutaraldehyde in 0.05 M phosphate buffer (pH 7.2) for 4 hr at room temperature. After being rinsed in three changes of buffer for 4 hr, the specimens are postfixed with 1% OsO_4 in the same buffer for 3 hr at 4°C.

ATTACHMENT OF CELLS TO SUBSTRATUM

The following procedure permits the study of ultrastructural details of cell adhesion and spreading during initial cell interaction with substratum (Grinnell *et al.*, 1976). Epon (substratum) is polymerized in flat embedding molds (10 × 4 × 3 mm) (Polysciences, Inc.) for 24 hr at 60°C. The substratum is rinsed for 1 hr at room temperature in the following adhesion salts to remove surface impurities: 0.8 mM $MgSO_4 \cdot 7H_2O$; 116 mM NaCl; 5.4 mM KCl; 10.6 mM Na_2HPO_4; 5.6 mM D-glucose; and 20 mM N-2-hydroxyethylpiperazine-N-2-ethane sulfonic acid (HEPES buffer, pH 7.0).

Cells grown in suspension culture in the logarithmic growth phase are collected by centrifugation at 500 g for 2 min. Cells are resuspended in 4 ml of adhesion salts containing fetal calf serum and placed in Falcon 10 × 35-mm Petri dishes. Each dish holds four epoxy substrates. About 1–2 × 10^6 cells are used in each experiment. Nonattached cells are removed by immersing the substrata in a beaker containing Dulbecco's phosphate-buffered saline (PBS).

The substrata with attached cells are placed in tissue vials and fixed with 3% glutaraldehyde in 0.1 M phosphate buffer (pH 7.4) for 15 min at room temperature and then for 2 hr at 4°C. The cells are first rinsed with 0.1 M phosphate buffer containing 0.2 M sucrose and then postfixed with 2% OsO_4 in buffer for 30 min. After dehydration, the epoxy substrata are placed longitudinally in BEEM capsules and embedded in fresh Epon.

BACTERIA (Ryter and Kellenberger, 1958)

Method 1 (General Method)

Solution A (tryptone medium)

Bacto-tryptone	1 g
NaCl	0.5 g
Distilled water to make	100 ml

Solution B (Veronal acetate buffer)
Sodium Veronal	2.94 g
Sodium acetate	1.94 g
Sodium chloride	3.40 g
Distilled water to make	100 ml

Solution C (Kellenberger buffer)
Veronal acetate buffer	5 ml
Distilled water	13 ml
HCl (0.1 N)	7 ml
CaCl$_2$ (1 M)	0.25 ml

The pH is adjusted to 6.0 with HCl. Freshly prepared buffer is recommended.

Solution D (fixative)
Kellenberger buffer	100 ml
OsO$_4$	0.5 g

Solution E (rinsing solution)
Kellenberger buffer	100 ml
Uranyl acetate	0.5 g

Prefixation is carried out by suspending bacteria in solution A and mixing a 30-ml aliquot of this suspension with 6 ml of solution D in a centrifuge tube. After centrifugation for 5 min at 1800 g, the pellet is resuspended in 1 ml of solution D and 0.1 ml of solution A for 16 hr at room temperature. The suspension is diluted with 8 ml of solution C and centrifuged for 5 min at 1800 g. The pellet is resuspended in 2% warm agar in solution A (0.03 ml), transferred as a drop on a glass slide, and allowed to solidify. The hardened agar is cut into small cubes (1 mm^3), treated with solution E for 2 hr at room temperature, dehydrated, and embedded.

Method 2 (Glutaraldehyde and OsO$_4$)

Mid-logarithmic cells are centrifuged at 12,000 g for 10 min and then suspended in a mixture of 5% acrolein and 1% glutaraldehyde in 0.1 M cacodylate buffer (pH 7.2) containing 5 mM CaCl$_2$ for 2–4 hr at room temperature. The cells are centrifuged and rinsed in the same buffer by centrifugation. The pellet is suspended in an equal volume of 2.5% agarose in 0.1 M cacodylate buffer held at 48°C. The resulting mixture is chilled, cut into small blocks, and rinsed twice with the same buffer. The cells are postfixed for 1 hr with 1% OsO$_4$ in 0.1 M cacodylate buffer containing 5 mM CaCl$_2$ and 500 µg ruthenium red/ml; ruthenium red is added immediately before use. The blocks are rinsed once in buffer and then twice in distilled water. They are treated with 0.5% uranyl acetate for 1 hr, rinsed twice in distilled water, and then dehydrated and embedded in Spurr resin.

Method 3 (for Extracellular Matrix)

Cells are concentrated by centrifugation and treated for 1 hr at room temperature with 0.15% ruthenium red in 0.1 M cacodylate buffer (pH 7.0). This procedure is followed by fixation for 1 hr at room temperature with a mixture of 3.6% glutaraldehyde and 0.15% ruthenium red in cacodylate buffer. The cells are rinsed twice in buffer and postfixed with a mixture of 1% OsO_4 and 0.15% ruthenium red in cacodylate buffer for 1 hr at 4°C. After being rinsed twice in buffer, the cells are encapsulated in 2% agar. The solidifed agar is cut into small pieces (1 mm³), which are treated with 0.5% uranyl acetate for 1 hr.

Method 4 (for Phage-Infected Bacteria)

Bacteria are sedimented, and the well-drained pellet is resuspended into a mixture of 0.1% uranyl acetate and 5% glutaraldehyde in Michaelis buffer at a final pH of 5.4 (the initial pH of the buffer is 5.9, which upon addition of a concentrated aqueous solution of uranyl acetate drops to 5.4) (Séchaud and Kellenberger, 1972). Fixation is completed overnight at room temperature.

About 1 ml of the fixed cells is spun down in a Micro-centrifuge tube used in a small, swinging, bucket-type tabletop centrifuge. The pellet (2–4 × 10⁸ bacteria) is suspended in 2% agar in Michaelis buffer (pH 5.7) by pouring 3–4 mm of agar in the tube and then mixing it with a microsyringe while holding the tube in a 45°C bath. The agar–bacteria mixture pulled back into the syringe makes a cylinder 6 mm long. The syringe is transferred to the refrigerator for some minutes; then the solidified cylinder is pushed out of the tube onto a glass slide and cut into small blocks, which contain a sufficiently high concentration of bacteria. These blocks are treated in a saturated aqueous solution of uranyl acetate for 2 hr at room temperature; dehydration and embedding follow.

Note: Bacteria fixed directly with OsO_4 show a contracted conformation of the nucleoplasm. This conformation occurs because OsO_4 destroys the permeability of the plasma membrane, resulting in the leakage of K^+ and enhanced permeability of Na^+. The entrance of Na^+ will set up a Donnan equilibrium, which induces uptake of water. On the other hand, prefixation with glutaraldehyde results in a dispersed state of the nucleoplasm. This phenomenon occurs because during glutaraldehyde fixation the impermeability of the plasma membrane to Na^+ is maintained although K^+ leaks out. The net loss of K^+ will result in a net loss of cations and hence a dispersed state of the nucleoplasm. Thus monovalent cations play an important role in determining the organization of the nucleoplasm during prefixation with OsO_4 or glutaraldehyde. Bacteria fixed with glutaraldehyde–uranyl acetate or glutaraldehyde–OsO_4 show contracted nucleoplasm. It is thought that at least most of the bacterial DNA *in vivo* is in a contracted conformation.

BACTERIAL COLONIES

The following method facilitates the study of nonhomogeneous bacterial colonies (e.g., mycoplasmas and L-colonies) (Bonnová and Rýc, 1976). A narrow strip of agar containing bacterial colonies is cut from a Petri dish (Fig. 7.1A). This strip is covered by a somewhat wider strip of cool agar of a lower concentration than the culture medium (Fig. 7.1B). The more solid agar with the cultivated bacterial colonies serves as a base for the subsequent cutting and trimming of agar blocks. A heated scalpel is applied to the sides of the agar strips to weld both layers together. The advantage of using a heated scalpel is that each bacterial colony with intact surface layers *in situ* is completely covered by agar. Each colony remains in the center of each block and is not heat-damaged. A razor blade is used to cut the agar block into small pieces, each containing a single colony.

With the aid of a heated razor blade, the blocks are trimmed to prism shape; the short side is less than 1 mm (Fig. 7.1C and D). The height of the prism should exceed the width to allow orientation. After dehydration and resin infiltration (Fig. 7.1E), the agar blocks are transferred to embedding capsules; the longer side of each block is horizontal in the capsule (Fig. 7.1F). After polymerization, the resin block is trimmed (Fig. 7.1G and H) for sectioning. Gradual enlargement of the embedded colony indicates sectioning toward the center of the colony.

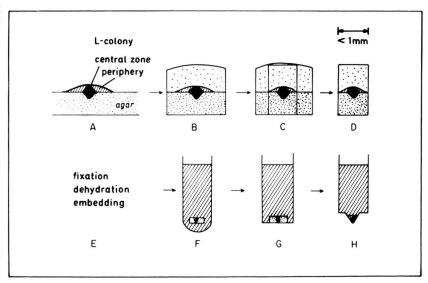

Fig. 7.1. Schematic diagram of embedding and orientation of bacterial colonies. (From Bonnová and Rýc, 1976.)

BACTERIAL MICROCOLONIES GROWN ON SOLID SURFACES

The following method utilizes Epon plates as a substrate for growth and adherence of microcolonies and avoids centrifugation, which might cause morphologic deformation (Kim *et al.*, 1977). Thin plates of Epon (1 mm thick) are cut into 1-cm squares, rinsed and soaked overnight in distilled water, and autoclaved for 15 min. These Epon pieces are transferred to a Petri dish containing liquid medium (Somerson *et al.*, 1973) in which PPLO bovine serum fraction (Difco) is substituted for horse serum (KC Biological, Inc.). Petri dishes are inoculated with *Miyagawanella pneumoniae*. After 24–48 hr of incubation at 37°C, the organisms grown on the surface of the Epon pieces are removed from the culture fluids. The adherent specimens are fixed, dehydrated, and embedded. Thin sections are cut vertically to the surface of the old Epon piece.

BONE (Undecalcified)

Method 1

A biopsy of human iliac crest bone is obtained by using an electric drilling machine. Slices about 1 mm thick are cut from the biopsy bone cylinder with a razor blade (Schulz, 1977). They are immediately fixed with 2.5% glutaraldehyde in cacodylate buffer (pH 7.4, 300 mosmols) for 4 hr. After being rinsed in buffer, specimens are postfixed with 1.3% OsO_4 in collidine buffer for 2 hr at 4°C. After a thorough rinse in the same buffer, specimens are dehydrated with ethanol (30 min at each step) followed by propylene oxide. Specimens are transferred to a 1 : 1 mixture of propylene oxide and Spurr embedding medium for 1 hr. About one-half of this mixture is discarded and then replenished by the fresh mixture. After 1 hr, specimens are transferred to pure Spurr medium and kept overnight at room temperature. (A light vacuum can be applied to facilitate infiltration.) Specimens are placed vertically in predried gelatin capsules and polymerized at 70°C for 8 hr or longer.

The embedding mixture is made up as follows:

ERL 4206	200 g
DER 736	100 g
NSA	450 g
DMAE	5 g

Method 2

Solution A

Formaldehyde (2%)	1 part
Glutaraldehyde (2%)	1 part

Solution B
OsO_4 1 g
Phosphate buffer (0.1 M, pH 7.2)   50 ml
Very small pieces of bone are fixed in 0.1 M phosphate-buffered solution A for 2 hr at room temperature. After being rinsed in the same buffer for 30 min at 4°C, the specimens are postfixed with solution B for 2 hr at 4°C. Dehydration and infiltration are carried out with graded series of acetone and Spurr embedding medium, respectively. Each step of infiltration should last 2 hr with continuous agitation. The specimens are left in the pure embedding medium overnight with agitation. Evacuation (preferably 10^{-3} torr) is then carried out at 60°C for 30 min.

BUFFY COAT

A special tube is made from Plexiglas (Russo, 1977). This tube separates into two parts, with 40% of the total volume remaining in the bottom section. A rubber O-ring makes the junction watertight (Fig. 7.2A). The inner diameter and capacity of this tube are smaller than those of standard 15-ml conical tube (Fig. 7.2B). The new tube requires only 1.5–2.2 ml of blood, permits easy removal of the buffy coat, and yields a thicker buffy coat.

After centrifugation of normal human blood at 1530 g for 10 min at 25°C, the buffy coat is located just above the junction (for pathologic blood samples, the blood should be diluted according to the hematocrit to ensure that the buffy coat falls above the junction). The plasma is removed and replaced with 3% buffered glutaraldehyde; after 5–10 min of fixation, the tube is separated. The lower portion of the tube containing the packed red blood cells is put aside. The excess fixative is aspirated from the upper portion with a Pasteur pipette.

Excess erythrocytes are gently loosened from the buffy coat with a wooden stick, and the intact buffy coat disk is carefully detached from the wall by pushing. The 2-mm-thick disk is sliced into thin segments, and the transverse width of the segments is reduced by removing the red blood cells from one side and the platelets and the coagulated plasma from the other. These segments are fixed for 2–3 hr with 2–3% buffered glutaraldehyde and then postfixed with 1% OsO_4 for 20 min.

CARTILAGE

The proximal ends of tibiae are removed under gentle anesthesia from rats, cut parallel to the long axis into slices 1 mm thick, and placed immediately in a solution of 2% glutaraldehyde in 50 mM cacodylate buffer (pH 7.4) containing 0.7% ruthenium hexamine trichloride (RHT); final osmolality is 330 mosmols.

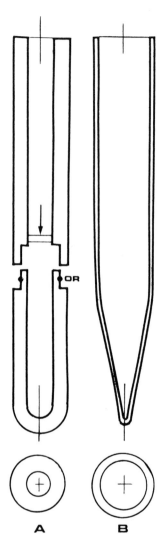

Fig. 7.2. Preparation of the buffy coat. Diagram compares the new tube (A) with a standard conical tube (B). The arrow indicates where the buffy coat is found with normal blood. OR, section of rubber O-ring. (From Russo, 1977. © 1977 The Williams & Wilkins Co., Baltimore.)

After the addition of RHT, the pH drops and stabilizes within the range of 6.7–6.75. The tissue blocks immersed in drops of fixative are further dissected with a razor blade into prismatic-shaped slices, with a side length of 1 mm and a height of 2–3 mm, each of which contains growth cartilage. Primary fixation is continued for 2–3 hr at room temperature.

After being rinsed for 5 min in cacodylate buffer (100 mM, pH 7.4) having an osmolality of 330 mosmols adjusted with NaCl, the specimens are postfixed for 2–3 hr with 1% OsO_4 in 100 mM buffer containing 0.7% RHT. The specimens are rinsed three times in the buffer and stored overnight in the same buffer at 4°C. Following dehydration, the specimens are infiltrated with a 1 : 1 mixture of propylene oxide and Epon with 0.6% accelerator for 1 hr, and then with Epon with 0.6% accelerator for 2 days in a glass desiccator all at ambient temperature. Polymerization of fresh Epon with 1.2% accelerator is carried out for 5 days at 60°C. For further details, see Hunziker *et al.* (1982).

CELL COLONIES GROWN ON SOFT AGAR

The following method is useful for the removal and preservation of individual cell colonies grown on soft agar (Safa and Tseng, 1983). Cell colonies (e.g., human neuroblastoma cells) plated in soft agar (0.5% under layer and 0.3% over layer) are fixed by flooding the entire dish with a mixture of 1% glutaraldehyde and 3% formaldehyde in 0.1 M cacodylate buffer (pH 7.4) for 1 hr at 4°C. After being rinsed in three changes of buffer, colonies are identified with an inverted microscope. Individual colonies are picked up by puncturing the agar with a 23-cm-long disposable Pasteur pipette. Small pieces of agar each containing a single colony are slid individually into the pipette by capillary action. Each piece is transferred to a 1.5-ml plastic Micro-centrifuge tube filled with 0.1 M cacodylate buffer. After postfixation with 1% OsO_4, the colonies are dehydrated in ethanol. At this stage, the colonies become darkened, thus facilitating the tracking of the sample. Low-speed centrifugation (2000 g for 2 min) between each step of dehydration keeps the colonies in the tip of the conical tube. After infiltration and embedding in a resin in the tube, the tips containing individual colonies are sawed off and mounted on a blank block for thin sectioning.

CELLS ATTACHED TO GLASS SURFACES

The procedure described here is useful for processing cells that are able to grow attached to glass surfaces, such as glass coverslips (Spindler, 1978). A glass plate is laid into a culture dish, and glass coverslips (18 × 18 mm) are placed on this plate. The cells (e.g., foraminiferans) are placed on the coverslips. After some hours, the cells become tightly attached. The coverslips with the attached cells are transferred to a small jar containing the fixative. After fixation with glutaraldehyde, the cells on the coverslips are rinsed in buffer and postfixed with OsO_4. After another rinse in the buffer, the cells are decalcified in EDTA, dehydrated, and infiltrated with embedding resin.

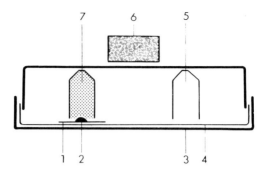

Fig. 7.3. Schematic diagram showing the method of embedding cells that are attached to a coverslip: 1, coverslip; 2, fixed cells; 3, Petri dish; 4, Parafilm; 5, empty BEEM capsule; 6, weight; 7, BEEM capsule filled with epoxy resin. (From Spindler, 1978.)

After remaining in pure resin for 12 hr at room temperature, each coverslip with attached cells is placed into its own Petri dish (10 cm in diameter), whose inner surface has been covered by Parafilm to prevent contact between the resin and the bottom of the dish. A BEEM capsule filled with resin is inverted over the specimen (Fig. 7.3). Two more empty capsules are placed in the same Petri dish to support the cover of the dish. A small weight (300 g) is placed on top of the cover to press the filled BEEM capsule tightly against the coverslip. With this method, no resin flows out of the inverted capsule, even when the cell to be embedded lies directly at the edge of the coverslip and the capsule protrudes over the edge.

After polymerization in an oven at 60°C, the coverslip (attached to the Parafilm) is cut out of the Parafilm with a scalpel and the capsule removed. The resin block with the attached coverslip is dipped for a few minutes into liquid air to burst away the coverslip. The embedded cell lies parallel to the surface of the resin block and only a few micrometers beneath the glossy surface of the block face. The block with the embedded cells can be easily trimmed and sectioned parallel to the surface.

CELLS IN CULTURE

Method 1 (e.g., Chick Embryo Spinal Cord Cells)

The culture is fixed for 10 min at 4°C in 3% glutaraldehyde containing 3% dextran and 3% glucose adjusted to pH 7.2 with 1 N NaOH. After a rinse in 0.25 M sucrose, the culture is postfixed for 1 hr at 4°C in 1% OsO_4 in 0.1 M phosphate buffer (pH 7.6).

Method 2 (e.g., Rat Liver Parenchyma and Human Glia Cells)

Glutaraldehyde (2%) in 0.1 M cacodylate buffer (pH 7.2) containing 0.1 M sucrose is recommended. The vehicle osmolality is 300 mosmols, and the total osmolality is 510 mosmols.

Method 3 (e.g., Limpet Blood Cells)

Cells are fixed in 2.5% vacuum-distilled glutaraldehyde (2.5%) in 0.1 M phosphate buffer (pH 8.2) for 3 min, spun down at 300 g for 5 min, and resuspended in fresh fixative for 20 min at room temperature. After being rinsed in artificial seawater, the cells are postfixed in 1% OsO_4 in the same buffer for 15 min at room temperature. They are again rinsed in artificial seawater and then suspended in 0.25% uranyl acetate in 0.1 M acetate buffer (pH 6.3) for 15 min. After centrifugation, two washings in the seawater, and a further centrifugation, the pellet is warmed to 55°C. At this temperature it is suspended in 2% agar and then centrifuged for 7 min, after which it is solidified by cooling with ice. For further details, see Gillett et al. (1975).

CELL LAYERS

Cells are cultivated in Falcon tissue culture dishes (Fig. 7.4) (Kuhn, 1981). Cell layers are fixed in situ with 2.5% glutaraldehyde in 0.1 M cacodylate buffer (pH 7.4) for 1 hr at room temperature. The fixation is started at room temperature immediately after the culture medium is withdrawn. A few minutes later, the fixation is continued at 4°C. After being rinsed overnight with 7% sucrose in the same buffer at 4°C, the culture layer is stained with 0.5% toluidine blue in the same buffer for 30 min at room temperature, rinsed in the buffer, and postfixed

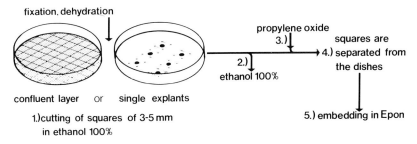

Fig. 7.4. Schematic representation of the method for separating squares of the cell layers from Falcon dishes. (From Kuhn, 1981.)

with 2% OsO_4 in the same buffer for 1 hr at 4°C. The toluidine blue staining facilitates the detection of the culture layer but does not alter the ultrastructure. Cells are treated with 0.1% uranyl acetate for 30 min, and then dehydrated in ethanol followed by propylene oxide.

Immediately before the change to propylene oxide, 3- to 5-mm squares are cut in the culture layer with a scalpel (Fig. 7.4). After the ethanol has been replaced with propylene oxide, the culture dishes are gently shaken. Within seconds the cutout culture squares are detached from the dish as the propylene oxide dissolves the plastic material. They are decanted into glass vessels, rinsed with propylene oxide, and embedded in Epon in flat molds. The culture squares are transferred from the glass vessels to the flat embedding molds with the aid of either wooden toothpicks, cut into small wedge-shaped spatulas, or glass beads (1.5–2 mm in diameter) produced on the extended end of a glass rod (2–3 mm in diameter).

CELL MONOLAYERS

Method 1

A 2.5 × 7.5-cm microscope slide is cleaned in a 1 : 5 mixture of nitric acid and sulfuric acid. A rather heavy coat of carbon is evaporated onto the slide in a conventional vacuum evaporator. The carbon film is stabilized by placing the slide in an oven at 180°C for 24 hr. A steel or glass ring, 5 mm deep, is sealed into the slide to form a shallow well. Cell monolayers are grown in the well. Fixation is carried out in 5.5% glutaraldehyde, prepared in Tyrode's solution (pH 7.3), for 5 min at room temperature. Postfixation is accomplished in 1% OsO_4, prepared in one-half isotonic Tyrodes's solution (pH 7.3), for 30 min at room temperature. Predried gelatin capsules are filled with a resin and inverted onto the monolayer. After polymerization, the capsule is pulled off the slide.

Method 2

Cells are grown on plastic coverslips (made of polymethylpentene), briefly rinsed in Ca^{2+}- and Mg^{2+}-free phosphate-buffered saline (pH 7.4), and fixed immediately with 2.5% glutaraldehyde in 0.1 M cacodylate buffer (pH 7.4) containing 0.12 M sucrose for 30 min at room temperature. After being rinsed for 10 min in the same buffer, the cells are postfixed with 1% OsO_4 in 0.04 M cacodylate buffer containing 0.14 M sucrose for 1 hr at 4°C. Following dehydration with acetone and infiltration with Epon, pieces of the coverslip are placed in the caps of BEEM capsules for polymerization. Coverslips can be easily separated from the polymerized resin by immersion in liquid nitrogen.

Polymethylpentene substrate is superior to glass because it does not require prior coatings (carbon) to assure resin separation. If desired, after the dehydration step, a portion of the same coverslip can be critical-point dried for scanning electron microscopy.

Method 3

Typically, 5×10^6 trypsinized cells are suspended in 50 ml of a suitable growth medium containing 20 mM HEPES buffer (pH 7.3 at 37°C) (Sargent *et al.*, 1981). The suspension is placed in a 50-ml siliconized Bellco spinner flask together with 0.5 g polystyrene Biosilon beads (160–300 μm in diameter). The flask is left unstirred overnight at 37°C to allow attachment of cells to beads. The flask is then stirred slowly over a period of days, the medium being changed as appropriate, and samples are taken periodically for light microscopy determination of the degree of cell growth.

The samples of beads are washed in three changes of phosphate-buffered saline (PBS) and fixed for 1 hr with 2.5% glutaraldehyde in PBS at room temperature. The samples are washed in three changes of PBS and heated in a water bath to 45°C. A 2% agar in PBS solution is prepared and maintained at 45°C. A few drops of 1% eosin are added to make the agar visible during processing. About equal volumes of agar solution and bead suspension are mixed, placed in a prewarmed 0.5-ml conical centrifuge tube, and centrifuged in a Micro-centrifuge as soon as possible. The agar block formed is gently separated from the tube by using a Pasteur pipette and flooding with 0.2 M cacodylate buffer (pH 7.2). The intact agar block is washed in two changes of cacodylate buffer and postfixed with 1% OsO_4 in buffer for 1 hr. The block is washed in buffer and then stained with 1% uranyl acetate for 45 min. The block is dehydrated with graded ethanol, treated with three 10-min changes of 1,2-epoxypropane to dissolve the beads, and embedded in Spurr resin.

Method 4 (Pig Aorta Endothelial Cells) (Beesley, 1978)

Lids of BEEM embedding capsules are sterilized in boiling water for 15 min. Endothelial cells are grown to confluence in these lids. Several lids are incubated together in closed Petri dishes stacked inside tightly sealed plastic boxes (Fig. 7.5). Adequate gassing of the cultures is assured by lifting the covers of the Petri dishes daily and gassing over the surface of the culture medium. Sufficient gas is allowed to escape inside the plastic box to replace the atmosphere with fresh gas. The lids of capsules are sufficiently translucent to allow the outlines of the growing cells to be viewed with an inverted microscope.

Lids containing the cell sheet (5–10 μm thick) are rinsed in phosphate-buff-

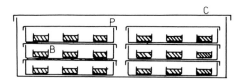

Fig. 7.5. Diagram showing the BEEM capsule lids (B) containing growing cells inside Petri dishes (P) which are stacked within larger plastic containers (C). (From Beesley, 1978. © 1978 The Williams & Wilkins Co., Baltimore.)

ered saline to remove serum. The monolayer is then fixed for 15 min with 2% glutaraldehyde in hypertonic sucrose–divalent cation salt solution at pH 7.2 (HEPES-buffered solutions of the chlorides of calcium 0.1 M and magnesium 0.02 M, containing 2–10% sucrose; the final osmolality is 500 mosmols). After a rinse in the buffer, the monolayer is postfixed for 15 min with 1% OsO_4 in 0.05 M cacodylate buffer (pH 7.2) containing 0.05 M NaCl and 0.02 M $CaCl_2$. The cells are rinsed with the buffer and then stained with 2% tannic acid in buffer for 15 min. After dehydration, the cells are left overnight in a 1 : 1 mixture of propylene oxide and Araldite or Epon. A BEEM capsule is filled with fresh Araldite and inverted onto the cells. After 18 hr at 60°C, the resin has been polymerized and the lid is easily peeled away, leaving the darkly stained cells embedded at the face of the Araldite block. Embedded cells can be viewed with an optical microscope so that selected areas may then be chosen for ultrathin sectioning.

Method 5 (Eppig *et al.*, 1976)

BEEM embedding capsules with lids open and facing up are placed in a vacuum evaporator and coated with a clearly visible carbon film. Coated capsules are sterilized with ethylene oxide in wrapped packages in a gas autoclave, and the lids are then closed inside a sterile transfer room. Cell suspensions are inoculated in 0.5 ml quantities into the capsules, which are sealed and inverted, so that the cells will settle on the circular lid surface. The inverted capsules are maintained sealed in 1-oz screwcapped jars that are gassed with 90% air and 10% CO_2 (Fig. 7.6A). Media changes are performed at 48-hr intervals over a 6-day period.

The tips of the capsules are cut off with a razor blade (Fig. 7.6B). The cells are rinsed by displacing the medium with PBS and then fixed with ice-cold 2.5% glutaraldehyde in 0.1 M sodium cacodylate buffer (pH 7.4) for 1 hr. After dehydration, the cells are infiltrated with a 1 : 1 mixture of propylene oxide and Epon for 3 hr. The capsules are inverted on a paper towel to drain out the infiltration mixture thoroughly, and fresh Epon is added and subsequently polymerized at 60°C. The capsules are filled completely if sections need to be cut parallel to the plane of the monolayer. If sections perpendicular to the plane of the monolayer are desired, the lid alone is filled with Epon. The capsules are carefully removed from the paper towel with a razor blade, and the blocks are trimmed to the region selected for sectioning. For perpendicular sectioning, the Epon disks are held with the Vise-grip accessory.

Method 6 (Human Brain Tumor Cells) (Kawamoto *et al.*, 1980)

Human brain tumor cells grown as monolayers in plastic T25 flasks are dispersed with 0.25% trypsin. About 400 cells are inoculated into four wells of a plastic Micro Test II tissue culture plate (Falcon 3040). Each well contains 0.3 ml of Eagle's minimal essential medium supplemented with 10% fetal bovine serum and an antibiotic–antimycotic mixture. Two adjacent rows of wells are left empty when different types of cells are grown in the same Micro Test plate. The plates are covered with a plastic lid (Falcon 1041). Incubation is carried out for 4 days at 37°C in a humidified atmosphere of 5% CO_2 in air (slow-growing cells may need longer times).

The cell monolayer is washed with Earle's balanced salt solution, the wells are filled with ice-cold 2.5% glutaraldehyde in 0.067 M phosphate buffer (pH 7.4), and the monolayer is fixed for 1 hr at room temperature. The cells are rinsed with the buffer and postfixed with 1% OsO_4 in chromate buffer (pH 7.4) for 30 min at room temperature. After dehydration with ethanol and embedding with Epon, the

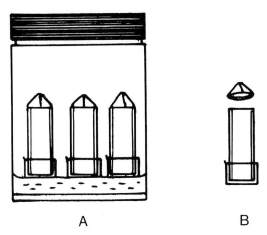

A   B

Fig. 7.6.   Diagram of inverted BEEM capsules used for cell culture. The capsules (20 mm in height) are shown (A) inside a gassed culture jar and (B) with tip removed for ease in performing fixation, dehydration, and epoxy-embedding. (From Eppig *et al.*, 1976.)

wells containing the embedded cells are cut out with an electric saw and the empty wells are discarded.

To separate the Epon blocks from the plastic growth surface, the wells are immersed in xylene for 14 hr at room temperature. The xylene containing the dissolved plastic is poured off and the Epon blocks are treated with fresh xylene for 30 min under continuous agitation with a magnetic stirring bar. After the xylene has been decanted, the slightly softened blocks are air-dried and then placed in the oven at 60°C for 2 hr. Xylene facilitates easy separation of cells embedded *in situ* from their plastic growth surface.

Method 7 (Shepard *et al.*, 1981)

The following method utilizes ruthenium red (RR) and *p*-phenylenediamine during processing to obtain intensely darkened cells. This treatment facilitates orientation and localization of monolayers within flat embedding molds when trimming and thin sectioning. Cells are grown in Lux Permanox Petri dishes. The tissue culture medium is rinsed away and replaced with 2% glutaraldehyde in 0.15 *M* cacodylate buffer (pH 7.2) containing 0.2% RR for 30 min at room temperature. After being rinsed in 0.15 *M* cacodylate buffer containing 0.2 *M* sucrose and 0.1% RR for 15 min (three changes), the specimens are postfixed

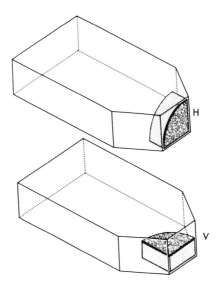

Fig. 7.7.   Diagram showing reembedding of monolayer fragments for horizontal (H) and vertical (V) sectioning, Fibroblasts of the monolayer are indicated by small circles. (From Shepard *et al.*, 1981. © 1981 The Williams & Wilkins Co., Baltimore.)

with 2% OsO_4 in 0.15 M cacodylate buffer containing 0.05% RR for 30 min at room temperature. The specimens are then rinsed in three changes of distilled water for a total of 5 min.

The specimens are dehydrated, starting with 30% ethanol according to standard procedures except that 70% ethanol contains 1% p-phenylenediamine. The latter is freshly prepared and filtered. For embedding, Spurr resin is added to a depth of 2–3 mm in the Petri dishes. Each dish is covered with a lid, and resin is polymerized at 60°C. If desired, monolayers can be studied and photographed before removal from the Petri dishes.

Selected parts of monolayers are circled with the tip of an 18-gauge needle under a dissecting microscope. Resin disks are removed with a $\frac{3}{16}$-in. bone biopsy trephine (Codman Instruments, Cat. No. 69-2500). The $\frac{3}{16}$-in. disks are quartered and reembedded in flat embedding molds. For horizontal sectioning of the monolayer, disks are pressed flat against the rectangular tip of the mold (Fig. 7.7H); elevation of the opposite end of the mold during polymerization prevents the monolayer from floating away from that position. Disks intended for vertical sectioning of the monolayer are placed at the tip of the mold with the cells facing up (Fig. 7.7V). Because of the 2- to 3-mm thickness of the resin, the monolayers will be centered in the tip of the mold. Vertical sectioning is generally preferred.

Method 8 (Neuronal Cultures) (Spoerri *et al.*, 1980)

The following *in situ* embedding technique is useful for studying individual cells. A plastic flask with the embedded cells can be stripped off easily without affecting the monolayer culture. Cells are grown in plastic flasks (e.g., Corning) in Eagle's minimum essential medium supplemented with 10% fetal calf serum. The cultures are fixed in situ for 15 min at 36°C by draining the growth medium and adding 6% glutaraldehyde in 0.2 M phosphate buffer (pH 7.2). Postfixation is then carried out with 2% OsO_4 in the same buffer for 30 min at room temperature.

After dehydration with ethanol (without propylene oxide), Epon mixture is poured into the flask to a depth of 8 mm (Fig. 7.8A). Epon mixture consists of 3 parts of Epon A and 2 parts of Epon B (Epon A: 62 ml Epon 812 + 100 ml DDSA; Epon B: 100 ml Epon 812 + 89 ml NMA + 1.5% DMP-30). The flask is transferred to a 37°C incubator overnight and is subsequently incubated at 60°C for final polymerization for 36 hr. The chosen area of growth is marked by scoring a circle around it (Fig. 7.8B), and this area of plastic (along with the cells) is cut away with a fine saw (Fig. 7.8C). This plastic piece is mounted on a Vise-type holder (Fig. 7.8D); the plastic now lies on top of the Epon block (Fig. 7.8E). The remnant of the plastic container is easily stripped off (Fig. 7.8F), and the Epon block with a relatively large surface is trimmed on the sides (Fig. 7.8G) and sectioned.

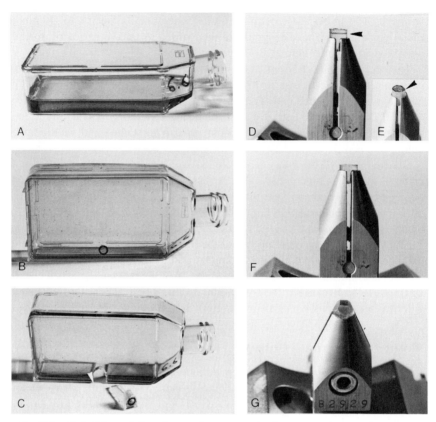

Fig. 7.8. Embedding technique for monolayer neuronal cultures grown in plastic flasks. (From Spoerri *et al.*, 1980.)

CELL SUSPENSIONS (Mereau *et al.*, 1982)

The following method is useful for processing many samples in a short time. One millimeter of culture is fixed with 1 ml of 1% glutaraldehyde in 0.1 *M* phosphate buffer (pH 7.4) for 20 min at room temperature. A filter disk (F in Fig. 7.9B) cut from a Millipore filter type RA (1.2 μm pore size) is damped with buffer, adjusted on the top of a glass tube (p in Fig. 7.9A, external and internal diameters 6 mm and 1 mm, respectively), and kept in place by vacuum. The cells are drawn up into a Pasteur pipette (Pp in Fig. 7.9A) with a peristaltic pump (pp in Fig. 7.9A). The pipette tip is placed 1 mm above the filter disk and facing the hole of the thick glass tube (Fig. 7.9B). By regulating the flow of the mixture with the peristaltic pump according to the suction, one can pack the cells into a pellet (pe in Fig. 7.9B).

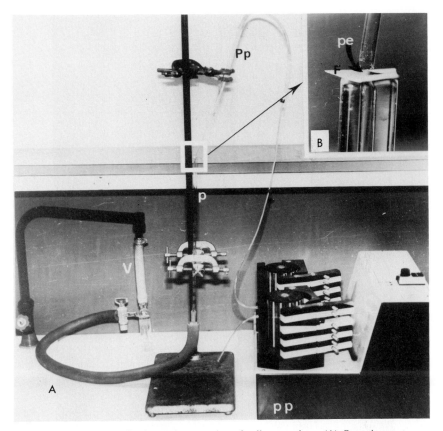

Fig. 7.9. Unit for collection and preparation of cell suspensions. (A) Operating system: p, pipette (0.2 ml, 1/100); pp, peristaltic pump; Pp, Pasteur pipette; V, vacuum. (B) Pellet after filtration: F, filter disk; pe, pellet. (From Mereau *et al.*, 1982.)

The filter disk with its adhering pellet is immediately immersed in and rinsed in three changes of the buffer. The disk is then immersed in 1% OsO_4 in 0.1 M phosphate buffer (pH 7.4) for 1 hr. *En bloc* staining with 1% uranyl acetate is carried out in 70% ethanol. During the second propylene oxide step, the pellet is separated from the filter disk by gently bending the latter.

CEREBROSPINAL FLUID CELLS (Lutke-Schipholt and Stadhouders, 1981)

This method allows the concentration of a small number of cells from free-floating cell suspension. A special conical "filtration-through-centrifugation"

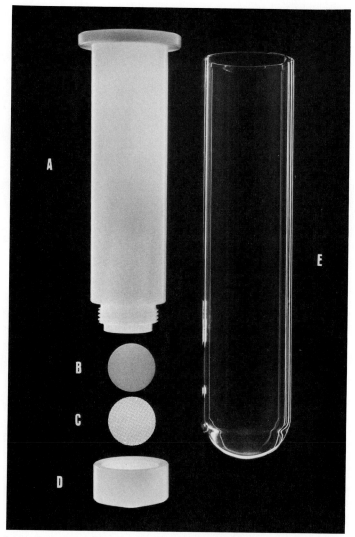

Fig. 7.10. Filtration unit used for concentration and preparation of small number of cells from free floating cell suspensions such as cerebrospinal fluid cells. (A) Conical Delrin tube. (B) Filter. (C) Support screen. (D) Tube cap. (E) Centrifugation tube. (From Lutke-Schipholt and Stadhouders, 1981. Reprinted by permission of Hemisphere Publishing Corporation.)

tube made of Delrin is used (Polaron Equip. Ltd.) (Fig. 7.10A). The bottom cap can be unscrewed to insert a Millipore filter together with a stainless steel filter support screen (Fig. 7.10B–D). The inner surface of the tube must be machined perfectly flat to prevent loss of cells. Moreover, the bottom surface of the tube and the inner face of the drilled cap must be very flat, so that the filter can be clamped without risk of leakage.

The filtration unit is placed in a standard centrifugation tube, which has a diameter small enough to make the upper rim of the filtration unit rest on the upper edge of the centrifugation tube (Fig. 7.10E). Typically, a glass tube with an outer diameter of 25 mm, a height of 95 mm, and a volume of 25 ml is satisfactory. Sufficient space should remain in the recipient tube to receive the filtered fluid.

The cerebrospinal fluid is obtained by lumbar puncture after careful cleaning of the skin. The fluid is collected in a 10-ml disposable polypropylene syringe and fixed immediately by mixing it with an equal volume of sterile 4% glutaraldehyde in 0.2 M cacodylate buffer (pH 7.3) for 6 hr at room temperature. To sterilize the glutaraldehyde, it is passed through a Millipore Millex filtration unit (diameter 25 mm, pore size 0.22 μm) directly into the syringe containing the fluid. After fixation the mixture is filtered by centrifugation force of 1500 g for 10 min, using the filtration unit with a filter having a pore size of 0.8 μm (MF-Millipore).

After filtration the tube cap is unscrewed and the filter is taken out and placed on a microscope slide. A drop of 2% agar solution that gels at low temperature (40°C) is placed on the central area of the filter, where the sedimented cells are present. The agar drop is sterilized by pressing it through a Millipore filter (pore size 0.22 μm). To gel the agar, the slide is placed in a refrigerator for 10 min. The center of the filter is cut out, refixed by placing it in 2% glutaraldehyde in 0.1 M cacodylate buffer (pH 7.3) for 2 hr, and rinsed in buffer. Postfixation is carried out with 2% OsO_4 in 0.1 M buffer for 1 hr. The filter–agar sandwich is stained *en bloc* with a 0.5% solution of uranyl acetate in buffer (pH 7.3) for 2 hr.

In the last step of dehydration in propylene oxide, the Millipore filter dissolves, leaving the cells embedded in the agar. After the agar block has been infiltrated with an embedding resin, the block is placed under slight pressure between two mica sheets and polymerized in an oven at 60°C for 48 hr. The mica sheets are peeled off, and the block is attached (cell side up) to a flat surface of a blank resin support block. For quantitative classification of cell types, sections should be cut in a plane parallel to the plane of the cell layer.

CHROMOSOMES

Method 1 (Polytene Chromosome Squashes)

The following method is useful for preparing polytene chromosome squashes (Wu and Waddell, 1982). This method provides close correlation between the light microscopic image and the TEM image. Slides are prepared by dipping a thoroughly cleaned microscope slide in 0.5 mg/ml bovine serum albumin in glass distilled water. Each slide is air-dried and used immediately. Salivary glands are dissected out from late third-instar larvae of *Drosophila melanogaster* in 45% acetic acid. Chromosome squashes are flattened by a combination of tapping with the eraser end of a heavy pencil and using adequate finger pressure. Only well-flattened chromosome squashes should be used.

Each slide with the chromosomes is frozen in liquid nitrogen, and the siliconized coverslip is flipped off with a razor blade. The slide is immediately dehydrated by soaking it in 100% methanol for 30 min and air-dried. Then the slide is soaked in Tris-buffered 2 XSSC (0.01 M Tris pH 7.0, 0.3 M NaCl, and 0.03 M Na-citrate) for 10 min, rinsed briefly with distilled water, and air dried. The slide is coated with 0.7% Parlodion in amyl acetate, air-dried, and examined with phase-contrast microscopy to select suitable areas with good chromosome spread. A suitable area is cut with a razor blade and a 100-mesh copper grid is placed over the selected chromosome squash under a phase-contrast microscope. The surface Parlodion film with the grid is *carefully* floated off on the surface of 0.3 M hydrofluoric acid. The grid is picked up on a small drum made of Saran Wrap over a small beaker, *gently* rinsed with distilled water, reexamined with phase-contrast microscopy to make sure that the chromosome squash has been properly transferred, and then viewed with the TEM. In this technique, conformational changes in nucleoproteins are inevitable because of the use of strong denaturing reagents.

Method 2 (Chromosomes from Human Lymphocytes) (Vincent et al., 1975)

The following procedure involves the blaze-drying of chromosomes from acid alcohol-fixed human lymphocytes. This procedure allows electron microscopy of specimens that have been fixed for routine clinical study. Human lymphocytes are obtained from 10-ml samples of heparinized whole blood. Buffy coats are prepared by centrifuging each sample at 300 g for 12 min at 24°C. Two 20-ml cultures are established from each buffy coat. Cells are cultured in McCoy's 5A medium (North American Biologicals, Inc.) supplemented with 20% fetal bovine serum (Microbiological Associates), 1% antibiotic/antimycotic 100 X mixture, and 1.5% reconstituted phytohemagglutinin M (Difco Lab.). Cultures are incubated for 70 hr at 37°C in 9% CO_2 atmosphere. Mitotic cells are arrested for 2 hr with 0.1 μg/ml of Colcemid (Ciba Pharmaceuticals) and then collected by centrifugation at 100 g for 10 min. The cell pellet is suspended in 0.075 M KCl for 10 min at 24°C, after which the cells are collected by centrifugation, fixed, and stored in a mixture (3 : 1) of methanol and acetic acid at −20°C.

Electron Microscope Grids

Formvar film is prepared by dipping slides into a 0.25% ethylene dichloride solution of Formvar 15/95 E (Shawinigan Resins Corp.). The dried film is removed and placed on a distilled-water surface, and grids are on the floating film. The grids and film are pressed down into the water and then out of the water with a microscope slide so as to position the grids between the film and the slide. The films are dried overnight and lightly carbonized in a vacuum evaporator.

Supporting Films for Metaphase Spreads

A 0.3% solution of Parlodion is prepared in a 1 : 1 mixture of ethanol and ethyl ether. The solution is applied to a freshly cleaved 5.1 × 3.7-cm mica sheet (Ladd Res. Ind.), and allowed to dry for 10 min. The film is heavily carbonized. Supporting films are prepared immediately before blaze-drying.

Blaze-Drying Technique

The bottom surface of a mica sheet containing a Parlodion film is wetted with several drops of hexylene glycol. The wetted sheet is placed on a block of dry ice, and the surface of the film is frosted by breathing once or twice across it. A drop of cell suspension in 3 : 1 methanol/acetic acid is placed on the frosted film, and the mica sheet is quickly brought close to a gas burner flame to ignite the fixative. After the flaming, the film is dried with cool air from a hair dryer and excess glycol is wiped from the mica sheet.

Transferring Metaphases to Grids

The metaphase spreads and their supporting film are separated from the mica by dipping the sheet into a bowl of distilled water at a 45° angle. A slide of Formvar-coated grids is oriented directly below the floating film and then slowly raised out of the water so that the film containing the metaphase spreads covers the grids. The slides are dried overnight in a vertical position.

Staining

For nonbanded chromosomes, slides of grids are stained with Giemsa for 5–15 min, rinsed in distilled water, and air-dried. The stain consists of 10% methanol, 70% distilled water, and 20% Giemsa.

Banding

Slides of grids are exposed to a 0.025% trypsin solution without antibiotics for 14 min at 24°C, rinsed in two changes each of 70% and 100% ethanol, air-dried, and stained for 2 min at 24°C in freshly prepared Giemsa stain. This stain consists of 1.5 ml of 0.1 M citric acid, 5 ml of Giemsa, and 50 ml of distilled water. The solution is brought to pH 7.0 with 0.1 M Na_2HPO_4. After the staining, the slides are rinsed in distilled water and air dried. Grids selected initially with the light microscope are examined with the electron microscope.

CILIATES

Method 1

Solution A

Phosphate buffer (0.15 M, pH 7.0)	4.8 ml
$MgSO_4$ (1 mM)	0.1 ml
Sucrose (0.1 M)	0.1 ml
OsO_4 (2%)	5.0 ml

Solution B: The ingredients of solution B are the same as those for solution A except that 25% glutaraldehyde replaces OsO_4.

Ciliates are fixed in solution A for 1.5 to 3 min at room temperature. An equal volume of solution B is added to the suspension and fixation is allowed to continue for 20–30 min. For embedding, Spurr medium is recommended.

Method 2

Phosphate buffer (0.15 M, pH 7.0)	7.2 ml
Sucrose (1 mM)	0.2 ml
Glutaraldehyde (2%)	0.2 ml

This solution is mixed with 10 ml of 2% OsO_4 immediately before use. After the culture medium is pipetted off, ciliates are fixed for 20–30 min at room temperature. For additional details, see Shigenaka *et al.* (1973).

CREAM CHEESE

In this method, specimens are taken from 1–2 cm under the surface of the cream cheese blocks (Kaláb, 1981). A small portion of the cheese (1 mm^3) is picked up on the tip of a stainless-steel needle and briefly dipped in a warm (35°C) 3% agar solution. After the agar covering the specimen has gelled within several seconds, the needle is withdrawn and the hole left in the specimen is filled with a droplet of agar. The specimen thus encased in agar is fixed with 1.4% glutaraldehyde in 0.1 M cacodylate buffer (pH 7.2) for 2 hr at 4°C. After being rinsed in the buffer, the specimen is postfixed with 2% OsO_4 in 0.05 M Veronal acetate buffer (pH 6.75) for 1 hr at 4°C.

CYSTS

Pelleted organisms are fixed with 5% glutaraldehyde in 0.5 M cacodylate buffer (pH 6.8) containing 1 mM $CaCl_2$ for 6 hr at room temperature. The specimens are centrifuged at 400 g for 10 min in 12-ml conical centrifuge tubes. After the supernatant is decanted, the pellet is embedded in agar. This embedment is accomplished by layering 2 ml of 2% agar at 45°C over the pellet, which is then lifted with a dental explorer into the molten agar. The centrifuge tube is plunged into an ice bath for 30 sec to solidify the agar.

The agar block containing the specimens is cut into small pieces and rinsed for 16 hr in 0.05 M cacodylate buffer (pH 6.8) containing 7.5% sucrose. Postfixation is carried out with 2% OsO_4 in 0.05 M cacodylate buffer containing sucrose for 6 hr at 4°C. After a rinse in distilled water for 2 hr, the specimens are treated with 1% uranyl acetate for 12–16 hr at 4°C. For embedding, Spurr mixture is recommended. Chiovetti (1978) has given an alternate method for embedding.

DIATOMS

Method 1

Cells are fixed with 3% glutaraldehyde in 0.1 M cacodylate buffer (pH 7.2) for 90 min at room temperature. After being rinsed in the buffer, they are postfixed with 2% OsO_4 in the same buffer for 12 hr at 4°C.

Method 2

The following procedure is useful for preserving extracellular polysaccharides. The cell pellet is transferred to a mixture of 1% glutaraldehyde and 1% alcian blue in culture medium adjusted to pH 6.5. After 5 sec, 1% OsO_4 in 0.1 M cacodylate buffer (pH 6.6) is added to yield a final OsO_4 concentration of 0.2%. The fixation is continued for 20 min and is followed by a rinse in distilled water and repeated centrifugation until the blue color no longer appears in the supernatant. Ultrathin sections are stained with the periodic acid–thiocarbohydrazide– silver proteinate technique (see p. 207).

ECHINODERMS (Dietrich and Fontaine, 1975)

Method 1

Calcified tissues are fixed for 2 hr at room temperature with 2.5% glutaraldehyde in 0.2 M Millonig's phosphate buffer (pH 7.6) containing 0.14 M NaCl

(960 mosmols). After a rinse in the same buffer, the specimens are postfixed with 2% OsO_4 in the same buffer without NaCl for 2 hr at room temperature. The specimens are rinsed three times (5 min each) with 0.3 M NaCl and then decalcified in a 1 : 1 mixture (pH 2.6) of 2% ascorbic acid and 0.3 M NaCl for 12–24 hr while being continuously stirred (e.g., angle-head rotor set at 3 rpm). A large volume of freshly prepared decalcifier (50 ml) should be used. Specimens are rinsed in several changes of glass-distilled water, dehydrated, and embedded as usual.

Method 2

Specimens are fixed with 3% glutaraldehyde in 0.1 M phosphate buffer (pH 7.8) for 3–4 hr at 4°C and then postfixed with 2% OsO_4 in the same buffer. The glutaraldehyde fixative should have an osmolality of 1000 mosmols (after the addition of sucrose).

EGGS AND ZYGOTES

For the fixative solution, 2.5 g of bovine serum albumin is dissolved in 30 ml of seawater. Then 2 g of paraformaldehyde powder is dissolved in 20 ml of seawater and the two solutions are mixed. Next, 7 g sucrose and 2 ml of acrolein are added to 30 ml of 0.2 M collidine buffer (pH 7.4). This solution is mixed with the previous mixture, and then 6 ml of 50% glutaraldehyde is slowly added. The mixture is brought to 100 ml with 0.2 M collidine buffer.

After insemination of oocytes, fertilized eggs are collected at intervals of a few minutes. The specimens are fixed in the preceding solution for 1 hr at room temperature. After being rinsed in seawater for 2 hr at room temperature, the specimens are postfixed with 1% OsO_4 in seawater for 90 min at 4°C.

EFFUSION FLUID

Aliquots of 4–50 ml of the effusion fluid are centrifuged at 1000 rpm for 5 min. The pellet is suspended in an appropriate volume of 8% bovine serum albumin (BSA) prepared in 0.05 M Tris-HCl buffer (pH 7.2). The suspension is centrifuged at 1000 rpm for 5 min, and about 2 drops of 25% glutaraldehyde are mixed to form a gel. The gel is extracted and cut into 1-mm³ cubes, and specimens are fixed with 2.5% glutaraldehyde in 0.1 M cacodylate buffer (pH 7.3) for 2 hr at room temperature. After a rinse in buffer containing 7.5% sucrose, the specimens are postfixed with 2% OsO_4 in the same buffer for 1 hr at

4°C. Following a rinse in the buffer containing sucrose, the specimens are dehydrated and embedded.

EMBRYOS (e.g., Chick)

Fixative solution

Glutaraldehyde (50%)	5 ml
Formaldehyde (8%)	25 ml
Cacodylate buffer (0.2 M, pH 7.6)	50 ml
$CaCl_2$ (anhydrous)	25 mg

Embryos *in situ* are fixed by pouring this solution (chilled) over them after removing the amnion. After 5 min, the embryo or a part of it is transferred to fresh solution for 30 min at 4°C. Following a brief rinse in cold buffer, the specimen is postfixed with 1% OsO_4 in the same buffer for 2 hr at 4°C.

EPIDERMAL HAIR OF SEEDS

Small pieces of a seed coat with attached hair are cut from a partially rehydrated seed and fixed with 4% glutaraldehyde in 0.1 M cacodylate buffer (pH 6.8) for 4 hr at room temperature. After a rinse in buffer, the specimens are postfixed with 1% OsO_4 in the same buffer.

EPITHELIUM (e.g., Cheek Pouch)

Segments of cheek pouch (hamster) are fixed with a mixture of 2.5% glutaraldehyde and 2% formaldehyde in 0.067 M cacodylate buffer (pH 7.4) (total osmolality is 1015 mosmols) for 30 min at room temperature. Tissue segments are then cut into 1-mm^3 blocks and fixed further for 30 min in the same fixative. After being rinsed in 0.1 M cacodylate buffer (pH 7.4, 180 mosmols), the specimens are postfixed with 2% OsO_4 in 0.1 M cacodylate buffer (total osmolality is 227 mosmols) for 90 min at 4°C.

ERYTHROCYTES (e.g., Avian)

Glutaraldehyde (2%) in 0.1 M cacodylate buffer (pH 6.85) containing sucrose is recommended. The vehicle osmolality is 320 mosmols. A slightly acidic pH enhances the stability of microtubules and microfilaments. The osmolality of avian blood plasma is 323 mosmols. The addition of 12% hexylene glycol to the buffer improves the preservation but is not essential. The addition of $MgCl_2$ helps preserve cell integrity, although it may cause fusion of the two nuclear membranes when used in a hypertonic buffer. For additional details, see Brown (1975).

EXAMINATION OF SPECIFIC PARTS
OF IDENTIFIED NEURONS (SOMATOSENSORY
CORTEX)

The following method is useful for ultrastructural examination of selected regions of a neuron that has been either Golgi-impregnated or filled with horse-radish peroxidase (HRP) (Kristt and Trythall, 1982). A well-impregnated neuron cell is selected in a thick section of cortical tissue (Fig. 7.11A). The cell is drawn with a camera lucida or photographed in a through-focus series and then drawn by tracing the composite print. A grid composed of 10-μm-square boxes is superimposed over the drawing (Fig. 7.11B). The tissue section containing the selected cell is flat-embedded in a resin, and then the resin block is trimmed as close to the cell as possible. A record is made of the distance from the edge of the block to the center of the soma, and this information is also indicated on the x axis of the grid (Fig. 7.11C). The depth of the cell from pis, or other point of reference, should also be noted along the y axis. Adjustments for changes in linear dimensions of the drawn cell, due to embedding, should be made on the grid.

The flat block containing the selected cell is rapidly glued edgewise (with cyanoacrylate) onto a polymerized resin blank or Lucite blank (Fig. 7.11D), that is, at 90° to the plane of drawing. Serial thick sections (1 μm) of the entire block containing the cell are cut. Each section is mounted on a glass slide and centered (Fig. 7.11E). A complete record of all tissue removed from the block must be kept because each thick section contains a known portion of the neuron, a portion determined by reference to the grid (Fig. 7.11B).

When certain regions of the cell are selected for study, each of the corresponding thick sections is mounted from its slide onto another resin block. This mounting is accomplished by placing a resin drop over the section and then inverting a resin blank over the section (Fig. 7.11G). The glass slide and the blank are placed overnight in an oven at 60°C. While the slide is still in the oven, the blank is gently and continuously tugged until it pops off with the adherent section (Fig. 7.11H). This block is trimmed very carefully and then thin-sectioned.

The thin sections are mounted onto a Formvar-coated specimen grid with a single hole (Fig. 7.11I). The coordinates for the neuronal elements that are to be

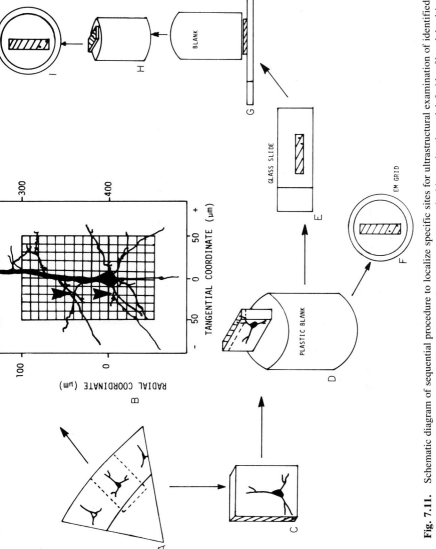

Fig. 7.11. Schematic diagram of sequential procedure to localize specific sites for ultrastructural examination of identified neurons. Steps A–I are identified in text. In (B), arrows mark two processes seen in thin section through left side of basal dendrite field at −14 μm. (From Kristt and Trythall, 1982.)

examined ultrastructurally on these thin sections are determined from the two-dimensional grid (Fig. 7.11B). Because extension of a process in the z axis will influence the accuracy of these two-dimensional coordinates, in some instances it may be desirable to estimate the z-axis dispersion of the processes. This estimate can be most conveniently done from the microscope's focus micrometer at the time the cell is photographed.

EXTRACELLULAR PROTOPLASTS IN SUSPENSION

The suspension is pipetted into small chambers of the "Lab-Tek" (Miles Laboratories, Inc.), which are partly filled with 1.5% agar in 0.1 M Na-cacody-late buffer (pH 7.0) containing 0.7 M NaCl (Fig. 7.12A) (Saikawa and Kobayashi, 1974). The specimens, which sediment on the agar plates in a few minutes, are fixed with the addition of a drop of 5% glutaraldehyde for 30 min and postfixed with OsO$_4$ for 1 hr (Fig. 7.12B). The solutions used for the fixation and rinsing are carefully pipetted out from the chamber by using fine glass capillary tubes (0.5 mm in diameter) to prevent loss of the specimen (Fig. 7.12C).

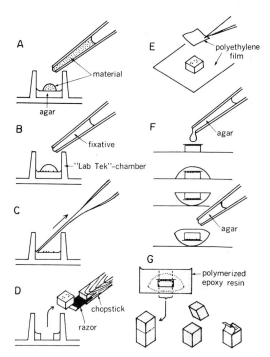

Fig. 7.12.   Schematic diagram of the preparation procedure for extracellular protoplast in suspension. (From Saikawa and Kobayashi, 1974.)

With the aid of a small spatula (Fig. 7.12D), the fixed specimens are trans-ferred as small blocks of agar (5 × 5 × 2 mm) to a glass surface and covered with thin polyethylene film (0.03 mm thick) (Fig. 7.12E); the film is slightly larger in size than the largest side of the agar block. One or two drops of agar are placed over the agar block and the film (Fig. 7.12F). The film-covered agar blocks are completely embedded in agar to avoid losing the specimen during subsequent treatments. The blocks are dehydrated and embedded in Epon in small polyethylene dishes (3 cm in diameter). The castings are cut with a fine saw along the ridges of the polyethylene film. Each polymerized block is broken in two parts at the border of the film (Fig. 7.12G). Ultrathin sections are cut after removing the film.

EYES

Fixation solution

Glutaraldehyde (25%)	3 ml
Formaldehyde (10%)	50 ml
NaCl	3 g
Sucrose	4.5 g
Phosphate buffer (0.1 M, pH 7.2) to make	100 ml

Specimens are fixed with this fixation solution for 1–2 hr at room temperature. After being rinsed briefly in 0.1 M buffer containing 8% sucrose, the specimens are postfixed with 1% OsO_4 in the same buffer without the sucrose for 1 hr at 4°C.

FETAL TISSUE (e.g., Pig Liver and Tooth Germs)

The recommended mixture contains 1.25% glutaraldehyde and 2% formalde-hyde in 0.1 M cacodylate buffer (pH 7.0) having an osmolality of 950 mosmols. Fixation is carried out for 90 min at 20°C. Postfixation is accomplished with 2% OsO_4 in the same buffer for 2 hr. For additional details, see Rømert and Mat-thiessen (1975).

FINE-NEEDLE ASPIRATION BIOPSY

Method 1 (Single Glomerulus) (Helin *et al.*, 1979)

Specimens of small size and of a fragmentary nature (e.g., single glomerulus) can be processed from fixation to polymerization in the same device. Specimens can be processed by either conventional or rapid methods for clinical diagnostics.

This device is constructed by connecting a BEEM capsule with a polyethylene adapter (Kodin Muovi, Helsinki, Finland) fitted to a syringe (Fig. 7.13a). A small piece of nylon sieve (100-μm-mesh opening) is attached to the capsule after the tip has been cut to make an open end of about 1 mm² (Fig. 7.13d).

Immediately after aspiration of the biopsy specimen, the aspiration syringe is filled with 4% glutaraldehyde in 0.1 M cacodylate buffer (pH 7.4) and fixed for 30 min at room temperature. The aspirate is transferred from the syringe to the processing device. The processing is carried out by allowing the solutions to drain through the sieve, which retains the specimens. The fixative is replaced with buffer, and rinsing is accomplished in 15 min. The buffer is replaced with 2% OsO₄ in buffer, and postfixation is accomplished in 30 min. After being rinsed in buffer, the specimens are stained *en bloc* with an aqueous solution of uranyl acetate for 5 min, a procedure followed by dehydration, infiltration, and embedding.

Method 2 (Akhtar *et al.*, 1980)

The aspirated sample is flushed out into a microtube (Beckman RIIC) containing 3% glutaraldehyde in 0.1 M phosphate buffer (pH 7.3) and fixed for 1 hr. The mixture is centrifuged at 1500 rpm for 10 min, and the pellet is rinsed with buffer for 10 min. The pellet is suspended in 1% OsO₄ in the same buffer, postfixed for 1 hr, and centrifuged at 1500 rpm for 5 min. After being rinsed in the buffer for 10 min, the pellet is stained *en bloc* with saturated aqueous solution of uranyl acetate for 1 hr. This procedure is followed by dehydration and embedding in resin in the microtube.

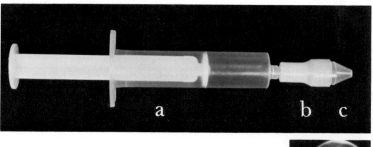

Fig. 7.13. Processing device, disposable plastic syringe: a, polyethylene adapter; b, BEEM capsule; c, with open tip; d, closed with a nylon sieve of 100-μm mesh opening. (From Helin *et al.*, 1979.)

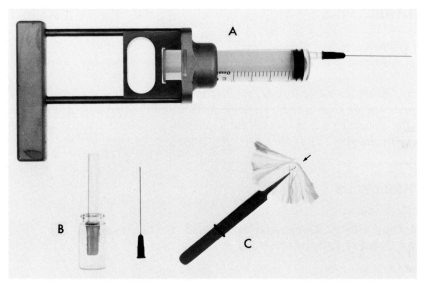

Fig. 7.14. (A) Instrument used for the aspirations. (B) The aspirated material is ejected into a glass tube sealed at one end with lens paper; the tube is placed in a beaker in which the material is processed till the embedding in Epon. (C) When the lens paper is opened up, the osmium-stained tumor tissue threads are easy to see (arrow) and handle for the final embedding in Epon. (From Kindblom, 1983.)

Method 3 (Soft Tissues and Bone Tumors) (Kindblom, 1983)

The following method is useful for light and electron microscopy of fine-needle aspiration biopsy samples in the preoperative diagnosis of soft tissues and bone tumors. The needles used for biopsy are 30–50 mm long and have an external diameter of 0.6–0.7 mm. During the aspiration, the needle is passed back and forth a few times for a length of about 10 mm. The pressure of the 20-ml syringes (Fig. 7.14A) is equalized before the needle is retracted from the tissue. The needle is detached from the syringe and a few milliliters of 2.5% glutaraldehyde in 0.1 M cacodylate buffer (pH 7.2) are aspirated into the syringe. The needle with the glutaraldehyde is immediately ejected with care into a glass tube sealed at one end with a double-layered ashless lens paper and unwaxed dental floss; then the tube is placed in a beaker containing glutaraldehyde for 30–60 min at room temperature (Fig. 7.14B). After being rinsed in buffer for 10 min, the specimen is postfixed with 1% OsO_4 in buffer for 30–60 min. This procedure is followed by dehydration and infiltration. Before embedding, when the double-layered ashless lens paper is opened, the osmium-stained tissue is seen adhered in fragments of up to 5 mm in length (Fig. 7.14C). These fragments are easy to handle for embedding.

FLAGELLATES

Fixation solution

Glutaraldehyde (25%)	20 ml
Acrolein	2 ml
Cacodylate buffer (0.1 *M*, pH 7.4) to make	100 ml

Specimens are fixed with this solution for 2 hr at 4°C. After a rinse in the same buffer containing 0.2 *M* sucrose, the specimens are postfixed with 2% OsO_4 in 0.1 *M* buffer for 2 hr at 4°C.

FREE CELLS

Method 1 (e.g., Cerebrospinal Fluid, Urine, Ascites, and Pleural Effusion) (Ito and Inaba, 1977)

The fluid is centrifuged at 2000 rpm for 20 min. All but the last drop of the supernatant is removed with a pipette (care being taken not to disturb the sedimented cells). About 3–5 ml of serum of the same patient is added, and the test tube is gently shaken to resuspend the cells. The suspension is centrifuged at 3000 rpm for 20–30 min. A pipette is used to discard the supernatant, and the test tube is placed at an angle in a beaker containing ice. With the use of a pipette, 3% glutaraldehyde (chilled) is introduced gently into the test tube, and the serum and cells are fixed for 30 min at 4°C (the test tube is not removed during fixation). The fixed and solidified serum containing the cells is removed from the test tube and cut into small pieces. Postfixation is accomplished with 1% OsO_4 for 1–2 hr at 4°C.

Method 2 (e.g., Leucocytes) (Shimizu *et al.*, 1978)

Collodion solution (collodion, 4 g; ether, 75 ml; ethanol, 25 ml) is poured into conical glass tubes that are rolled so that the solution adheres to their inside surface. The excess solution is removed from each tube. The tube is air-dried under an electric fan while being rolled to ensure a collodion film of uniform thickness.

Leucocytes drawn from their buffy coat are suspended in plasma or bovine serum albumin, transferred to the tubes, and centrifuged at 1500 *g* for 5 min. The supernatant is discarded, and 1.5% glutaraldehyde in 0.067 *M* cacodylate buffer (pH 7.4) is gently layered over the pellet. After 20 hr, the cell sediment is solidified into the same shape as the tip of the conical tube. When the collodion film is removed from the tube, the pellet also comes off the tube. The film is gently torn off and separated from the pellet under a dissecting microscope. The pellet is cut into 4–12 pieces with a razor blade and transferred to 1% OsO_4 solution in 0.05 M cacodylate buffer (pH 7.4) containing 5% sucrose at 4°C. Centrifugation is unnecessary for any of the dehydration steps because the pellet pieces remain intact. The pellet consists of several layers, which, in sequence starting from the tip of the tube, are as follows: red blood cells, neutrophil leucocytes, monocytes, lymphocytes, immature leucocytes, and platelets. This sequence allows selection of desired cell types in the conical pellet.

FUNGI

Method 1 (General Method)

Hyphae and spores are fixed for 3 hr with 4% glutaraldehyde in 0.1 M cacodylate buffer (pH 7.0) containing 0.01 M $CaCl_2$ and 3% Triton X100. After pelleting, the specimens are further fixed in a fresh solution of glutaraldehyde without the detergent for 15 hr (overnight) in the refrigerator. The specimens are rinsed twice in collidine buffer (pH 7.4) and postfixed with 1% OsO_4 in collidine buffer for 12 hr at 4°C.

Method 2 (*Neurospora*)

A compact pellet of hyphae is formed by removing the water in a Buchner funnel under vacuum. Segments measuring 2 mm on each side are cut and fixed with 1–3% glutaraldehyde in 0.1 M cacodylate buffer (pH 7.2) for 1 hr at room temperature. A brief vacuum treatment at the beginning of fixation removes air bubbles trapped in the specimen pellet. The specimens are rinsed in the same buffer and postfixed with 1% OsO_4 for 1 hr at 4°C.

Method 3 (Filamentous Fungi)

Glass, screwcapped vials containing "tissue carriers" (Ted Pella, Inc.) are used to substitute three to five membrane supports each in 8–10 ml of anhydrous substitution fluid (e.g., 2.5% OsO_4 in acetone) at −85°C. Specimens are grown over 5 × 5-mm squares of cellulose membranes immobilized by quenching them in molten fluoroform (Freon 23) and then transferred to these vials for several days; the substitution fluid is replaced twice during this process by the tissue carriers. The specimens are exposed to anhydrous chemical fixatives in acetone as follows: 2% glutaraldehyde plus 2% OsO_4 at −85°C for 12 hr, at −20°C for 12 hr, and then at 0°C for 45 min; two changes of 2% OsO_4 at 0°C for 45 min each; 2% OsO_4 at 21°C for 2 hr.

Water is distilled from 70% aqueous glutaraldehyde at 70 mmHg under a nitrogen atmosphere. The still pressure is slowly reduced to 10 mmHg, and anhydrous glutaraldehyde is distilled at a vapor pressure of 97°C. The distillate is maintained at 35°C and diluted with anhydrous acetone (dried over a molelcular sieve) as soon as the distillation has been completed; otherwise the 100% glutaraldehyde will polymerize. (The elimination of glutaraldehyde from this procedure should result in good preservation of most cell components.)

The specimens are infiltrated with Epon–Araldite as follows: 5, 10, 15, and 25% resin mixtures in acetone for 1 hr each; 40 and 55% resin mixtures for 90 min each; 70 and 85% resin mixtures for 2 hr each; two changes of 100% resin for 8 hr each. The membrane supports and adhering hyphae remain in the same vial for substitution through 100% resin and are placed on clear-glass microscope slides before polymerization at 70°C. Well-frozen cells are selected by using 100× oil and phase-contrast optics; they are then targeted and remounted for sectioning.

Thin sections are stained by immersion in 2% aqueous uranyl acetate for 30 min and then in lead citrate for 10 min. Uranyl acetate is prepared by using glass-distilled water, which is Millipore-filtered and mixed, 9 : 1, with Millipore-filtered isobutyl alcohol. The mixture is agitated on a vortex mixer for 3 min and used immediately.

Method 4 (Glass Bead Treatment) (Modified from Olson and Eden, 1977; Olson, 1978)

The fungal and yeast cell wall (especially in case of spores) can be a barrier to fixation and infiltration by the embedding resin. This problem can be alleviated in some cases by mechanically breaking or removing the cell wall by glass bead treatment after preliminary fixation. The cells (0.1–0.3 ml packed volume) are fixed with a 1 : 1 mixture of 2% glutaraldehyde and 2% formaldehyde in 0.1 M cacodylate buffer (pH 7.2) for 2–4 hr at room temperature. After being rinsed three times for 10 min with the same buffer, the cells are pelleted in a conical centrifuge tube (15 × 100 mm). Then they are mixed with glass beads (0.45- to 0.5-mm or 1-mm beads) in a ratio of 1 part cells to 5 parts beads. The mixture should have the consistency of a moist paste. If the mixture has lower viscosity, more beads are added.

The cell wall is mechanically broken by rapid mixing on a vortex mixer. The centrifuge tube is placed against the side of the rotating head of the mixer. The centrifuge tube is placed against the side of the rotating head of the mixer, not in the depression at the center of the head. With a mixer having a speed of 2750 rpm and a radial eccentricity of 4.5 mm, the time required to break the cell wall ranges from 1 to 30 sec, depending on the cell type. Mixing is usually continued until 50% of the cells are partially or totally free of the cell wall.

The glass bead–cell mixture is suspended in distilled water, beads are allowed to settle, and the cell suspension is decanted. The pellet is refixed with the aldehyde mixture for 1 hr and then postfixed with 1% OsO_4. The cells are rinsed and stained *en bloc* with a saturated solution of uranyl acetate, after which they are pelleted by centrifugation in a warm (45°C) 1% agar solution (1 ml). After the agar has solidified, 30% ethanol is forced between the glass and the agar to free the pellet from the tube. Following dehydration, the agar pellet is sliced into three layers in 100% ethanol (top, middle, and bottom). A small portion of each layer is removed, spread in ethanol on a glass slide, and quickly examined with the light microscope. The portion of the pellet containing the largest number of cells where the cell wall has been broken or removed is cut into 1-mm^3 pieces and infiltrated with the embedding resin.

This glass bead technique is also useful for preparing cell-free homogenates, if the cells have not been prefixed with aldehydes.

HeLa CELLS

Method 1

Solution

Glutaraldehyde (8%)	10 ml
OsO_4 (2%)	10 ml

This solution is prepared in 0.1 M cacodylate buffer (pH 7.25) and mixed immediately before use at 4°C, the mixture having an osmolality of 607 mosmols. Cultures are grown in Falcon T-30 flasks at 37°C, using Eagle's minimal essential medium with 10% fetal calf serum as the fluid phase and 5% CO_2 and air as the gas phase. The cultures are allowed to grow for 2 weeks until a monolayer has formed. Fixation is carried out in the preceding solution for 1 hr at 4°C. The cells are prestained with 0.5% uranyl acetate in Veronal acetate buffer before dehydration. For additional details, see Chang (1972).

Method 2

Solution A

Glutaraldehyde (25%)	4 ml
Acrolein (25%)	4 ml
Phosphate buffer (0.1 M, pH 7.3) to make	100ml
$CaCl_2$	a trace

The osmolality should be 393 mosmols.

Solution B

OsO_4	1 g
Phosphate buffer (0.1 M, pH 7.3)	50 ml

The osmolality should be adjusted to 366 mosmols with glucose.

Cells are fixed in solution A for 10 min and then pelleted. After being fixed in the same solution for another 15 min, the pellet is postfixed with solution B for 1 hr at 4°C.

LEAVES

Fixative solution

Glutaraldehyde (25%)	15 ml
Formaldehyde	25 ml
Cacodylate buffer (0.2 *M*, pH 7.2)	18 ml
CaCl$_2$	25 mg

A metal ring (3 mm high and 12 mm in diameter) is sealed with lanolin onto the upper surface of an intact leaf. No lanolin should be present within the ring. With the use of a hypodermic needle, the ring is filled with the preceding solution and then covered with a glass coverslip. This solution is changed after 45 min with fresh solution, and fixation is allowed to continue for another 45 min at room temperature. The leaf tissue in contact with the fixative is cut out and sliced into 0.5-mm strips. These strips are placed in fresh solution for 30 min at 4°C. After being rinsed in the buffer, the strips are postfixed with 2% OsO$_4$ for 2 hr.

LIPOSOME PREPARATION

Method 1 (Kim and Martin, 1981)

The following procedure is useful for preparing cell-size unilaminar liposomes with high captured volume and defined size distribution. Liposome vesicles of 9.2 μm average diameter can be prepared. Amphipathic lipids totaling 3.33 μmol plus small amounts of neutral lipids (e.g., triolein and cholesterol oleate) are combined in a 1-2 dram vial (1.4 × 4.5 cm, screw cap lined with aluminum foil). Chloroform is added to bring the volume to 1 ml. The material (e.g., glucose, collagen, or ferritin) to be trapped is dissolved in 1 ml of 150 m*M* aqueous sucrose solution and then added to the chloroform–lipid solution.

The vial is immediately shaken with a standardized mechanical shaker for 45 sec to produce a water-in-chloroform emulsion. When the aqueous phase is being added to chloroform, it is necessary to avoid producing a chloroform-in-water emulsion; therefore the aqueous phase is added in three divided portions with manual swirling after the addition of each aliquot. The emulsion, after mechanical shaking, is divided into two equal portions, and each is transferred into a new vial containing the "mother liquor."

The "mother liquor" is prepared by adding 0.5 ml of lipid solution in diethyl ether (3.33 μmol total of amphipathic lipids plus a small amount of neutral lipid/ml ether) to each of two new vials containing 2.5 ml of 200 mM sucrose solution and then shaking the vial mechanically for 15 sec.

After the water-in-chloroform has been transferred into the "mother liquor," the vial is immediately shaken mechanically for 10 sec to produce microscopic chloroform spherules, most of which contain smaller water droplets within.

The chloroform and ether must now be evaporated to form liposomes. The chloroform spherule suspension in the "mother liquor" is layered on the bottom of a 250-ml filtration flask (bottom diameter 8 cm). A stream of nitrogen gas at 1.5 l/min is introduced into the flask through a piece of glass tubing protruding 5 cm into the mouth of the flask. The flask is gently swirled every 15–30 sec to keep the chloroform spherules suspended. Throughout the solvent evaporation (which takes about 45 min) the flask is kept in a 37°C water bath. Complete evaporation of chloroform and ether is indicated by a marked decrease in turbidity. Remaining traces of ether and chloroform may be removed by incubating the liposome preparation at 100°C in a loosely capped vial for 10 min.

Method 2

Varying proportions (1–4 ml) of a 0.16 M KCl solution adjusted with 1/10 volume of Veronal buffer to pH 7.35 and ethanolic solution (1–4 ml) of lecithin (1.19 mM) are mixed in a test tube to give a total volume of 4 ml (Goto and Sato, 1980). This mixture is sonicated for 15 sec at room temperature at power level 4 in a Branson sonifier Model W-185. The tube is sealed with aluminum foil that is covered with Parafilm and left at 50°C for 30 min. The tube is opened and 0.1 ml of solution is transferred to a second tube and diluted with 0.16 M KCl solution (pH 7.35) to give a final ethanol concentration of 0.8%. About 0.5 ml of 1% La(NO$_3$)$_3$ solution is added to this tube and is followed by 0.5 ml of freshly prepared and filtered 1% KMnO$_4$. After thorough mixing the mixture is centrifuged at 3000 rpm for 1 min. The resulting pellet is washed with distilled water three times by centrifugation, dehydrated, and embedded in a resin. Thin sections are mounted on collodion-coated grids and examined without poststaining. This method yields cell-sized, single-layered liposomes that show semipermeability and trilamellar properties characteristic of biologic membranes.

LIQUID SPECIMENS (e.g., Milk and Fruit Juices)

Method 1 (Milk)

The following microencapsulation method is useful for processing suspensions and emulsions. Microcapsules are made by dipping a stainless-steel rod (0.5 mm in diameter) in a warm 4% agar solution. When the rod is removed (within a few

minutes) from the solution, it is covered with a thin layer of agar (Fig. 7.15, 1).
(M. Kaláb, 1983 personal communication). This layer is allowed to solidify,
forming an agar capsule. The capsule is cut with a razor blade (Fig. 7.15, 2). The
agar capsule is filled by dipping its lower end into the liquid specimen; the liquid
is drawn in by aspirating it with the rod functioning as a piston (Fig. 7.15, 4).
The trapping of air bubbles in the capsule should be avoided. The lower end of
the capsule is blotted and then sealed by dipping it into the hot agar solution. This
procedure is done most conveniently with a drop of the agar solution on a small
spatula. The capsule is placed on a glass slide, and its upper end is cut from the
rod with a razor blade. This end is also sealed with hot agar solution. This
capsule encasing the liquid specimen is fixed, dehydrated, and embedded in a
resin in the same way as is a tissue block.

Method 2 (Orange Juice)

In a method similar to the preceding one, a Pasteur pipette is filled with a
liquid (e.g., orange juice) and then dipped in a 4% agar solution (Fig. 7.16A)
(Jewell, 1981). The pipette carrying the liquid is removed and the agar is allowed
to solidify. The microcapsule formed is manipulated off the pipette by using the
thumb and forefinger (Fig. 7.16C), The microcapsule is filled by discharge from
the pipette, completely removed from the pipette, and sealed with agar. This

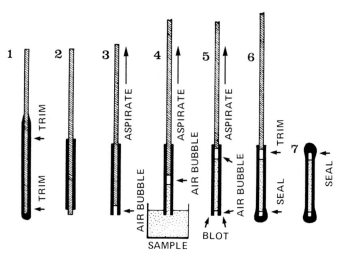

Fig. 7.15. Microencapsulation method. (1) A glass rod is coated with agar gel (solid black line)
and the gel is trimmed. (2) An agar gel tube is formed around the glass rod. (3) The air bubble is
aspirated. (4) The glass rod is used as a piston to aspirate a liquid specimen. (5) The lower end of the
agar gel tube is blotted with filter paper. (6) The clean lower end of the agar gel tube is sealed with
agar and the upper end is cut off the tube. (7) The upper end of the tube is also sealed with agar, and
the encapsulated specimen is immediately placed in a fixative. (From Kaláb, 1981.)

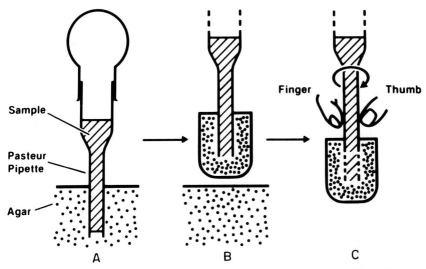

Fig. 7.16. Microencapsulation method. (A) Pasteur pipette full of juice dipped in molten agar. (B) Pipette is removed and agar allowed to set. (C) Microcapsule is manipulated off pipette by use of finger and thumb. The microcapsule is filled by discharge from the pipette and then fully removed and sealed with agar. (From Jewell, 1981.)

microcapsule is fixed with 3% glutaraldehyde in 0.1 M cacodylate buffer (pH 7.2) for 4 hr at 4°C, rinsed in buffer, and postfixed with 1% OsO_4 for 24 hr at 4°C.

Alternatively, if fat globules are studied, milk is added to an equal volume of 2.5% glutaraldehyde in 0.1 M cacodylate buffer (pH 7.2) and the mixture is centrifuged. The resulting cream layer is dispersed in buffer and then retrieved by centrifugation. The compacted cream is postfixed with 1% OsO_4 for 1 hr. After it has been rinsed and suspended in distilled water, the specimen is collected on a Nucleopore filter (200 nm pore size) and covered with a thin layer of 2% agar. After solidification, the agar gel is stained *en bloc* with uranyl acetate, cut into small pieces, and dehydrated and embedded.

Method 3 (Colostrum)

Milk cells are isolated by diluting the colostrum with an equal volume of PBS and then centrifuging at 650 g for 20 min at 4°C (Lee and Outteridge, 1981). The cell pellet is washed twice in 10 ml of PBS, centrifuged at 650 g for 20 min, and resuspended in a small volume (about five times the volume of the cell pellet) of PBS. This sample is fixed for 1 hr at room temperature by adding an equal volume of 4% glutaraldehyde in 0.1 M phosphate buffer (pH 7.2) containing 2% sucrose. The suspension is centrifuged at 650 g for 20 min. After the supernatant is discarded, the pellet is broken into small pieces of about 0.1–0.2 mm in diameter and then postfixed with 1% OsO_4 in 0.1 M Veronal acetate buffer (pH 7.2) for 1 hr at room temperature. The specimens are stained *en bloc* with 2% uranyl acetate for 30 min. The pellet is dehydrated and embedded according to standard procedures. Milk components (e.g., neutrophils, macrophages, casein micelles, fat droplets, and lymphocytes) can be easily recognized in ultrathin sections.

LIVERWORT

Specimens are transferred to a mixture of 1% glutaraldehyde and 1% formaldehyde in 0.05 M PIPES buffer (pH 6.8). After 10 sec, 1% OsO_4 in the same buffer is added, and the fixation is continued for 20 min at room temperature (Lehmann and Schultz, 1982).

LOBSTER

Method 1

Soft tissue pieces are fixed with 3% glutaraldehyde in 0.1 M sodium cacodylate buffer (pH 7.4) containing 12% glucose for 3 hr at room temperature. After a thorough rinse in 0.1 M cacodylate buffer containing 24% sucrose, the specimens are postfixed with 1% OsO_4 in the same buffer. The specimens are then rinsed in a 0.1 M cacodylate buffer series containing 24, 12, and 6% sucrose, after which they are dehydrated in acetone. The specimens are infiltrated overnight in a 1 : 1 mixture of Spurr resin and acetone and then embedded.

Method 2

Neuromuscular tissue blocks are fixed for 2 hr at room temperature with a mixture of 1% glutaraldehyde and 2% formaldehyde in 0.1 M phosphate buffer (pH 7.4) containing 3% NaCl and 4% sucrose. After a rinse for 1 hr in 0.1 M phosphate buffer containing 8% sucrose, the specimens are postfixed for 1 hr with 1% buffered OsO_4 containing 4% sucrose.

LYMPHOCYTES (PERIPHERAL, HUMAN) (Nunes *et al.*, 1979)

Two milliliters of whole blood obtained by venipuncture are gently mixed with 0.2–0.4 ml of heparin. Glass or plastic heparinized tubes (0.9 mm in internal diameter and 75 mm in length) are immediately filled by capillarity, sealed with plasticine, and centrifuged for 4 min on a standard microhematocrit centrifuge. Eight to 10 tubes are used for each blood sample. After centrifugation, the tubes are scored with a diamond pencil and broken at the buffy coat–plasma boundary. The plasticine-sealed end is pressed into a dental wax plate to extrude the buffy coat. The broken end is immersed in a horizontal position into a drop of fixative placed on dental wax or Parafilm. In less than 5 min the buffy coat is hard enough to be separated easily from the erythrocyte layer. This separation can be helped with a razor blade under binocular microscope control (Fig. 7.17).

Alternatively, the tube is sectioned at the erythrocyte–buffy coat boundary, and the buffy coat is blown into a drop of fixative solution with a microrubber bulb; however, care must be taken to avoid gross contamination of the fixative with the plasma. When plastic tubes are used, they are cut with a scalpel blade on a hard surface at the level of the segment containing the buffy coat, which is then placed in the fixative; in 10 min the pellet can be easily dislodged from the plastic ring with a metal needle.

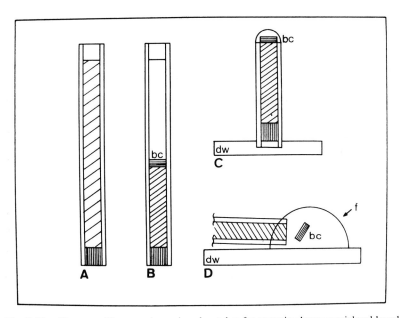

Fig. 7.17. Diagram of the procedure using glass tubes for preparing human peripheral lympho-cytes. (A) Plasticine-sealed microhematocrit tube filled with heparinized blood. (B) Centrifuged tube showing a microbuffy coat (bc). (C) Erythrocyte-containing broken end of the tube, pressed into the dental wax plate (dw) with extrusion of the microbuffy coat. (D) The broken end of the tube horizontally immersed in a drop of fixative (f), allowing the separation of the microbuffy coat. (From Nunes *et al.*, 1979. © 1979 The Williams & Wilkins Co., Baltimore.)

The microbuffy coats obtained by both procedures are cylindrical, being 0.9 mm in diameter and 0.5 mm high. They are fixed with 2.5% glutaraldehyde in 0.1 M cacodylate buffer (pH 7.3) and postfixed with 1% OsO_4 in the same buffer. This procedure is followed by *en bloc* staining with 0.5% uranyl acetate in distilled water. The coats are then dehydrated and embedded in flat plastic molds, with the base of each coat facing the base of the mold.

MARINE INVERTEBRATE TISSUES (Eisenman and Alfert, 1982)

Specimens are fixed for 10 min with a mixture of 0.05% OsO_4 and 4% glutaraldehyde in 0.2 M Na-cacodylate buffer (pH 7.2) containing 0.35 M sucrose. This mixture is prepared immediately before use. The specimens are fixed for 1 hr with 4% glutaraldehyde in 0.2 M Na-cacodylate buffer (pH 7.2) containing 0.1 M NaCl and 0.35 M sucrose. After two rinses (5–10 min each) in 0.2 M Na-cacodylate buffer (ph 7.2) containing 0.3 M NaCl, the specimens are postfixed for 1 hr with 1% OsO_4 in Na-cacodylate buffer (pH 7.2) containing 0.3 M NaCl.

MICROORGANISMS (Rittenburg *et al.*, 1979)

Cells are fixed with an equal volume of 4% glutaraldehyde in 0.1 M PIPES buffer (pH 7.2) for 30 min. About 2 ml of the suspension of fixed cells is drawn into a 3-ml syringe, which is attached to a filter holder containing a 13-mm-diameter polycarbonate membrane filter (Fig. 7.18) (Unipore, Bio-Rad Labs., Richmond, CA). The syringe contents are passed through the filter until a slight resistance is felt. (The pore size of the filter can be varied according to specimen size; pore diameters of 0.8 and 3 μm are used for collecting the bacterial and protozoan cells, respectively).

A fresh syringe is attached to the filter holder, and cells are washed by passing 3 ml of buffer through the filter, taking 1 min to complete the washing. A fresh syringe containing 2 ml of 2% OsO_4 solution in the same buffer is attached to the filter holder. About 1 ml of this solution is slowly passed through the filter. The syringe is left attached to the filter holder, and postfixation is accomplished in 25 min at 4°C. The remaining 1 ml of OsO_4 is slowly passed through the filter. A fresh syringe is attached to the filter holder and 4 ml of distilled water is passed through the filter to wash the cells.

With a fresh syringe, 2 ml of 2% aqueous uranyl acetate solution is passed through the filter, taking 5 min to complete the process. About 1 ml of each of 25, 50, and 95% acetone is passed through the filter. Dehydration is completed

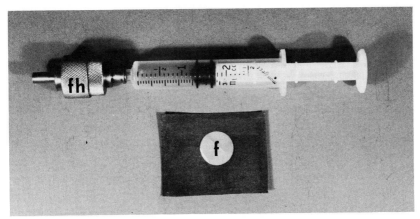

Fig. 7.18. Syringe-filter unit. The filter holder (fh) must be of metal to withstand the acetone treatments. The polycarbonate membrane filter (f) is also shown. (From Rittenburg *et al.*, 1979. © 1979 The Williams & Wilkins Co., Baltimore.)

by passing 3 ml of 100% acetone through the filter. The filter membrane is removed from the holder and soaked in a 1 : 1 mixture of acetone and Spurr resin for 30 min at room temperature. The filter membrane is transferred to a 1 : 2 mixture of acetone and resin for 15 min and then to two changes of 10 min each of pure embedding resin. The membrane is cut into small pieces and embedded in fresh resin. The membrane pieces can either be placed at the tip of the capsule manually or be gently centrifuged to the tip.

MICROTUBULES

Tissue specimens are fixed with a mixture of 1% glutaraldehyde and 1% formaldehyde in 0.1 M phosphate buffer (pH 7.3) containing 1% PTA for 2 hr at room temperature. The specimens are rinsed with the buffer containing 7% sucrose and then postfixed with buffered 2% OsO_4 for 1 hr at room temperature. The specimens are treated with 0.5% aqueous uranyl acetate for 20 min before dehydration.

MICROTUBULES AND MICROFILAMENTS

Method 1 (Plant and Animal Tissues)

Root tissues are fixed for 2 hr at 22°C with 1% glutaraldehyde in 50 mM phosphate buffer (pH 6.8) containing 0.2% tannic acid. After a brief rinse in the buffer, the specimens are postfixed with 1% OsO_4 in the same buffer for 1 hr at the same temperature. The diameter of microfilaments enlarges with increased concentration of tannic acid. The addition of various ions, use of low temperatures, and excessive hydrostatic pressure are undesirable for satisfactory preservation of these structures.

Animal tissues are fixed for 1 hr at room temperature with 2% glutaraldehyde in 0.05 M phosphate buffer (pH 7.2) containing 4–8% tannic acid and 0.015 M $CaCl_2$. After a brief rinse in the buffer, the specimens are postfixed with OsO_4. Tannic acid is dissolved in a buffer with moderate heating; on cooling, a brown precipitate is formed, which can be cleared by centrifugation. The resultant solution is clear and brownish. Glutaraldehyde is added just before fixation.

Method 2 (Pituitary Gland)

The following method is modified from Warchol et al. (1974) and is useful for measuring microtubules and microfilaments in the pituitary gland. Very small tissue specimens are fixed in specimen vials containing 2.5% glutaraldehyde in 0.1 M cacodylate buffer (pH 7.0) (400 mosmols) and placed on a rotator for 1–2 hr at room temperature. The specimen vials are rotated in a refrigerator for 60 min and then transferred to crushed ice for 10 min. One part of ice-cold OsO_4 in the buffer containing 6% sucrose (400 mosmols) is added at 4°C to 2 parts of the preceding fixative containing the specimens. The specimen vials are placed on a rotator and fixed for 30 min at 4°C. The fixation mixture is decanted and replaced with a 1 : 2 mixture of glutaraldehyde and OsO_4 (4°C). After 30 min fixation using a rotator at 4°C, the mixture is replaced with 1% OsO_4 (4°C) in the same buffer for 1–2 hr at the same temperature. After being rinsed twice with ice-cold distilled water, the specimens are treated with 0.5% uranyl acetate in 0.1 M Veronal acetate buffer (pH 5.8) for 8 hr at 4°C.

MITOTIC APPARATUS (MAMMALIAN CELLS)

This method is useful for preserving microtubules and membranes in the mitotic apparatus. The duration of two fixations is critical. HeLa cells are grown on tube slips in monolayer culture in Eagle's minimal essential medium (EMEM), supplemented with 10% calf serum, 1% glutamine, and 1% nonessential amino acids. The number of cells in mitosis is increased by a thymidine block (2 mM). After careful rinsing in warm Hanks' balanced salt solution (HBSS), the cells are fixed with 2.5% glutaraldehyde in EMEM for 15 min at 37°C. This mixture is prepared immediately before use. At room temperature the specimens are rinsed again in HBSS for 10 min and then postfixed with a saturated aqueous solution of $KMnO_4$ for 5 min. They are then rinsed in 25% ethanol until the rinsing solution remains colorless, and dehydration with acetone follows.

PHAGE PARTICLES (Lickfield and Menge, 1980)

Infected cells are prefixed with 0.2% OsO_4 in 0.41 M Ryter–Kellenberger (RK) buffer (Michaelis buffer and 0.01 M $CaCl_2$, pH 6.1) for 15 min at room temperature. About 1 ml of 1% OsO_4 solution is added to a 5-ml sample and immediately vigorously shaken. Samples are centrifuged at 4000 g at room temperature for 5 min. The pellet is thoroughly mixed with 1 ml of 1% OsO_4 solution and 0.1 ml of LB broth (amino acid source) and then kept overnight at room temperature. About 5 ml of RK buffer is added, and centrifugation at 4000 g for 5 min follows. The supernatant is discarded, residues are carefully removed with filter paper, and the pellet is warmed in a water bath to 47°C.

One drop of warm (47°C), melted agar in RK buffer is added to the pellet and thoroughly mixed with it. After the agar has cooled to room temperature, a small strip of the specimen in a drop of 5% uranyl acetate in RK buffer is cut to cubes measuring no more than 1 mm³. These specimens are transferred to 5% uranyl acetate in RK buffer for postfixation for 2 hr at room temperature. They are dehydrated with acetone: 30, 50, 70, 95, and 100% (twice), each for 15 min. Absolute acetone can be obtained by using a molecular sieve (0.3 nm) as a drying agent. Infiltration is accomplished with acetone–Epon mixtures of 3 : 1, 1 : 1, and 1 : 3 for 30 min each, followed by Epon (twice) for 45 min each. Polymerization is completed at 45°C for 12 hr and then at 65°C for 36 hr. Thin sections are poststained with lead citrate.

PLANT TISSUES (WOODY)

Tissue specimens (1–2 mm) are immersed in 6% glutaraldehyde in 0.05 M cacodylate buffer (pH 7.1) and aspirated under low vacuum at room temperature for 4 hr. The specimens are transferred to fresh fixative, and fixation is continued for another 4 hr under high vacuum. After several rinses in buffer for 1 hr, the specimens are postfixed in buffered 2% OsO_4 overnight in a refrigerator at 4°C. The specimens are rinsed in buffer for 1 hr and then dehydrated.

PLANT TISSUES (SOFT)

Specimens should be prefixed for 1 hr with 5% glutaraldehyde in 0.08 M PIPES buffer (pH 6.8), and postfixed for 1 hr with 2% OsO_4 in 0.18 M PIPES buffer, the osmolalities being 800, 600, and 680 mosmols, respectively. All treatments are carried out at room temperature with gentle agitation. The OsO_4

solution in PIPES buffer should be prepared immediately before use, because a brownish color develops after 1 hr.

PLATELETS (HUMAN BLOOD)

Method 1 (Mattson *et al.*, 1977)

About 9 ml of whole blood is mixed with 1 ml of 3.8% sodium citrate. After centrifugation at 200 g for 5 min at room temperature, the citrated platelet-rich plasma is withdrawn with a syringe and transferred to a centrifuge tube. To this plasma is added an equal volume of 0.1% glutaraldehyde in 0.2 M cacodylate buffer (pH 7.0) having a total osmolality of 407 mosmols. After gentle mixing by inversion, the platelets are centrifuged at 900 g for 10 min. The supernatant is carefully poured off, and the platelet button in the tube is overlayered with 3% glutaraldehyde in the same buffer. The tube is capped and allowed to remain at room temperature for 2 hr.

The platelet button is rinsed in the buffer and then postfixed for 30 min with 1% OsO_4 in the buffer having an osmolality of 331 mosmols. After being rinsed twice (14 min each time) with the buffer, the pellet is dislodged, transferred to a glass screwcapped bottle, and dehydrated and embedded.

Method 2

Blood is collected in a plastic test tube containing 3.8% trisodium citrate anticoagulant in a ratio of 9 parts anticoagulant to 1 part coagulant. The sample is mixed by gentle inversion and then centrifuged at 750 rpm for 8 min. Citrated platelet-rich plasma (CPRP) is transferred to a clean plastic test tube and kept at room temperature to avoid cold-induced morphologic changes (White and Krivit, 1967).

One volume of CPRP is poured in a slow, steady stream into 10 volumes of freshly prepared 0.2% glutaraldehyde in 0.1 M cacodylate buffer, which is continuously agitated. Equal volumes of fixative and CPRP are used. The addition of the fixative stops the physiologic reaction immediately. This initial fixation is accomplished in 30 min at room temperature.

Preparation for the TEM

CPRP in glutaraldehyde is centrifuged at 3100 rpm for 10 min, and the glutaraldehyde is decanted. The intact pellet is fixed with 3% glutaraldehyde for 2 hr at room temperature; care should be taken not to break the pellet. The pellet is gently dislodged and transferred to a glass Petri dish containing 3% glutaraldehyde. It is cut into 1-mm^3 pieces, which are placed into a vial, washed twice in the buffer, and postfixed in 1% OsO$_4$ for 1 hr at 4°C.

Preparation for the SEM

An equal volume of 6% glutaraldehyde is added to the CPRP in 0.2% glutaraldehyde and fixed for 24 hr at room temperature. The platelets are centrifuged at 3100 rpm for 10 min, and the supernatant is discarded. The platelets are gently resuspended in 5 ml of distilled water, washed for 10 min, centrifuged at 3100 rpm for 10 min; the supernatant wash is discarded. This process is repeated three more times and is followed by dehydration.

Method 3 (McLean *et al.*, 1982)

The following method avoids repeated centrifugation steps and the concomitant resuspensions of the pellet, and it can be completed in 1 day. Whole blood from normal volunteers is drawn by clean venipuncture into plastic syringes and immediately transferred into plastic test tubes that contain 3.8% sodium citrate (1 volume citrate to 9 volumes blood). The citrated blood is centrifuged at 120 g for 10 min at room temperature. The platelet-rich plasma is transferred to clean plastic tubes that are capped and kept at room temperature.

One volume of platelet-rich plasma is added to 45 volumes of 2.5% glutaraldehyde in 0.1 M phosphate buffer (pH 7.4) in a plastic test tube. The tube is tightly capped and the platelets are fixed for at least 1 hr at room temperature with continuous gentle mixing by inversion. The sample is gently transferred to a plastic conical centrifuge tube and centrifuged at 1000 g for 10 min. The supernatant is carefully removed and discarded. About 1 ml of 1% liquified agarose at 45–56°C is layered over the pellet and the tube is rapidly centrifuged to prevent resuspension of the platelets. After the agarose has been allowed to harden, the agarose block containing the platelets is removed from the test tube. The platelet pellet is clearly visible at the tip of the block.

The pellet is separated from the remaining agarose and cut into 1-mm^3 pieces. They are placed in vials, washed twice in 0.1 M phosphate buffer, and postfixed with 1% OsO$_4$ for 1 hr at 4°C. Further processing follows standard procedures.

POLLEN GRAINS (FOSSIL MATERIAL)

Pollen grains are macerated from mature seed fern specimens (e.g., *Dolerotheca sclerotica*) using 5% HCl (Taylor and Rothwell, 1982). The grains from immature pollen organs are removed from cellulose acetate peels in the following manner. Desired portions of sporangia that show evidence of pollen are placed in a depression slide and the adhering mineral material is dissolved in several changes of 5% HCl and then rinsed in distilled water. The peel sections are placed in 100% acetone and dissolved. All grains (macerated from sporangia or dissolved from acetate peels) are fixed with 2% aqueous $KMnO_4$ for 30 min at 60°C on a Fluopore filter in the dark. The filters with the grains are washed in several changes of distilled water. The samples are placed in vials and treated with 2% aqueous uranyl acetate for 1 hr at 4°C in the dark, dehydrated, and embedded in a low-viscosity resin.

POLLEN WALLS

During conventional fixation a considerable part of the soluble material on the surface and in cavities of the pollen wall and in the anther loculus between the tapetum and the pollen grains is extracted. These materials are satisfactorily preserved with the following procedures (Dunbar, 1981).

As will be apparent from the following discussion, all three methods outlined here need to be used separately for preserving various components of a pollen grain. In one method, pollen grains are fixed with a mixture of 2% glutaraldehyde and 0.5% tannic acid in 0.1 M phosphate buffer (pH 7.2) for 1–2 hr at room temperature. After being rinsed three times in buffer, the specimens are postfixed with 0.2% OsO_4 in the same buffer for 1 hr.

Alternatively, the specimens are fixed with a mixture of 7% glutaraldehyde and 0.5% alcian blue in 0.1 M cacodylate buffer (pH 7.0) for 1–2 hr at room temperature and then postfixed with 0.2% OsO_4 in the same buffer without previous rinsing. Dehydration is accomplished in an acetone series from 30 to 95% and is followed by five changes in 100% acetone. Infiltration and embedding are carried out with Spurr resin.

The inclusions from the cytoplasm in the intine and cell organelles of the pollen cytoplasm are better preserved with the tannic acid method than with the alcian blue method, while other cellular substances in cavities of the pollen wall and outside the wall are better preserved with the following method; these materials are extracted with the tannic acid–glutaraldehyde method.

Pollen grains are fixed with a mixture of 1% glutaraldehyde and 0.1% alcian blue in 90% acetone for 1–2 hr at room temperature. Without rinsing, the specimens are transferred to 90% acetone containing two crystals of OsO_4 for 15 min. The specimens are transferred directly to 100% acetone; five changes in 100% acetone are recommended. Standard dehydration is undesirable.

Losses of water-soluble surface materials from the pollen grain walls during fixation can also be prevented by adding cetylpyridinium chloride (CPC) to glutaraldehyde (Grote *et al.*, 1983). Parts of the mature pollens collected from flowers are fixed for 2 hr at room temperature with 6.25% glutaraldehyde in 0.071 M cacodylate buffer (pH 7.4) containing 0.5% CPC. The specimens are frozen in a drop of this fixative solution upon the dish of a cryomicrotome, and sections 10–20 μm thick are cut at −18°C. The sections are brought to room temperature and fixed for an additional period of 2 hr in the same fixative. After being rinsed for 1 hr in 0.1 M cacodylate buffer (pH 7.4) containing 7.5% sucrose, the specimens are postfixed for 2 hr at room temperature with 2% OsO_4 in 0.1 M phosphate buffer (pH 7.4). No poststaining is required. The exine of pollens is covered with a layer of electron-dense material, which is absent in the conventionally fixed specimens.

POTATO TUBER

Unequal hardness of plant tissues can impede their uniform infiltration with a resin. Starch grains in the tissue are generally poorly embedded, and therefore cutting of ultrathin sections is difficult. The inclusion of dibutyl phthalate in a mixture of Epon and Araldite is useful for embedding difficult plant tissues such as potato tubers (Allen and Friend, 1983). The embedding mixture consists of Epon (3 parts), Araldite (3 parts), DDSA (8 parts), and the accelerator DMP-30. Dibutyl phthalate is added in the ratio of 3 drops/5 ml of the resin mixture.

The specimens are fixed with 3% glutaraldehyde in 0.05 M potassium phosphate buffer (pH 6.8) for 90 min at room temperature. After being rinsed for 1 hr in several changes of the buffer, the specimens are postfixed with 2% OsO_4 in the buffer for 2 hr. They are dehydrated first with ethanol and then with propylene oxide. Finally, they are serially infiltrated with 25, 50, 75, and 100% resin, with evacuation for 5 min following each step.

PREPARATION AND REMOVAL OF SELECTED CELLS FROM CYTOLOGIC SMEAR PREPARATIONS (Mather *et al.*, 1981)

Body cavity smears are prepared on glass slides and fixed with 2.5% glutaraldehyde in 0.1 M cacodylate buffer (pH 7.4) for 2–4 hr at room temperature. After being rinsed several times in buffer, the smears are stained by the Papanicolaou method. They are mounted by using Polymount and a coverslip and then examined with the light microscope. Single cells or cell groups to be removed are recorded by photography, and a diamond marker is employed to mark on the underside of the slide the selected cells in the center of the circle. The coverslip is removed with xylene, and the smear is rehydrated through alcohols before being destained with 1% acid alcohol. The smears are rinsed in distilled water and buffer and then postfixed with 1% OsO_4 for 30 min.

After dehydration through graded alcohols to 100% acetone, the smear is

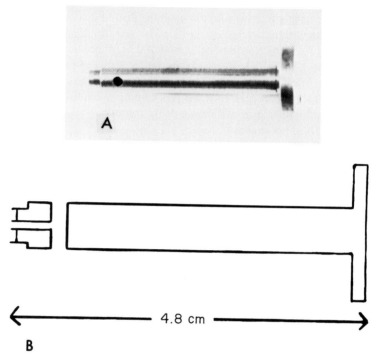

Fig. 7.20. (A) Modified wad punch. (B) Diagram showing the modification of the wad punch. (From Mather *et al.*, 1981.)

impregnated with a resin. A modified wad punch (Fig. 7.20) is used to make paper markers, which are the exact size of the gelatin capsule to be used (size 3 capsule, 5.5 mm diameter) and have a hole 1 mm in diameter at the center. A wad punch, usually used for punching holes in paper, consists of a hollow metal tube with a sharpened end. The modification consists of tooling the sharpened end to punch out the proper-sized paper marker and hole.

Excess resin is removed and a drop of fresh resin is placed over the marked area. The paper marker is centered over the selected area; its position can be checked by light microscopy. A gelatin capsule is filled with resin and laid on top of the marker. The slide and capsule are placed overnight in an oven at 60°C to polymerize. The slide and capsule are placed in a freezer at −20°C for a few minutes, after which the slide and capsule separate without difficulty. The part of the resin block that contains the selected cells is identified by the hole (1 mm in diameter) in the center of the paper marker, and the rest of the resin can be trimmed away and a trapezium formed. Ultrathin sections must be cut with care because the selected cells are very close to the surface of the block.

PRESERVATION OF THE LARVA–SUBSTRATE
RELATIONSHIP (Bergquist and Green, 1977)

Marine larvae (e.g., *Halichondria* and *Microciona*) are collected as they are released, and then are transferred by pipette into 250-ml plastic Nalgene or tripour beakers containing 150 ml of fresh seawater. The beakers have a 3-mm-thick layer of hardened Epon on the bottom. The larvae are allowed to settle for 1–5 days, with fresh seawater added every day. Preparation of settled larvae is achieved by decanting the seawater and then introducing successive solutions into the beaker, decanting, and then adding the next solution. Each solution is added in sufficient volume to cover the larvae completely.

Specimens are fixed with 4% glutaraldehyde in 0.1 *M* cacodylate buffer (pH 7.2) and then postfixed with 1% OsO_4. After dehydration, a second layer of Epon (identical in composition to the layer forming the bottom of the beaker) is poured over the specimens. The beaker is rotated on an angle for 2 hr before polymerization. The final result is an Epon bilayer (about 6 mm thick) in which the larvae are adhering undisturbed to the bottom layer and are embedded in the top layer. These bilayered disks can be removed by flexing the sides of the beakers. Blocks of an appropriate size are cut from the disk with a jeweler's saw and are oriented in such a way that sections can be cut through the larvae at right angles to the substrate.

By orienting the interface parallel to the knife edge when sectioning, one can overcome any problems caused by differences in consistency between the two layers. Uneven embedding can be improved by embedding under vacuum or increasing the rotation time during embedding. To achieve settlement of difficult larvae on Epon, it may be necessary to roughen the surface with fine emery paper or to allow a conditioning algal film to develop before introducing the larvae.

PROTOPLASTS (LETTUCE LEAF) (Seed, 1980)

About 1 ml of 8% glutaraldehyde is added drop by drop down the side of a Petri dish containing 7 ml of protoplast suspension (approximate concentration 5 $\times$ 10^4 protoplast/ml) over the course of 1 hr at room temperature. The dish is gently swirled after the addition of each droplet; the final concentration of glutaraldehyde is 1%. The suspension is subsequently transferred to a centrifuge tube and centrifuged at 35 *g* for 4 min. The protoplast pellet is resuspended in 4 ml of 1.5% glutaraldehyde in 0.1 *M* cacodylate buffer (pH 7.4), which is allowed to run slowly down the side of the tube as it is gently rotated. The preparation is fixed for 2 hr at 4°C.

The sample is then centrifuged, and the protoplasts are resuspended in 4 ml of 3% glutaraldehyde in buffer and incubated a further 4 hr at 4°C. The protoplasts are subsequently rinsed twice in buffer and incubated in buffer for 17 hr at 4°C. Following this procedure, the buffer is removed and the protoplasts are postfixed with 4 ml of 1% OsO_4 in buffer for 2 hr at 4°C. After protoplasts have been rinsed in buffer, the pellet is suspended in 3 drops of a warm solution of 1% agarose or agar in distilled water, which is then poured onto a glass slide, allowed to solidify, and divided into 3 × 3-mm segments. These segments are transferred to vials and very gradually dehydrated and embedded in a resin. If needed, the preparation can be incubated under vacuum before polymerization so as to facilitate penetration by pure resin mixture.

ROOTS

Roots are cut into segments (less than 1 mm on a side) in 3% glutaraldehyde in 0.1 M PIPES buffer (pH 8.0) having an osmolality of 750 mosmols. After being fixed for 1 hr at room temperature and then rinsed in the buffer for 30 min, the specimens are postfixed with 2% OsO_4 in the same buffer for 1 hr at 4°C.

ROSETTED CELLS (Payne and Satterfield, 1980)

Rosettes are prepared by mixing a suspension of white blood cells (1 × 10⁶ cells/ml) with an equal volume of a suspension of red blood cells (1–5 × 10⁷ cells/ml). The volume of the final rosetted preparation varies between 0.2 and 4 ml. An equal volume of 3% glutaraldehyde in 0.1 M phosphate buffer (pH 7.2) is added to the cell suspension and gently mixed by drawing the suspension up and down in a glass pipette. Any pellet present is resuspended. The preparation is allowed to prefix in the diluted glutaraldehyde for 1 hr at room temperature. The suspension is gently centrifuged at 1000 rpm in a Sorvall GLC-1 swinging bucket clinical centrifuge for 5 min. The pellet is resuspended in a small aliquot of human plasma and transferred to a large (17 × 100 mm) round-bottomed plastic centrifuge tube (No. 2059 tube; Falcon, Oxnard, Calif.). The suspension is again centrifuged at 1000 rpm in the clinical centrifuge for 5 min.

The supernatant is removed and replaced with cold 3% glutaraldehyde in 0.1 M phosphate buffer. The pellet is fixed for 1 hr at 4°C. The fixative is decanted and replaced with the buffer. After two changes of buffer (15 min each), the pellet is postfixed for 90 min in 1% OsO_4 in buffer. Following dehydration with ethanol, the pellet is loosened from the bottom of the tube with a thin wedge-shaped wooden stick. The ethanol is replaced with a 1 : 1 mixture of 100% ethanol and Spurr mixture and allowed to infiltrate for 2–4 hr.

The pellet is broken up into smaller pieces, transferred to a new vial of pure resin after being blotted on filter paper, and infiltrated for 20 hr at room temperature. These pieces are placed in BEEM capsules containing fresh resin and are polymerized at 70°C.

RUST FUNGI

Blocks (1 mm³) of diseased plant tissue are dissected in the same fixative as given for potato tubers (p. 343) (Coffey *et al.*, 1972). Postfixation, dehydration, and embedding follow standard procedures for plant tissues given on p. 339.

SEEDS

Fixative solution
Glutaraldehyde (8%)	25 ml
Formaldehyde (8%)	25 ml
Collidine buffer (0.05 *M*, pH 7.3)	100 ml

Specimens are fixed in this solution for 5 min at room temperature and then transferred to fresh solution at 4°C for another 2 hr. After a rinse in the same buffer for 1–2 hr, the specimens are postfixed with 1% OsO_4 in the buffer for 2–12 hr at 4°C.

SEEDS (DRY) (Horowitz, 1981)

Dry seeds are imbibed in 1% glutaraldehyde in 0.1 *M* cacodylate buffer (pH 7.0) for 48 hr. Cotyledons are separated from the embryonic axis and testa and then chopped into 1-mm³ cubes. These specimens are placed in 5% glutaralde-hyde in the same buffer plus 2% sucrose first under vacuum for 1 hr and then at atmospheric pressure for another hour. The specimens are rinsed overnight with the same buffer containing sucrose, postfixed with 2% OsO_4 in buffer for 1 hr, and then rinsed several times with distilled water.

After dehydration, the specimens are placed in 100% acetone for 6 hr. They are then placed in a 1 : 1 mixture of acetone and Spurr resin without hardener and allowed to remain under vacuum for 1 hr. Twice the amount of resin is added, and the specimens are left under vacuum for another hour, after which they are transferred to pure resin without hardener and shaken gently for 36 hr. Fresh resin plus hardener is poured into gelatin capsules, and the specimens are placed in them. Polymerization is completed at 70°C for 8 hr. Thin sections are stained by placing a drop of 1% $KMnO_4$ solution onto the section side of the grid for 2 min. After being rinsed with distilled water, the sections are stained with lead citrate.

SEEDS (WATER-IMPERMEABLE COAT) (Paul and Egley, 1983)

The following method allows the study of semithin sections of whole seeds with the light microscope and thin sections of selected areas with the electron microscope. Large seeds (e.g., okra) are glued onto plastic cylinders (Ladd Cat. No. 23656) with a rapid-bonding methacrylate cement (Duro Super Glue). After drying, the block is clamped in a specimen holder and the seed is sawed with a jeweler's saw (Fig. 7.21), leaving a solid rim of seed coat to provide stability during subsequent processing. Two sides of the seed can be sawed off to facilitate fixative penetration and to predetermine cross or longitudinal sections.

The specimen is fixed for 12–48 hr with 2–5% glutaraldehyde in 0.2 M cacodylate buffer (pH 7.0). After being rinsed for several hours in buffer, the specimen is postfixed overnight at 4°C with 1% OsO_4. It is then rinsed in distilled water for 3–4 hr, dehydrated, and embedded over 2–3 days in Spurr medium. At the final embedding step, the resin containing the specimen is poured into an aluminum weighing dish (Fisher Cat. No. 8 732 5A). The specimen is oriented, and the dish is transferred to a vacuum oven at 70°C. After polymerization, the specimen is sawed out of the hardened block and glued to a plastic cylinder for trimming and sectioning.

Fig. 7.21. Okra seed glued to plastic cylinder clamped in block-trimming chuck. One side of the seed has been sawed away to permit penetration of fixative. (From Paul and Egley, 1983.)

Fig. 7.22. Top: specimen supports for thin-layer preparations of free-floating cells; left, glass coverslip with scored orientation grid; right, combined support consisting of a 50-µm circular Mylar polyester slip glued to a glass coverslip with Epon, and scored with an orientation grid. Bottom: Epon block placed in a slide-sized Plexiglas holder for examination on the micrscope stage. In the middle, the support has been removed from the block; the relief of the orientation grid is visible on the block face (arrow). (From Weber, 1977. © 1977 The Williams & Wilkins Co., Baltimore.)

SINGLE CELLS FROM SUSPENSIONS (Weber, 1977)

The following procedure is useful for electron microscopy of individual cells from suspensions such as blood and peritoneal exudate. The cell suspension is collected in a glass tube containing an appropriate aldehyde fixative solution. Fixation, rinsing, and concentration of cells are carried out by low-speed centrifugation. Alternatively, spontaneous sedimentation may be preferable to protect delicate specimens.

Thin-Layering and Immobilization

The rinsed suspension is sedimented, and the pellet is resuspended in a solution of 5% bovine serum albumin (BSA) in 0.1 M Tris-HCl buffer (pH 7.5). This suspension should yield a single layer during subsequent sedimentation: the appropriate amount of BSA is determined empirically. Small drops (5 ml) of the suspension are placed on separate specimen supports, on an area previously scored with an orientation grid (Fig. 7.22). The preparations are kept in a humidity chamber for 30 min and then transferred to a large Petri dish containing cellulose soaked in freshly prepared 20% acrolein. In the tightly closed dish, cells are exposed to acrolein vapors for 30 min at room temperature. The BSA is cross-linked as shown by slight opacification.

Specimen Supports

Slips cut from Mylar polyester sheets type A (Du Pont) or glass coverslips are used for thin-layer preparations. It is advisable to use 50- or 75-μm Mylar sheets and to glue them to glass coverslips with a resin. The coverslips have been coated with Teflon. (A sufficient number of coverslips are spread out and sprayed with Teflon. Excess Teflon is removed by wiping each coverslip with a cloth in order to restore its transparency for light microscopy. For this purpose, the coverslip is placed on another coverslip "glued" onto paraffin in a glass dish of appropriate size.) The orientation grid (2.5 × 2.5 mm) is scored in the specimen support with a diamond scriber or a fine dissecting knife.

Selection of Cells for Sectioning

The specimen is placed face down on a drop of buffered 20% glycerine on a microscope slide and viewed under a light microscope, using interference or phase contrast. The cell to be sectioned and its location in the orientation grid are recorded with the aid of three photomicrographs using 40×, 10×, and 2.5× magnifications. Mylar slip specimens should be viewed with the microscope slide inverted on the stage. Alternatively, instead of photomicrographs, drawings are equally good as a reference. Following the light microscopy, the specimen is removed from the slide, rinsed in buffer, postfixed in OsO_4, and embedded in Spurr medium.

Embedding

Specimens are dehydrated in acetone and embedded in Spurr medium. After the preparations have been placed on the embedding capsule containing the final resin monomer, the supports should be covered with a glass slip or a piece of a glass slide.

Relocating a Selected Cell in the Embedded Thin Layer

Glass supports are separated from the polymerized block by immersing them in liquid nitrogen; Mylar slips can be removed without previous cooling. The block is placed vertically in a holder, allowing the light to pass through the block; the holder is located on the stage of a transmission light microscope, and the embedded thin layer is examined through the block face by using bright-field

illumination or interference contrast. The position of the cell selected and photographed prior to embedding can be relocated with ease with the aid of the
reference photomicrographs and the orientation grid, which appears as a relief on
the block face. The appropriate area of the face is trimmed under the binocular
microscope of the ultramicrotome.

Although the use of low-power objectives is sufficient for relocation of a cell
after embedding, higher magnifications may be used for examining the embedded thin layer. Thus it is possible in many cases to identify cells in the embedded
preparations when examining the latter at 40×; that is, a particular cell of
characteristic appearance can be selected in the polymerized block without the
preembedding light microscopy procedure.

SKIN

Method 1 (Fish) (Hawkes, 1974)

Specimens (1 mm³) are fixed for 1 hr at room temperature with a mixture of
0.75% glutaraldehyde, 3% formaldehyde, and 1% acrolein in 0.1 M cacodylate
buffer (pH 7.2) containing 0.02% $CaCl_2 \cdot H_2O$, 0.02 M s-collidine, 5.5%
sucrose, and 2% DMSO. After a rinse in 0.1 M cacodylate buffer (pH 7.2)
containing 0.02% $CaCl_2 \cdot H_2O$ and 5.5% sucrose, the specimens are postfixed
with 2% OsO_4 in buffer for 1 hr. Thin sections are stained first with lead citrate,
then with uranyl acetate, and again with lead citrate.

Method 2 (Frog)

Fixative solution
Formaldehyde (8%)	25 ml
Glutaraldehyde (50%)	5 ml
Cacodylate buffer (0.1 M, pH 7.3) to make	50 ml

A skin biopsy is removed and cut into very small pieces (less than 1 mm³), which
are fixed in the above solution for 90 min at room temperature. After being
rinsed in the buffer for 10 min, the specimens are postfixed with 1% OsO_4 in the
same buffer for 2 hr at 4°C.

SLIME MOLD

Fixative solution
Formaldehyde (2%)	25 ml
Glutaraldehyde (12%)	5 ml
Phosphate buffer (0.2 M, pH 7.2) to make	50 ml

Agar-grown cells are exposed to OsO_4 vapors for 30 min at room temperature. Cells and spores from the agar plate are washed in the buffer for 20 min and centrifuged at low speeds. They are embedded in 1.5% agar, and small cubes of agar are fixed in the above solution for 45 min at room temperature. After being washed in the same buffer, the cubes are postfixed in 1% OsO_4 in the same buffer for 45 min at room temperature. The cubes are treated with aqueous 0.5% uranyl acetate for 5 min.

SMALL SPECIMENS

Method 1 (Small Invertebrates and Biopsy Specimens) (Carson et al., 1982)

The following method is useful not only for processing small specimens (e.g., small invertebrates and biopsy specimens) without inadvertent loss during processing, but also for obtaining a specific orientation during embedding and sectioning. The soft and flexible lens paper conforms to the contour of the specimen and remains firmly attached to the specimens throughout processing. This paper also acts as the reference point during embedding, so that specimens can be embedded in a specific orientation for sectioning. The cyanoacrylic adhesive has the advantage of being equally effective in wet conditions. The presence of paper and adhesive does not seem to affect the sectioning.

Nasal scrapings taken with a curette from the inferior nasal turbinate are blotted briefly on a filter paper to remove excess fluid and mucus. These scrapings are transferred with a dissecting pin to a small piece of lens paper covered with a thin film of cyanoacrylic adhesive (Eastman 910). Care should be taken to keep the epithelial surface facing up, away from the adhesive. The lens paper with the adhesive is prepared 2 min before specimens are taken. The paper with the adhesive is prepared by placing one drop of adhesive on the paper, and the adhesive is then spread evenly with an applicator stick.

Immediately following the specimen transfer, the lens paper with the specimens is flooded with 2.5% glutaraldehyde in 0.1 M phosphate buffer (pH 7.4) with the aid of a pipette. The adhesive is allowed to set for 10 min. The specimen along with the lens paper is cut into smaller pieces and fixed with the glutaraldehyde solution for 1 hr, rinsed in buffer, postfixed with 1% OsO_4, stained *en bloc* in aqueous uranyl acetate, dehydrated, and embedded in a resin.

Method 2 (Eggs, Embryos, and Small Isolated Organs) (Junquera and Went, 1981)

The following method is useful for collecting, fixing, dehydrating, staining, and embedding small specimens such as eggs, embryos, and small isolated organs of different origins. The system consists of a standard glass filtration apparatus connected by suitable length of tubing to a vacuum membrane pump (Fig. 7.23). For specimens larger than 30 μm, filters that can be used for processing consist of a nylon-tissue disk with a diameter fitting the glass filtration apparatus and with opening sizes starting from 10 μm.

Specimens are transferred to the cylinder of the apparatus, and an appropriate vacuum is used to suck away the suspension, which is immediately replaced with the fixative. The specimens collected on the filter disk are fixed for an adequate time. The operation is repeated for each subsequent solution according to the chosen treatment. Care should be taken not to overload the apparatus: obstruction of the filter complicates the control of the suction forces, thereby increasing the risk of drying or damaging the specimens.

After infiltration with 30% resin, the filter disk along with the specimens is removed from the apparatus and placed in a small dish containing 30% resin. Embedding is completed by transferring the disk into resin mixtures of higher concentrations. Alternatively, the specimen disk is removed from the apparatus after *en bloc* staining, and infiltration and embedding are completed.

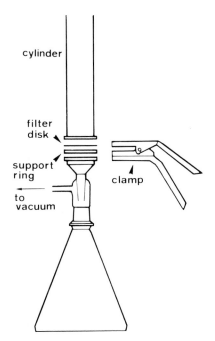

Fig. 7.23. Diagram of the glass filtration apparatus. (From Junquera and Went, 1981.)

SPERMATIDS

Solution A

Glutaraldehyde (25%)	25 ml
Cacodylate buffer (0.1 M, pH 7.2)	75 ml

Solution B

Sucrose	7.5 g
Cacodylate buffer (0.1 M, pH 7.2)	100 ml

Solution C

Potassium bichromate (5%)	15 ml
KOH (2.5 M)	2 ml
Distilled water	2 ml
Cacodylate buffer (0.1 M, pH 7.2)	10 ml

Solution D

OsO_4	1 g
Cacodylate buffer (0.1 M, pH 7.2)	100 ml
Sucrose	0.4 M

Small pieces of the testis are fixed in solution A for 1 hr at 4°C. After a rinse in buffer, the specimens are stored in solution B for 30 hr, transferred to solution C for 30 min, rinsed in solution B, and finally postfixed in solution D for 80 min, all at 4°C.

SPERMATOZOA

Method 1

Fixative solution

Glutaraldehyde (8%)	25 ml
Formaldehyde (10%)	25 ml
Picric acid	0.02 g
Phosphate buffer (0.1 M, pH 7.2) to make	100 ml

Two drops of semen are collected directly into 20 ml of this solution at room temperature. Proteinaceous seminal clumps, formed rapidly owing to coagulation, are discarded. The remaining nearly pure cellular suspension is centrifuged at 1000 rpm for 10 min (the total duration of fixation is 25 min). The supernatant is discarded and the pellet of highly concentrated spermatozoa is washed in phosphate buffer for 15 min. The pellet is postfixed with 1% OsO_4 in 0.1 M phosphate buffer (pH 7.2) for 15 min at room temperature.

Method 2

The semen is allowed to liquify and then is fixed for 2 hr at room temperature by adding more than an equal volume of 3% glutaraldehyde in 0.1 M cacodylate buffer (pH 7.3) containing 2.5 mM $CaCl_2$ (Ryder and Mackenzie, 1981). The specimens are centrifuged at 1000 g for 4 min, and the pellet is suspended in 10 ml of fresh buffer for 24 hr at 4°C. The specimens are postfixed with 1% OsO_4 in the same buffer for 2 hr at room temperature. They are then centrifuged at 1000 g for 4 min, and the pellet is suspended and rinsed in 50% ethanol for 10 min. After centrifugation at 1000 g for 4 min, the pellet is well mixed in 0.1–0.5 ml of 50% ethanol. The suspension is drawn up into a Pasteur pipette and then slowly dispensed drop by drop onto the surface of a stack of fast-filter papers (Whatman 41). The liquid is completely absorbed into the filter paper between drops. Thus a pile of cellular material is created on the surface of the filter paper (Fig. 7.24A). This material is removed with a fine spatula (Fig. 7.24B), placed on a 2–4% agar plate, and covered with molten agar at 60°C (Fig. 7.24C). After solidifying in a refrigerator, the surplus agar is trimmed off around the specimen (Fig. 7.24D), and the agar–specimen block is dehydrated and embedded in a resin.

Method 3

About 5 ml of sperm–Ringer solution is added to 5 ml of semen in a centrifuge tube. This tube is placed in a water jacket at 24°C, and the tube and jacket are transferred to a cold room for 5 hr at 2°C. The tube is centrifuged at 600 g for 15 min at 2°C. The sediment (spermatozoa) is resuspended in 3 ml of sperm–Ringer solution. This washing procedure is repeated three to five times. The final sediment is suspended in 5 ml of saline (0.154 M) containing streptomycin sulfate (50 μg/ml).

Equal volumes of the preceding suspension and Karnovsky's fixative (1.33 M formaldehyde, 0.5 M glutaraldehyde, and 4 mM $CaCl_2$ in 0.15 M sodium cacodylate buffer, pH 7.4) are mixed in a centrifuge tube and left overnight for fixation at 2°C. The tube is centrifuged at 20,000 g for 15 min, and the supernatant is discarded. An equal volume of 0.2 M cacodylate buffer is added to the sediment, and the tube is left overnight at 2°C. The tube is centrifuged at 20,000 g for 15 min and the supernatant is discarded. The solid pellet is cut into small pieces, which are postfixed for 2 hr at 4°C with OsO_4 (79 mM in 0.2 M collidine buffer, pH 7.4) containing 81 mM $CaCl_2$.

SPINDLE MICROTUBULES (ANAPHASE OF GRASSHOPPER SPERMATOCYTES)

Agar-treated glutaraldehyde is superior to conventional glutaraldehyde for preserving spindle microtubules in anaphase of grasshopper spermatocytes (Nicklas *et al.*, 1982). A 0.5% solution of agar is prepared in distilled water and set aside to gel. A 24% solution of glutaraldehyde is prepared from 70% glutaral-

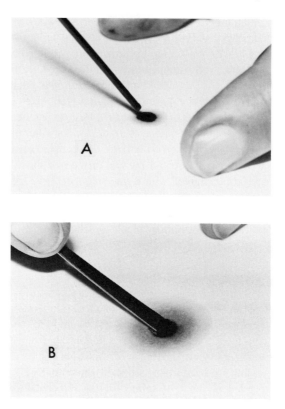

Fig. 7.24. Stages in the preparation of semen. (A) After fixation a pile of cellular components is formed dropwise on the surface of a fast filter paper. (B) Cellular material is removed using a fine spatula. (C) Once on the agar plate, the specimen is covered with molten agar. (D) Excess agar is

dehyde available in sealed glass ampoules. Then 23.3 ml of 24% glutaraldehyde is added to 50 ml of 0.5% agar. The two components are broken up into a slurry with a glass rod and then mixed on a rotary shaker at 200 rpm for 30 min at 4°C. The glutaraldehyde is separated from the agar by vacuum filtration through coarse filter paper in a Büchner funnel. The concentration of the recovered, treated glutaraldehyde is calculated from its optical density at 280 nm, using the optical density of the initial 24% glutaraldehyde as a standard. A typical yield is 22 ml of glutaraldehyde at a concentration of 11%. The treated glutaraldehyde is stored at 4°C, and aliquots are appropriately diluted with a buffer/NaCl stock solution each day to give buffered glutaraldehyde at the final concentration desired. Fresh agar-treated glutaraldehyde is prepared each week. Fixation with this glutaraldehyde is carried out according to a lengthy procedure, which allows studies of living cells to be combined with electron microscopic studies of cells (Nicklas *et al.*, 1979).

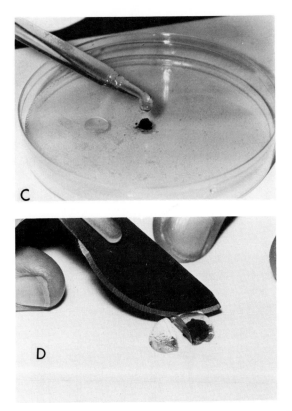

trimmed away and the resulting block is processed in the conventional manner. (From Ryder and Mackenzie, 1981.)

SPONGES

Method 1 (General Method) (Simpson and Vaccaro, 1974)

Sponges are flooded with an ice-cold mixture of 1 part 2.5% glutaraldehyde and 2 parts of 1% OsO_4 in 0.1 M cacodylate buffer (pH 7.4). After 5 min, the initial fixative is decanted and replaced with fresh solution. After 30 min, the dishes are decanted and rinsed with two changes of Tyrode's solution (Difco 555–72) and treated with uranyl acetate (0.25% in 0.1 M Veronal acetate buffer, pH 6.3) for 30 min. After two changes of Tyrode's solution, the sponges are dehydrated and embedded. If necessary, silicon can be removed from thin sections by floating the grids (face down) on a 2.5% aqueous solution of hydrofluoric acid for 45 min to 2 hr.

Method 2 (Freshwater or Marine Sponges) (Lethias et al., 1983)

Freshwater sponges are fixed with 0.4% glutaraldehyde in 0.02 M cacodylate buffer (pH 7.4) for 1 hr at 4°C. The osmolality of the solution is 80 mosmols. After several rinses in 0.04 M buffer, the specimens are postfixed with 1% OsO_4 in the buffer for 1 hr at 4°C. Marine sponges are fixed for 1 hr at 4°C with 0.4% glutaraldehyde in 0.1 M cacodylate buffer (pH 7.4) containing 0.3 M NaCl. The osmolality of this solution is 1100 mosmols. After several washings in 0.2 M buffer containing 0.35 M NaCl, the specimens are postfixed for 1 hr at 4°C with 1% OsO_4 in 0.1 M buffer containing 0.3 M NaCl.

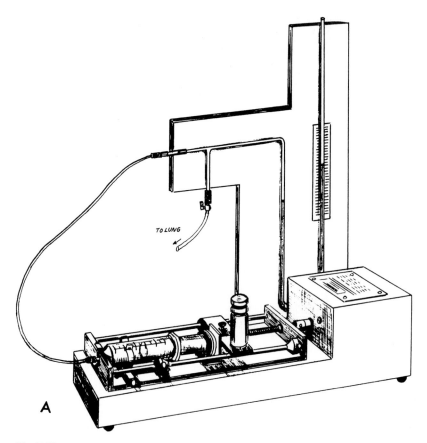

TO LUNG

A

Fig. 7.25. (A) Illustration of Harvard infusion/withdrawal pump used to insufflate OsO_4 vapors into the degassed lung. (B) Diagrammatic representation of the arrangement of equipment used for vascular perfusion fixation of the lungs and maintaining pulmonary pressure. The upper portion of

SPORES

Method 1

Spores are harvested by centrifugation at 3000 g and rinsed twice with distilled water. The spores are resuspended in molten 2% agar and solidified in the refrigerator. The agar is cut into 1-mm cubes and fixed with 15% formaldehyde in 0.1 M phosphate buffer (pH 7.2) for 2 hr at room temperature. The cubes are transferred to 6% glutaraldehyde in the same buffer for 2 hr. After a rinse in buffer, the cubes are postfixed with 1% OsO$_4$ in the same buffer for 2 hr. The addition of dibutyl phthalate to Spurr resin allows the cutting of thin sections of the ascus and spore walls of Ascomycetes (Merkus *et al.*, 1974).

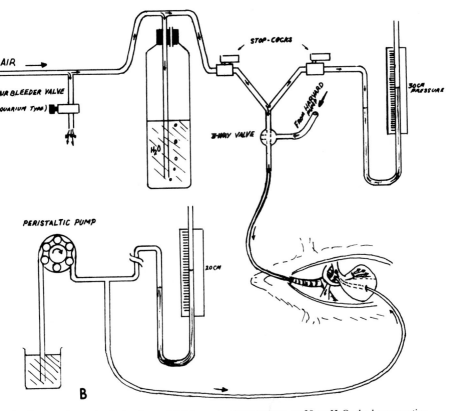

the diagram shows the device for maintaining pulmonary pressure at 30 cmH$_2$O; the lower portion shows the device for the perfusion of solutions directly into the pulmonary artery that are drained from the left atrium. (From Callas, 1982.)

Method 2 (Dormant Spores of *Bacillus subtilis*)

If fixative penetration is a problem, this method may be helpful. Dormant spores of *Bacillus subtilis* are rinsed five times with deionized water by centrifugation. Cells are prefixed with 5% glutaraldehyde in 0.1 M sodium cacodylate buffer (pH 7.2) for 5–8 hr at 4°C and centrifuged at 1500 g. After being washed five times with 0.05 M buffer, the pellet is fixed with 1% OsO_4 in buffer for 3 days at 4°C and then washed by centrifugation. The postfixation is accomplished with 2% $KMnO_4$ in deionized water for 3 hr at 4°C. The cells are rinsed repeatedly with 0.05 M buffer by centrifugation until the purple color of $KMnO_4$ in the supernatant is absent. The cells are suspended in molten 2% agar, and then the solidified agar is cut into 1-mm cubes for dehydration.

Method 3

Spores are fixed for 2–4 hr at room temperature with 4% glutaraldehyde in 0.1 M cacodylate buffer (pH 7.1) containing 0.01 M $CaCl_2$ and 3% Triton X100. The spores are pelleted and refixed overnight in this fixative without detergent at 4°C. The pellet is rinsed first in 0.2 M cacodylate buffer and then in collidine buffer, and is finally postfixed with 1% OsO_4 in collidine buffer for 12 hr at 4°C. Spores of different species and developmental stages may require changes in the duration of fixation.

SURFACTANT OVER ALVEOLAR SURFACE (Callas, 1982)

Osmium tetroxide vapor insufflation into the airway combined with vascular perfusion preserves the alveolar lining layer. This method preserves the surfactant as a continuous layer over the alveolar surface. Rats are anesthetized by an intraperitoneal injection of pentobarbital (35 mg/kg body weight). A tracheal cannula is inserted; it is attached to a reservoir of 100% oxygen, which the animal breathes for 5 min. A midline abdominal incision is made, and 0.5 ml of heparin is injected into the inferior vena cava and allowed to circulate for 5 min before the diaphragm is opened, allowing the lungs to collapse and degas.

After the lungs have been degassed, the pulmonary artery is cannulated. The tracheal cannula is connected to a calibrated motor driven syringe (Harvard infusion/withdrawal pump) that inflates the lungs at a rate of 24 ml/min (Fig. 7.25A). Transpulmonary pressure is monitored on a U-tube water manometer. Before infusion, 0.5 g of OsO_4 crystals is placed in a 35-ml glass syringe connected to the pump. Gentle heating of the syringe over an open flame (41°C) allows vaporization of the OsO_4. The lungs are reinflated with the OsO_4 vapors to a transpulmonary pressure of 30 cmH$_2$O, which is maintained for the duration of the infusion procedure (Fig. 7.25B).

The pulmonary arterial cannula is attached to a peristaltic pump (Cole-Parmer, Chicago), and the perfusion pressure, maintained at 20 cmH$_2$O, is measured with a U-tube water manometer (Fig. 7.25B). A 1% heparin-procaine solution in normal saline (0.9%) solution is introduced into the pulmonary artery. Blood washed from the lungs is drained by the transection of the left atrium. Lungs are perfused with 3% glutaraldehyde in 0.1 M phosphate buffer (pH 7.4, 550 mosmols) at a rate of 150 ml/30 min at 20 cmH$_2$O. About 1% OsO$_4$ in phosphate buffer is perfused at a rate of 50 ml/10 min at 20 cmH$_2$O in a fume hood. This step is followed by sequential perfusions at 20 cmH$_2$O with normal saline solution at a rate of 50 ml/10 min. The lungs are perfused with 50 ml of ethanol solutions of increasing concentration (10 min each). After perfusion with 95% ethanol (50 ml/10 min), the lungs are removed, cut into 1-mm slices, and processed for embedding in a resin.

TEETH

Rats are perfused with lactated Ringer's solution for 1 min through the left ventricle, and then perfused for 10 min with a mixture of 2% acrolein, 2.5% glutaraldehyde, and 3% formaldehyde in 0.06 M cacodylate buffer (pH 7.3) containing 0.05% CaCl$_2$ (Nanci et al., 1983). The mandibular incisors are dissected and washed in 0.1 M cacodylate buffer containing 0.05% CaCl$_2$. Postfixation follows for 2 hr at room temperature with 2% aqueous OsO$_4$. Thin sections are cut with a diamond knife and quickly removed from the distilled water in the knife trough to prevent crystallite dissolution. Poststaining with aqueous heavy metal salt solutions also causes dissolution of enamel crystallites. It should be noted that enamel crystallites are intrinsically electron opaque and therefore do not need poststaining to be viewed. Crystallites in osmicated enamel are more resistant to demineralization and electron beam damage. When grid-mounted sections are demineralized for 1 hr at room temperature with 1% formic acid in 10% sodium citrate (pH 5.0), all intrinsic electron opacity is removed. To visualize the organic matrix alone, sections are decalcified and then poststained with uranyl acetate followed by lead citrate.

TETRAHYMENA

Pelleted specimens are fixed for 2 hr at 4°C in a mixture of glutaraldehyde (0.5%) and OsO$_4$ (1%) buffered with 0.5 M cacodylate buffer (pH 7.2) having an osmolality of 120 mosmols. After a rinse in buffer, the pellet is postfixed with 1% OsO_4 for 1 hr at 4°C.

TISSUE CONTAINING HARD MINERAL FIBERS

Tissue specimens are fixed for 2 hr at 4°C with 3% glutaraldehyde in 0.1 M Sörensen's phosphate buffer (pH 7.2), rinsed in buffer, and then postfixed for 1 hr with buffered 1% OsO_4. After dehydration, the specimens are embedded in a resin. A major problem with sectioning tissues containing mineral fibers is that the fibers invariably tear or fall out of the sections. This problem can be minimized by cutting sections 0.3–0.5 μm with a diamond knife (8° cutting angle) at a speed of 1 mm/sec and using a scanning transmission electron microscope (STEM) at 200 kV. The sections should be stained by submersion in uranyl acetate in 70% ethanol for 15 min and then in lead citrate for 3–8 min.

TISSUE PREPARED ANHYDROUSLY IN ORGANIC REAGENTS

The following method avoids the use of aqueous solvents because specimens are exposed only to organic reagents (Landis *et al.*, 1977). Aqueous solvents, for example, induce transformations of the calcium phosphate solid mineral phase of bone tissue. Bones are dissected in 100% ethylene glycol on an ice-filled Petri dish. About 1-mm³ bone pieces are placed into loosely capped scintillation vials containing 3–4 ml of 100% ethylene glycol at room temperature. The vials are transferred to a vacuum desiccator connected to a mechanical pump. The desiccator may be placed on a shaker or rotator, in which the specimens are agitated throughout processing.

After 3 hr, the pump is disconnected, but the vacuum is maintained, and the vials are left on the shaker at 4 or 20°C for 24 hr. The glycol is replaced with Cellosolve at atmospheric pressure at 4 or 20°C. The Cellosolve is changed after 12 hr. At the end of an additional 12 hr, the Cellosolve is replaced with a 1 : 1 mixture of propylene oxide and LX 112 (Epon) in which the tissues remain for 1 week at room temperature. For Spurr embedment, Cellosolve is replaced in the following manner: two 1-hr changes of 100% ethanol, 1 hr in ethanol–Spurr (1 : 1), 1 hr in ethanol–Spurr (1 : 3), and overnight in 100% Spurr, all in a shaker or rotator at room temperature.

During sectioning, the diamond knife trough contains ethylene glycol. Sectioning speeds of 0.5–2.0 mm/sec are desirable with as small a block face as possible. Grids holding the sections are dried first by using filter paper to remove excess glycol and then by placing them under vacuum for a few hours. The grid is floated (section side down) on a drop of 8% uranyl acetate in glycol for 10–60 min at room temperature. After it has been dried by being touched with filter paper, the grid is placed in a vacuum desiccator to remove residual glycol.

The structure of dry tissues (e.g., seeds, spores, and lichens) changes on coming in contact with aqueous solutions (Hallam, 1976). A hydration effect may precede the fixation. In the following method specimens do not come in contact with aqueous solutions. Tissues are fixed with 6% solution of anhydrous glutaraldehyde in dimethyl sulfoxide (DMSO) for 24 hr and then treated with chloroform for 4 hr to remove excess glutaraldehyde. Postfixation is accomplished with 2% OsO_4 in chloroform for 2 hr. Water is removed from 50% glutaraldehyde by treating it with molecular sieves. Chloroform should be warmed to facilitate the dissolution of OsO_4, a procedure that must be carried out in a fume hood. Infiltration is accomplished in a 1 : 1 mixture of acetone and Spurr resin for 4 days before infiltration with pure resin for 3–4 weeks. Embedded tissues are very difficult to section; a diamond knife should be used, and the trough should be filled with 70% solution of DMSO in water.

A modification of the preceding method has been used to demonstrate mineral granules in osteoblast mitochondria and collagenous matrix (Manston and Katchburian, 1984). Such electron-opaque granules are less prominent and fewer in number in similar specimens prepared by conventional aqueous procedures. Bone fragments are immersed in 6% glutaraldehyde in DMSO for 18 hr at room temperature. The specimens are immersed in pure DMSO for 1 hr and then treated with a graded series of DMSO–ethanol mixtures up to pure ethanol. After two changes of propylene oxide, the specimens are embedded in a resin. Thin sections floated on ethylene glycol are collected on the grids.

TRACHEAL EPITHELIUM TISSUE

Tracheal epithelium (cat) is fixed for 2 hr at room temperature with half-strength Karnovsky's (1965) fixative in cacodylate buffer (pH 7.4) having a total osmolality of 965 mosmols (Tandler *et al.*, 1983). After being rinsed in buffered sucrose (final osmolality of 300 mosmols), the specimens are postfixed for 2 hr with 2% OsO_4 in the same buffer (final total osmolality of 365 mosmols). The specimens are rinsed in distilled water and then soaked overnight in cold aqueous 0.25% uranyl acetate.

VIROSOMES (Almeida *et al.*, 1975)

Liposomes are prepared from a 9 : 1 mixture of lecithin and dicetylphosphate. After this mixture has dried, phosphate-buffered saline solution is added to give a lipid concentration of 1 g/dl and the mixture is sonicated for 90 min in a Megason water bath at a frequency of 50 KHz. Most of the liposomes range in size from 20 to 100 nm. Viral subunits (prepared by detergent disruption of purified influenza virus) are isolated by density gradient centrifugation and concentrated by vacuum dialysis. These subunits are added to the liposome preparation to give a final composition of 200 μg/ml viral subunit and 0.5 g/dl lipid, and the mixture is sonicated for an additional 15 min. The name *virosome* was introduced because the influenza virus subunits relocate on the liposomes to give an appearance similar to that of the original virus.

VIRUSES (PLANT TISSUES)

Method 1

This method is useful for preserving labile, viral crystalline inclusions (e.g., wheat streak and barley stripe mosaic viruses) in plant tissues (Langenberg, 1982a). Tobacco mosaic virus inclusions are not preserved with this method.

Infected leaf tissue pieces are vacuum infiltrated with a solution of 0.1 M $KHPO_4$ and 0.05 M sodium citrate buffer (pH 7.4) and then placed on cold 0–25% (w/v), freshly prepared glycerol–sucrose gradients in buffer and allowed to equilibrate overnight at 5°C. The gradient is prepared by layering successively 5 ml of 6% glycerol–sucrose, 3% glycerol–sucrose, and buffer only. Buffer-infiltrated tissues may settle at the bottom in 5–6 hr. However, overnight equilibration is preferred. Specimens are transferred to a mixture of cold buffered solution of 25% glycerol–sucrose (15 ml) and 25% glutaraldehyde (10 ml) for 24 hr at 4°C. After being rinsed three times (15 min each) in cold buffer, specimens are postfixed with cold buffered 0.1% OsO_4 for 15 min, dehydrated, and embedded.

Method 2

This method is recommended for distinguishing particles of small polyhedral plant viruses containing single-stranded RNA from ribosomes in thin sections of infected cells (Hatta and Francki, 1981). The method is based on the assumption that the rRNA is digested and the RNA within the virus particles is little affected. However, this method is somewhat less satisfactory for some viruses whose encapsidated RNAs are susceptible to RNase when the virus has been fixed with aldehydes. Tissue pieces (2 mm³) from infected leaves or roots are fixed for 16 hr at 4°C with a mixture of 4% formaldehyde and 2% glutaraldehyde in 0.1 M cacodylate buffer (pH 7.0). After being rinsed with several changes of SSC buffer (0.15 M NaCl and 0.015 M sodium citrate) (pH 7.0) for 6 hr at room temperature, the specimens are incubated in pancreatic RNase (0.1–2.0 μg/ml) (Type IIA, Sigma Chemical Co.) in SSC buffer for 16 hr at 25°C. Before use, stock solutions of RNase are heated at 100°C for 10 min to eliminate the activities of any other contaminating nucleases. The specimens are postfixed with 1% OsO_4 for 1 hr at 4°C, dehydrated, and embedded.

WHOLE-CELL PREPARATIONS

Method 1 (Aizu et al., 1981)

A 5-mm paper punch is used to pierce several holes into round Thermonox plastic coverslips (25 mm in diameter, Lux Sci. Co.) (Fig. 7.26A). These coverslips are coated with Formvar film and then lightly carbon-coated on both sides and sterilized by ultraviolet irradiation for 10–15 min. After the coverslips have been placed in a 35-mm plastic Petri dish, cells (1×10^6 cell/dish) are plated on them. When ready for fixation, the coverslip is removed from the dish, briefly rinsed in phosphate-buffered saline, and immersed in 2.5% glutaraldehyde in 0.1 M cacodylate buffer for 15 min at room temperature. A brief rinse with 0.2 M cacodylate buffer is followed by postfixation with 1% OsO_4 in 0.1 M cacodylate buffer for 15 min.

The coverslip is washed in 30% acetone for 30 sec and then stained with 0.2% uranyl acetate in 30% acetone for 10 min at room temperature. After dehydration, the coverslip is transferred to a stainless-steel basket and critical-point dried (Hayat, 1978). For viewing with the SEM, dried cells are coated with carbon, using a vacuum evaporator.

A small quantity of 0.2% Neoprene in toluene is applied to heavy-gauge slot grids (1×2-mm hole) which are laid on the Formvar-coated round holes of the coverslip (Fig. 7.26B). The Neoprene is spread between the grid and Formvar film, but is not allowed to overflow into the slot of the grid. Using a fine forceps, the grids (containing the cells and the Formvar film) are removed from the coverslip. These grids can be stored in a desiccator before examination with the TEM.

Method 2 (Marek and Kelley, 1983)

This method is simple and avoids the problem of delicate stripping of Formvar films from coverslips (Aizu et al., 1981). Moreover, this method allows cells to contract under ATP-induced contraction of cell models. Glass coverslips (12 mm round) are cleaned overnight in 1 N nitric acid, rinsed in distilled water, treated with EDTA in water (1 g/l), and then rinsed in distilled water. These coverslips are stored in 100% ethanol. Individual coverslips are wiped dry with lens paper

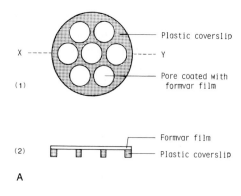

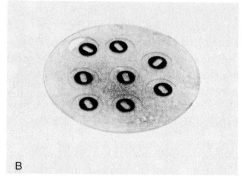

Fig. 7.26. (A) Diagram of the plastic coverslip coated with Formvar film. 1, top view. The plastic coverslip (25 mm in diameter) has seven pores, each 5 mm in diameter, which are covered with Formvar film. 2, transverse section at the *X–Y* level in (1). The coverslip is put with the Formvar-coated side upward so as not to make direct contact with the bottom of the Petri dish. (B) Photograph showing the plastic coverslip on which heavy slot grids are laid. (From Aizu *et al.*, 1981.)

and placed in a Denton Vacuum Evaporator. Two inches of palladium wire (99.95%, 0.008 in. diameter) are evaporated onto the coverslips in a vacuum of 1 $\times$ 10^{-5} torr at a distance of 5 in. The coverslips are then sterilized by being immersed overnight in 70% ethanol and dried under UV radiation in a Bioquest laminar flow hood. Alternatively, the coverslips can be dry-autoclaved. These coverslips are stored in Petri dishes until used.

Confluenced cell cultures are trypsinized by treatment with a solution of 0.45 mg/ml trypsin, 1 mM EDTA, and isotonic glucose–KCl–NaCl. The cells are plated at a 1 : 7 dilution on palladium-coated coverslips in culture dishes (L60 $\times$ 15 mm). After about 2 days, cells are extracted in a solution of 0.5% Triton X100, 4 M glylcerine, 0.1 M PIPES buffer (pH 7.2), 2 mM MgCl$_2$, and 50 mM KCl for 5 min at room temperature. The cells are fixed with a mixture of 2.7%

glutaraldehyde and 0.2% tannic acid in 0.1 M cacodylate buffer (pH 7.2) for 30 min at room temperature. After a rinse in the buffer, the preparations are dehydrated in ethanol and critical-point dried. They are then coated with a thin layer of carbon in an evaporator. Pieces of the palladium–cell preparation are transferred through several rinses of distilled water with a wire loop. These specimens are picked up on Formvar-coated grids (100–150 mesh).

Method 3 (Myoblasts) (Pudney and Singer, 1980)

Muscle tissue from human biopsy is finely minced and placed in a culture dish containing Eagle's minimum essential medium (EMEM) minus Ca^{2+} and Mg^{2+} to which 20% fetal calf serum has been added. Cells are cultured for 3 weeks, with the medium being changed every week. After 3 weeks, the cells have grown sufficiently for subpassage and are detached by means of a mixture of 0.5% trypsin and 5 mM EDTA in EMEM and plated at a density of 10^5 per plate.

Titanium grids are placed on a 5% Formvar film that has been released from a clean slide onto distilled water. The film and its grids are picked up on a glass coverslip, dried, and lightly coated with carbon. They are sterilized by immersion in 70% ethanol, dried, and placed in a standard Falcon plastic tissue-culture dish. Alternatively, grids on the coverslip can be sterilized with a short dose of UV light in a tissue-culture hood.

Cultured cells are rinsed with EMEM and then for 5 min with a mixture of 0.05% trypsin and 5 mM EDTA in EMEM, minus Ca^{2+} and Mg^{2+}, to remove them from the culture dish. The cell suspension is centrifuged at 700 g for 3 min and resuspended in EMEM containing 10% horse serum and 1% embryo extract. A suspension of myoblasts is plated onto the grids in 5 drops of medium and allowed to incubate for 2 hr. More medium is added so as to cover the grids, and the dishes are gently agitated to remove unattached cells from the grids. After the incubation has continued for 24 hr, the cells on the grids are viewed under a phase-contrast microscope to determine their condition.

The coverslips carrying the grids are removed from the medium, rinsed in hypotonic buffer (10 mM NaCl, 1.5 mM MgCl$_2$, 10 mM Tris-HCl, pH 7.4), and placed in 1% solution of Triton X100 in the same buffer for 2 min at room temperature. Once the coverslips have been gently rinsed in the buffer, the cells are fixed by immersing the preparations in 5% glutaraldehyde in 0.2 M collidine buffer (pH 7.3) for 10 min. Postfixation with OsO_4 is undesirable, because it will damage the integrity of actin. The specimens are briefly washed in the buffer and then in distilled water and 30% acetone. Dehydration is continued by passing them through a graded series of acetone, 70–100% (2 min each). Grids are removed from the coverslips and critical-point dried from CO_2. The cells are lightly coated with carbon for transmission electron microscopy. The same specimens can be observed with the SEM after a gold–palladium coating.

The preceding procedure is useful for any type of cells, provided they can grow on a carbon substrate. The filaments retain their three-dimensional configuration when maintained in a buffer for up to 60 min following extraction. This means that the extracted cell preparations are open to experimental manipulations such as immunocytochemistry, either at the light or electron microscope level.

WOOD

Solution A
Glutaraldehyde (25%)   12 ml
Phosphate buffer (0.025 *M*, pH 6.9) to make   100 ml
Solution B
Phosphate buffer (0.025 *M*, pH 6.9)   100 ml
Sucrose (0.2 *M*)   5 ml
CaCl$_2$ (0.001 *M*)   10 ml
Small pieces (less than 1 mm on each side) of the tissue are fixed in solution A for 2 hr at room temperature; the fixation is continued in the same solution for another 22 hr at 4°C. The specimens are washed in solution B for 4 hr and postfixed with 2% OsO$_4$ in 0.05 *M* phosphate buffer (pH 6.9) for 2 hr at 4°C. The specimens are dehydrated after being rinsed in solution B for 10 min.

WOOL FOLLICLES

Wool follicles obtained from a sheep are fixed for 4 hr at room temperature with a mixture of 5% glutaraldehyde and 4% formaldehyde in 0.1 *M* cacodylate buffer (pH 7.2). After being rinsed in 0.1 *M* cacodylate buffer (pH 7.2) containing 7.5% sucrose and then in 0.05 *M* phosphate buffer (pH 6.0) containing 7.5% sucrose, specimens are postfixed for 1 hr at 4°C with 8 m*M* OsO$_4$ in phosphate buffer. Specimens are treated for 30 min at 4°C with 1% tannic acid in phosphate buffer (pH 7.0) and then stained *en bloc* with 2% uranyl acetate in 0.1 *M* acetate buffer (pH5.2) for 1 hr at room temperature. Dehydration is carried out in hexylene glycol, and embedding in a resin follows.

YEAST

Method 1

Cells are pelleted by centrifugation at 1000 *g* and then rinsed twice in 0.02 *M* collidine buffer (pH 7.4) containing 0.2 *M* sucrose and 0.001 *M* CaCl$_2$; this rinsing is done by resuspension and recentrifugation. The cells are resuspended in 50% DMSO containing 2% acrolein and 2% glutaraldehyde buffered with 0.02 M collidine buffer for 7 hr at 4°C. The cells are rinsed twice (30 min each time) in the first solution and postfixed for 7 hr at 4°C with 2% OsO_4 in Veronal acetate buffer (pH 7.4). The cells are rinsed twice (10 min each time) in the first solution and then dehydrated.

Method 2

The pellets of yeast cells (e.g., *Candida*) are suspended in a few drops of rat serum. After centrifugation, the supernatant is discarded, and 2% glutaraldehyde in 0.1 M cacodylate buffer (pH 7.3) is gently added without disturbing the pellet. The pellet is allowed to solidify for 20 min at room temperature. To achieve an immediate fixation, the pellet is frozen on a drop of the fixative placed on a cold microtome table. Pellet sections 8 μm thick are cut and immediately dropped into the fixative, where they remain for 20 min. After being rinsed for 15 min in 0.1 M cacodylate buffer containing 0.2 M sucrose, the sections are postfixed for 30 min with 1% OsO_4 in 0.1 M cacodylate buffer. Following a brief rinse in buffer, the sections are treated with 0.5% uranyl acetate in Veronal acetate buffer (pH 5.2) for 10 min, dehydrated, and embedded in a resin.

Method 3

Cells in agar are placed in 0.1–0.5 M cysteine solution in phosphate buffer (pH 7.4) for 15–30 min. After a rinse in the same buffer, the cells are fixed in 6% glutaraldehyde in the same buffer for 2 hr at room temperature. Postfixation is accomplished in 2% OsO_4 in the same buffer for 1–2 hr and then the cells are stained *en bloc* for 2 hr at room temperature in 3% uranyl acetate in 30% ethanol. Cysteine treatment facilitates the penetration of the fixatives into the cells, presumably by weakening the cell wall.

Method 4 (*Saccharomyces cerevisiae*)

The following method is recommended for fixing yeast cells during mitosis or conjugation (modified from Byers and Goetsch, 1975). The cells are fixed for 30 min at 20°C with 2% glutaraldehyde either in 0.1 M cacodylate buffer (pH 6.8) containing 5 mM $CaCl_2$ or in 0.04 M phosphate buffer (pH 6.5) containing 1 mM $MgCl_2$. The cells are refixed for 16 hr at 4°C in the same fixative solution. Postfixation is accomplished with 2% OsO_4 in 0.1 M potassium acetate (pH 6.1) for 1 hr at 4°C. After a rinse in distilled water, the cells are treated *en bloc* with 2% aqueous uranyl acetate for 1 hr at 20°C before dehydration.

Internal structure is preserved better if cell walls are removed after the initial fixation. For this purpose, cells in stationary phase or meiosis are treated with a mixture of 0.1 M β-mercaptoethanol in 0.02 M EDTA and 0.2 M tris(hydroxymethyl)aminomethane hydrochloride (pH 8.1) for 10 min at 20°C before fixation. Mitotic cells do not require this pretreatment. After glutaraldehyde fixation, walls are removed by incubation of the rinsed cells in a 1 : 20 (v/v) mixture of glusulase solution (Endo Labs.) and 0.2 M phosphate–citrate buffer (pH 5.8) for 30 min at room temperature. Osmication follows as previously described.

Method 5 (*Candida albicans*)

Cells are fixed overnight at 4°C with 2% glutaraldehyde in 0.01 M phosphate buffer (pH 5.5). The cells are rinsed three times with buffer and then postfixed with 1% OsO_4 in the same buffer for 4–6 hr. Dehydration is accomplished through graded series of ethanol, starting with 70% for 30 min at each step, followed by two changes in propylene oxide (15 min each). Cells are gradually infiltrated with Spurr mixture, 1 hr being allowed at each step. Cells are kept in pure Spurr mixture for 4–6 hr at room temperature and then polymerized. Each solution change is accomplished by centrifugation. By using glutaraldehyde–OsO_4–tannic acid successively one can intensify cell wall structure.

Method 6 (Cell Wall)

The following method yields excellent staining of cell walls. First, 1% acrolein and 1% tris(1-aziridyl)phosphine oxide (TAPO) (K & K Labs., Plainview, N.Y.) are mixed in 0.1 M phosphate buffer (pH 7.2) and kept at room temperature for at least 50 min. Then, after centrifugation, the pellet is fixed for 10 min at 4°C in this mixture. The pellet is postfixed for 20 hr with 4% unbuffered OsO_4 and subsequently stained overnight with 0.5% uranyl acetate in Michaelis buffer (pH 5.8). After dehydration and embedding, ultrathin sections are post-stained with uranyl acetate and lead citrate.

Method 7 (Mitotic Spindle)

Arrested cells (*Saccharomyces cerevisiae*) are treated for 3 min with DET solution containing 0.1 M dithiothreitol, 0.02 M EDTA (disodium salt), and 0.2 M tris(hydroxymethyl)methyl ammonium chloride (pH 8.1). After being rinsed in PC buffer (0.2 M phosphate–citrate buffer, pH 5.8) at 36°C, the specimens are fixed for 30 min with 3% glutaraldehyde in 0.1 M Na-cacodylate buffer (pH 6.8) containing 5 mM $CaCl_2$, and then with fresh fixative for 16 hr at 4°C. The specimens are rinsed once in cacodylate buffer and three times in PC buffer and then incubated for 2 hr in 4% β-glucuronidase type H-2 in PC buffer. After a rinse in cacodylate buffer, the cells are postfixed with 2% OsO_4 in cacodylate buffer for 1 hr, rinsed in distilled water, and stained *en bloc* for 1 hr with 2% aqueous uranyl acetate.

Appendix

Commonly Used Chemicals for Stock Solutions of Specified Molarities

Chemical	Formula	Molecular weight
Acetic acid	$CH_3 \cdot CO_2H$	60
Barium chloride	$BaCl_2 \cdot 2H_2O$	244
Boric acid	H_3BO_3	62
Calcium acetate	$Ca(CH_3 \cdot CO_2)_2$	158
Calcium chloride	$CaCl_2 \cdot 6H_2O$	219
Citric acid	$C_6H_8O_7 \cdot H_2O$	210
s-Collidine	$2,4,6-(CH_3)_3(C_5H_2N)$	121.18
Copper sulfate	$CuSo_4 \cdot 5H_2O$	250
Hydrochloric acid	HCl	36.465
Glucose	$C_6H_{12}O_6$	180
Glycine	$NH_2 \cdot CH_2CO_2H$	75
Lead nitrate	$Pb(NO_3)_2$	331
Magnesium chloride	$MgCl_2 \cdot 6H_2O$	203
Magnesium nitrate	$Mg(NO_3)_2 \cdot 6H_2O$	256
Magnesium sulfate	$MgSO_4 \cdot 7H_2O$	246
Manganese chloride	$MnCl_2 \cdot 4H_2O$	198
Potassium carbonate	K_2CO_3	138
Potassium chloride	KCl	74.5
Potassium cyanide	KCN	65
Potassium ferricyanide	$K_3Fe(CN)_6$	329
Potassium sodium tartrate	$NaKC_4H_4O_6 \cdot 4H_2O$	282

(continued)

Chemicals for Stock Solutions *(Continued)*

Chemical	Formula	Molecular weight
Silver nitrate	$AgNO_3$	170
Sodium acetate	$CH_3 \cdot CO_2Na \cdot 3H_2O$	136
Sodium bicarbonate	$NaHCO_3$	84
Sodium cacodylate (trihydrate)	$Na(CH_3)_2AsO_2 \cdot 3H_2O$	214
Sodium chloride	$NaCl$	58.5
Sodium citrate	$Na_3C_6H_5O_7 \cdot 2H_2O$	294
Sodium dichromate	$Na_2Cr_2O_7 \cdot 2H_2O$	298
Sodium dihydrogen orthophosphate	$NaH_2PO_4 \cdot 2H_2O$	156
Sodium hydroxide	$NaOH$	40
Sodium phosphate (monobasic)	$NaH_2PO_4 \cdot H_2O$	138
Sodium phosphate (dibasic)	$NaH_2PO_4 \cdot 2H_2O$	141.98
Sodium sulfide	$Na_2S \cdot 9H_2O$	240
Sodium tetraborate	$Na_2B_4O_7 \cdot 10H_2O$	381
Succinic acid	$C_4H_6O_4$	118
Sucrose	$C_{12}H_{22}O_{11}$	342
Tris(hydroxymethyl)methylamine	$NH_2C(CH_2OH)_3$	121
Uranyl acetate	$UO_2(CH_3 \cdot COO)_2 \cdot 2H_2O$	424

COMMONLY USED SALTS AND THEIR PHYSICOCHEMICAL PROPERTIES

Sodium cacodylate:
    Anhydrous $Na(CH_3)_2AsO_2$ Unstable; readily absorbs 3 molecules of water
    Trihydrate $Na(CH_3)_2AsO_2 \cdot 3H_2O$ Stable
Disodium hydrogen phosphate:
    Anhydrous Na_2HPO_4 Unstable; absorbs 2–7 molecules of water
    Dihydrate $Na_2HPO_4 \cdot 2H_2O$ Stable
    Heptahydrate $Na_2HPO_4 \cdot 7H_2O$ Relatively stable
    Dodecahydrate $Na_2HPO_4 \cdot 12H_2O$ Unstable; loses water
Sodium dihydrogen phosphate:
    Anhydrous NaH_2PO_4 Unstable; absorbs water
    Monohydrate $NaH_2PO_4 \cdot H_2O$ Unstable; absorbs water
    Dihydrate $NaH_2PO_4 \cdot 2H_2O$ Stable
Sodium acetate:
    Anhydrous CH_3COONa Unstable; readily absorbs 3 molecules of water
    Trihydrate $CH_3COONa \cdot 3H_2O$ Relatively stable; water content may vary
Calcium chloride ($CaCl_2$):
    Anhydrous   Unstable; very hygroscopic
    Dihydrate $CaCl_2 \cdot 2H_2O$ Unstable; absorbs water
    Hexahydrate $CaCl_2 \cdot 6H_2O$ Unstable; water content may vary

Source: Kalimo and Pelliniemi (1977).

EARLE'S BALANCED SALT SOLUTION

	Components		References													
			A		B		C		D		E		F		G	
ID number		Formula weight	g/l	M	g/l	M	g/l	M	g/l	M	g/l	M	g/l	M	g/l	M
1	NaCl	58.44	6.800	0.1164	6.800	0.1164	6.800	0.1164	6.800	0.1164	6.800	0.1164	6.800	0.1164	6.800	0.1164
2	KCl	74.56	0.400	0.0054	0.400	0.0054	0.400	0.0054	0.400	0.0054	0.400	0.0054	0.400	0.0054	0.400	0.0054
3-a	$CaCl_2$	110.99	0.200	0.0018			0.200	0.0018	0.200	0.0018	0.200	0.0018	0.200	0.0018	0.200	0.0018
3-b	$\cdot 2H_2O$	147.01														
3-c	$\cdot 6H_2O$	219.08			0.393	0.0018[hc]										
4-a	$MgSO_4$	120.37	0.100	0.0008												
4-b	$\cdot 7H_2O$	264.48			0.200	0.0008[hc]	0.100	0.0004[hd]	0.200	0.0008[hc]	0.200	0.0008[hc]	0.001	0.0004[hd]	0.200	0.0008[hc]
5-a	NaH_2PO_4	119.98	0.125	0.0010			0.125	0.0009[hd]					0.125	0.0009[hd]		
5-b	$\cdot H_2O$	137.99							0.140	0.0010[hc]	0.140	0.0010[hc]			0.140	0.0009[hd]
5-c	$\cdot 2H_2O$	153.99	0.140		0.140	0.0009[hd]										
6	Glucose	180.16	1.000	0.0056	As desired		1.000	0.0056	1.000	0.0056	1.000	0.0056	1.000	0.0056	1.000	0.0056
7	$NaHCO_3$	84.01	2.200	0.0262	2.240	0.0267[hd]	2.200	0.0262	2.200	0.0262	2.200	0.0262	2.200	0.0262	2.200	0.0262
8	Phenol red		None		0.015[s]		0.050[s]		0.01[s]		0.01–0.05[s]		0.05[s]		0.02[s]	

Source: Burke and Croxall (1983).

Note: A = Earle (1934, 1943); B = Adams (1980); C = Bashor (1979); D = Kuchler (1977); E = Parker (1961); F = Paul (1975); G = Whitaker (1973).
[hc] Hydration state (h) of component varies from that given in cited paper but is adjusted in gram weight to give correct (c) molarity.
[hd] Hydration state (h) of component varies from that given in cited paper and results in different (d) molarity.
[s] Substitution: identifies a component that does not appear in the cited paper.

HANKS' BALANCED SALT SOLUTION (1948, 1949)

References

ID number	Components	Formula weight	A g/l	A M	B g/l	B M	C g/l	C M	D g/l	D M	E g/l	E M	F g/l	F M	G g/l	G M
1	$NaCl$	58.44	8.000	0.1369	8.000	0.1369	8.000	0.1369	8.000	0.1369	8.000	0.1369	8.000	0.1369	8.000	0.1369
2	KCl	74.56	0.400	0.0054	0.400	0.0054	0.400	0.0054	0.400	0.0054	0.400	0.0054	0.400	0.0054	0.400	0.0054
3-a	$MgSO_4$	120.37														
3-b	$\cdot 7H_2O$	246.48	0.200	0.0008	0.200	0.0008	0.200	0.0008	0.200	0.0020[hd]	0.200	0.0008	0.100	0.0004[wd]	0.200	0.0008
4-a	$MgCl_2$	95.21														
4-b	$\cdot 2H_2O$	131.25														
4-c	$\cdot 6H_2O$	203.31														
5-a	$CaCl_2$	110.99	0.200	0.0018	0.140	0.0013			0.140	0.0013	0.140	0.0013			0.200	0.0018
5-b	$\cdot 2H_2O$	147.02											0.100	0.0005[s]		
5-c	$\cdot 6H_2O$	219.08					0.276	0.0013								
6-a	NaH_2PO_4	119.98														
6-b	$\cdot H_2O$	137.99														
6-c	$\cdot 2H_2O$	153.99														
7-a	Na_2HPO_4	141.96														
7-b	$\cdot 2H_2O$	177.96	0.060	0.0003	0.060	0.0003							0.060	0.0003	0.060	0.0004
7-c	$\cdot 7H_2O$	268.07					0.090	0.0003	0.090	0.0003						
7-d	$\cdot 12H_2O$	358.20									0.120	0.0003				
8-a	KH_2PO_4	136.09	0.060	0.0004	0.060	0.0004	0.060	0.0004	0.060	0.0004	0.060	0.0004	0.060	0.0004	0.060	0.0004
8-b	$\cdot 2H_2O$	172.13														
9	Glucose	180.16	2.000	0.011	1.000	0.0056	As desired		1.000	0.0056	1.000	0.0056	1.000	0.0056	1.000	0.0056
10	$NaHCO_3$	84.01	1.400	0.0167	0.350	0.0042	0.168 / 0.015[wd]		0.350 / 0.010[wd]	0.0042	0.350 / 0.01–0.05[wd]	0.0042	0.350	0.0042	0.350	0.0042
11	Phenol red		0.020		0.020						0.020		0.020		0.020	

Source: Burke and Croxall (1983).

Note: A = Hanks (1948); B = Hanks and Wallace (1949); C = Adams (1980); D = Kuchler (1977); E = Parker (1961); F = Paul (1975); G = Penso and Balducci (1963).

[hd] Hydration state (h) of component varies from that given in cited paper and results in different (d) molarity.

[s] Substitution: identifies a component that does not appear in the cited paper.

[wd] Weight (w) given, which varies from that of cited paper and results in different (d) molarity.

PHOSPHATE-BUFFERED SALINE (Dulbecco and Vogt, 1954)

Components ID number	Formula weight	A g/l	A M	B g/l	B M	C g/l	C M	D g/l	D M	E g/l	E M	F g/l	F M	G g/l	G M
1 NaCl	58.44	8.000	0.1369	10.00	0.1715[wd]	8.000	0.1369	8.000	0.1369	8.000	0.1369	8.000	0.1369	8.000	0.1369
2 KCl	74.56	0.200	0.0027	0.25	0.0034[wd]	0.200	0.0027	0.200	0.0027	0.200	0.0027	0.200	0.0027	0.200	0.0027
3-a Na_2HPO_4	141.96	1.150	0.0081	1.44	0.0101[wd]			1.150	0.0081						
3-b $\cdot H_2O$	159.96													1.420	0.0089[hd]
3-c $\cdot 2H_2O$	177.96									1.150	0.0064[hd]				
3-d $\cdot 7H_2O$	268.07														
3-e $\cdot 12H_2O$	358.20					2.310	0.0064[hd]					2.900	0.0081[hc]		
4-a KH_2PO_4	136.09	0.200	0.0015	0.25	0.0018[wd]	0.200	0.0015	0.200	0.0015	0.200	0.0015	0.200	0.0015	0.200	0.0015
4-b $\cdot 2H_2O$	172.13														
5-a $CaCl_2$	110.99	0.100	0.0009			0.100	0.0009	0.100	0.0009	0.100	0.0009	0.100	0.0009	0.100	0.0009
5-b $\cdot 2H_2O$	147.02			0.10	0.0007[hd]										
5-c $\cdot 6H_2O$	219.08														
6-a $MgCl_2$	95.21														
6-b $\cdot 2H_2O$	131.25														
6-c $\cdot 6H_2O$	203.31	0.100	0.0005	0.10	0.0005	0.100	0.0005	Omitted[o]		0.100	0.0005	0.100	0.0005	0.100	0.0005
7 Phenol red		0.0	0.0	0.0	0.0	0.0	0.0	0.0		0.0		0.0		0.2[s]	

Source: Burke and Croxall (1983).

Note: A = Dulbecco and Vogt (1954); B = Adams (1980); C = Bashor (1979); D = Kuchler (1977); E = Paul (1975); F = Schmidt (1964); G = Whitaker (1973).

[hc]Hydration state (h) of component varies from that given in cited paper but is adjusted in gram weight to give correct (c) molarity.

[hd]Hydration state (h) of component varies from that given in cited paper and results in different (d) molarity.

[o]Omission of component that does not appear in the cited paper.

[s]Substitution: identifies a component that was listed in cited paper.

[wd]Weight (w) given, which varies from that of cited paper and results in different (d) molarity.

BALANCED SALT SOLUTIONS

Tyrode

Solution A

Distilled water	70 ml
NaCl	20 g
KCl	0.5 g
$CaCl_2 \cdot 6H_2O$	0.5 g
$MgCl_2 \cdot 6H_2O$	0.25 g
Distilled water to make	100 ml

Solution B

Distilled water	90 ml
$NaHCO_3$	5.0 g
NaH_2PO_4	0.25 g
Distilled water to make	100 ml

Final solution

Distilled water	940 ml
Solution A	40 ml
Solution B	20 ml
Glucose	1 g

Amphibian Ringer

Distilled water	180 ml
NaCl	1.3 g
KCl	0.04 g
$CaCl_2 \cdot 6H_2O$	0.04 g
Distilled water to make	200 ml

Mammalian Ringer

Distilled water	180 ml
NaCl	1.8 g
KCl	0.84 g
$CaCl_2 \cdot 6H_2O$	0.48 g
$NaHCO_3$	0.1 g
Glucose	0.1 g
$MgCl_2 \cdot 6H_2O$	0.05 g

Phosphate-Buffered Physiological Saline

1 M NaCl	12 ml
0.2 M KCl	2 ml
0.2 M CaCl$_2$	1 ml
0.1 M MgSO$_4$·7H$_2$O	1 ml
0.2 M Phosphate buffer (pH 6.9)	10 ml
Distilled water to make	100 ml

Standard Cacodylate Washing Solution

0.2 M Sodium cacodylate	24 ml
0.2 M Cacodylic acid	10 ml
0.2 M Calcium acetate	1 ml
Isotonic sodium sulfate to make	100 ml

If cacodylic acid is not available, the following solution will suffice:

0.2 M Sodium cacodylate	34 ml
0.1 M Nitric or sulfuric acid	20 ml
0.2 M Calcium acetate or nitrate	1 ml
0.2 M Sodium sulfate	38 ml
Distilled water to make	100 ml

KREBS–HENSELEIT SALINE (KHS)

Stock solution A:	NaCl	138.456 g/l
Stock solution B:	KCl	35.42 g/l
	MgSO$_4$·7H$_2$O	29.414 g/l
	KH$_2$PO$_4$	16.247 g/l
Stock solution C:	NaHCO$_3$	42 g/l
Stock solution D:	CaCl$_2$·6H$_2$O	55.326 g/l
	or	
	CaCl$_2$ (anhydrous)	28.19 g/l

Final solution

Stock solution A	50 ml
Stock solution B	10 ml
Stock solution C	50 ml
Distilled water to make	990 ml

The final solution is bubbled with 95% O_2 and 5% CO_2 for 10–15 min. Then 10 ml of stock solution D is added, and bubbling is continued. The pH is adjusted to

7.4. For preparing calcium-free KHS, stock solution D is substituted with NaCl (22.147 g/l).

LOCKE SOLUTION

NaCl	9.0 g
KCl	0.4 g
$CaCl_2$	0.2 g
$NaHCO_3$	0.3 g
Glucose	0.5 g

These ingredients are dissolved in 1 l of distilled water.

PHYSIOLOGIC SALT SOLUTIONS

Specimen	NaCl	KCl	$CaCl \cdot 2H_2O$	$MgCl_2 \cdot 6H_2O$	$NaHCO_3$	Additions
Amphibia	6.50	0.14	0.16	—	0.20	0.012 g Na_2HPO_4
Artificial seawater	19.0	0.53	0.49	5.10	—	1.5 g $MgSO_4$
Cockroach	10.93	1.57	1.10	0.36	—	—
Crustacea (crayfish)	12.00	0.40	1.99	0.53	0.20	—
Crustacea (marine)	30.65	0.99	1.82	5.04	to pH 7	—
Drosophila	7.50	0.35	0.28		0.20	—
Insects	9.00	0.20	0.27	—	to pH 7.2	4 g dextrose
Invertebrates	6.8	0.20	0.20	0.10	0.12	0.2 g Na_2HPO_4 + 0.77 g glucose·H_2O
Mammals (Krebs)	6.92	0.35	0.37	—	2.10	1.18 ml 1 M $MgSO_4$ + 0.16 g KH_2PO_2
Mammals (Locke)	9.00	0.42	0.32	—	0.20	2 g dextrose
Mammals (Tyrode)	8.00	0.20	0.27	0.21	1.00	0.06 g $Na_2HPO_4 \cdot H_2O$
Mollusks (Helix)	7.74	0.89	2.64	5.62	2.05	—
Mollusks (marine)	30.97	0.80	1.91	—	—	phosphate buffer to pH 7.2

Note: Values given are in grams per liter. To avoid precipitations with $NaHCO_3$, bicarbonate should be added after all other salts have been dissolved.

MOLARITIES OF SOLUTIONS ISOTONIC WITH
MAMMALIAN BLOOD
(Δ = $-0.56°C$, 300 mosmols)

$NaCl$, KCl, or NaH_2PO_4	0.16 M
$CaCl_2$	0.12 M
$MgCl_2$	0.11 M
Na_2HPO_4	0.13 M
Sucrose	0.29 M
Glucose	0.30 M

ARTIFICIAL SEA WATER

One liter of artificial seawater is prepared by dissolving 28.3 g of NaCl in 900 ml of distilled water. Appropriate quantities of 1 M stock solutions of the following components are added and the mixture brought to 1 l with distilled water. The pH is 7.7 and density is 1.026.

Component	Concentration (g/l)
$MgSO_4$	3.43
$MgCl_2$	2.40
$CaCl_2$	1.22
KCl	0.76
$NaHCO_3$	0.21
$NaBr$	0.082
H_3BO_4	0.062
$NaSi_4O_9$	0.0098
Al_2Cl_6	0.0066
H_3PO_4	0.0049
$LiNO_3$	0.0035
NH_4OH	0.0018

Instant Seawater

One pound of Aquamarine salts (Aquatrol, Inc.) is dissolved in about 12 l of distilled water by constant stirring for 1 hr at room temperature. The solution is filtered through glass wool to remove a small residue of undissolved materials.

References

Abrunhosa, R. (1972). Microperfusion fixation of embryos for ultrastructural studies. *J. Ultrastruct. Res.* **41**, 176.

Achong, B. G., and Epstein, M. A. (1965). A method for preparing microsamples of suspended cells for light and electron microscopy. *J. R. Microsc. Soc.* **84**, 107.

Adams, C. W. M., Bayliss, O. B., Hallpike, J. F., and Turner, D. R. (1971). Histochemistry of myelin. XIII. Anionic staining of myelin basic proteins for histology, electrophoresis, and electron microscopy. *J. Neurochem.* **18**, 389.

Adams, R. L. P. (1980). Media formulations. *In* "Laboratory Techniques in Biochemistry and Molecular Biology, Vol. 8, Cell Culture for Biochemists," pp. 246–270. Elsevier/North-Holland, New York.

Aggarwal, S. K. (1976). Platinum/pyrimidine complexes for electron microscopic cytochemistry of deoxyribonucleic acid. *J. Histochem. Cytochem.* **24**, 984.

Ainsworth, S. K., Ito, S., and Karnovsky, M. J. (1972). Alkaline bismuth reagent for high resolution ultrastructural demonstration of periodic-reactive sites. *J. Histochem. Cytochem.* **20**, 995.

Aizu, S., Itoh, T., and Yamamoto, T. Y. (1981). A simple method for whole-cell preparation in electron microscopy. *J. Microsc. (Oxford)* **124**, 183.

Akhtar, M., Ali, M. A., Owen, E., and Bakry, M. (1980). A simple method for processing fine-needle aspiration biopsy specimens for electron microscopy. *J. Clin. Pathol.* **33**, 1214.

Albersheim, P., and Killias, U. (1963). The use of bismuth as an electron stain for nucleic acids. *J. Cell Biol.* **17**, 93.

Albert, E. N., and Fleischer, E. (1970). A new electron-dense stain for elastic tissue. *J. Histochem. Cytochem.* **18**, 697.

Allen, C. F., Cressman, M. W. J., and Johnson, H. B. (1955). Tetraiodophthalic anhydride. *Org. Synth. Collect.* **3**, 796.

Allen, F. H. E., and Friend, J. (1983). Embedding difficult plant tissue—a technical note. *Proc. R. Microsc. Soc.* **18**, 290.

Allizard, F., and Zylberberg, L. (1982). A technical improvement for sectioning laminated fibrous tissues for electron microscopic studies. *Stain Technol.* **57**, 335.

Almeida, J. D., Edwards, D. C., Brand, C. M., and Heath, T. D. (1975). Formation of virosomes from influenza subunits and liposomes. *Lancet* **ii**, 899.

Almeida, J. D. (1980). Practical aspects of diagnostic electron microscopy. *Yale J. Biol. Med.* **53**, 5.

Altman, F. P., and Barrnett, R. J. (1975). The ultrastructural localization of enzyme activity in unfixed tissue sections. *Histochemistry* **44**, 179.

Altmann, P. L., and Dittmer, D. S., eds. (1973). "Biological Data Handbook," 2nd ed. Fed. Am. Soc. Exp. Biol., Bethesda, Maryland.

Angermüller, S., and Fahimi, H. D. (1982). Imidazole-buffered osmium tetroxide: an excellent stain for visualization of lipids in transmission electron microscopy. *Histochem. J.* **14**, 823.

Angold, R. (1980). Staining lipids in resin sections. *Proc. R. Microsc. Soc.* **15**, 170.

Ashton, F. T., and Pepe, F. A. (1981). The myosin filament. VIII. Preservation of subfilament organization. *J. Microsc. (Oxford)* **123**, 93.

Bachhuber, K., and Frösch, D. (1983). Melamine resins, a new class of water-soluble embedding media for electron microscopy. *J. Microsc. (Oxford)* **130**, 1.

Bachmann, L., and Schmitt, W. W. (1971). Improved cryofixation applicable to freeze-etching. *Proc. Natl. Acad. Sci. U.S.A.* **68**, 2149.

Barajas, L., Wang, P., Powers, K., and Nishio, S. (1981). Identification of renal neuroeffector junctions by electron microscopy of reembedded light microscopic autoradiograms of semithin sections. *J. Ultrastruct. Res.* **77**, 379.

Barnard, T. (1980). Ultrastructural effects of the high molecular weight cryoprotectants dextran and polyvinyl pyrrolidone on liver and brown adipose tissue in vitro. *J. Microsc. (Oxford)* **120**, 93.

Barnard, T. (1982). Thin frozen-dried cryosections and biological x-ray microanalysis. *J. Microsc. (Oxford)* **126**, 317.

Barnes, C. D., and Etherington, L. D. (1973). "Drug Doses and Laboratory Animals." Univ. of California Press, Berkeley.

Bashor, M. M. (1979). Basic methods of dispersion and disruption of tissues. *In* "General Cell Culture Techniques" (W. B. Jacoby and I. H. Pastan, eds.), Methods in Enzymology, Vol. 58, pp. 119–131. Academic Press, New York.

Baumeister, W., and Seredynski, J. (1976). Preparation of perforated films with predeterminable hole size distributions. *Micron* **7**, 49.

Baumeister, W., and Hahn, M. (1978). Specimen supports. *In* Principles and Techniques of Electron Microscopy: Biological Applications" (M. A. Hayat, ed.), Vol. 8. Van Nostrand-Reinhold, New York.

Beards, G. M. (1982). A method for the purification of rotaviruses and adenoviruses from feces. *J. Virol. Methods* **4**, 343.

Beesley, J. E. (1978). A new technique for preparing cell monolayers for electron microscopy. *Stain Technol.* **53**, 48.

Behrman, E. J. (1984). The chemistry of osmium tetroxide fixation. *Scanning Electron Microsc.* p. 1.

Bell, G. M., and Roscoe, D. H. (1982). Simple and rapid structure determination of virions in bacteriophage plaques. *Micron* **13**, 41.

Bennett, H. S., Wyrick, A. D., Lee, S. W., and McNeil, J. H. (1976). Science and art of preparing tissues embedded in plastic for light microscopy, with special reference to glycol methacrylate, glass knives, and simple stains. *Stain Technol.* **51**, 71.

Bergh Weerman, M. A. v.d., and Dingemans, K. P. (1984). Rapid deparaffinization for electron microscopy. *Ultrastruct. Pathol.* **7**, 55.

Bergquist, P. R., and Green, C. R. (1977). A method for preserving larva substrate relationships during preparation for electron microscopy. *Biol. Cell.* **28**, 85.

Berkowitz, L. R., Fiorello, O., Kruger, L., and Maxwell, D. S. (1968). Selective staining of nervous tissue for light microscopy following preparation for electron microscopy. *J. Histochem. Cytochem.* **16,** 808.

Bernhard, W. (1969). A new staining procedure for electron cytology. *J. Ultrastruct. Res.* **27,** 250.

Bernhard, W., and Avrameas, S. (1971). Ultrastructural visualization of cellular carbohydrate components by means of concanavalin A. *Exp. Cell Res.* **64,** 232.

Berthold, C. H. (1968). A study on the fixation of large mature feline myelinated ventral lumbar spinal-root fibers. *Acta Soc. Med. Ups., Suppl.* No. 9, 1.

Birch-Andersen, A., Ferguson, D. J. P., and Pontefract, R. D. (1976). A technique for obtaining thin sections of coccidian oocytes. *Acta Pathol. Microbiol. Scand.* **84,** 235.

Björkman, N., and Hellström, B. (1965). Lead–ammonium acetate, a staining method for electron microscopy free of contamination by carbonate. *Stain Technol.* **40,** 169.

Björkman, N., Bantzer, V., Hasselager, E., Holm, H., and Kjaersgaard, P. (1981). Perfusion *in vivo* of the porcine placenta. Fixation for EM. *Placenta* **2,** 287.

Bloom, E. E., and Aghajanian, G. K. (1968). Fine structural and cytochemical analysis of the staining of synaptic junctions with phosphotungstic acid. *J. Ultrastruct. Res.* **22,** 361.

Bohman, S.-O. (1974). The ultrastructure of the rat renal medulla as observed after improved fixation methods. *J. Ultrastruct. Res.* **47,** 329.

Bonnová, E., and Rýc, M. (1976). A method of orientation of embedded bacterial colonies prepared for ultrathin sections. *Fol. Biol. (Prague)* **22,** 366.

Boshier, D. P., Holoway, H., and Kitchin, L. F. (1984). A comparison of standard lipid staining techniques used in electron microscopic studies of mammalian tissues. *Stain Technol.* **59,** 83.

Bosman, F. T., and Go., P. M. N. Y. H. (1981). Polyethylene glycol embedded tissue sections for immunoelectron microscopy. *Histochemistry* **73,** 195.

Bowes, D., Bullock, G. R., and Winsey, N. J. P. (1970). A method for fixing rabbit and rat hind limb skeletal muscle by perfusion. *Proc. Int. Congr. Electron Microsc., 7th, Grenoble, Fr.* **1,** 397.

Braak, H., and Braak, E. (1982). A simple procedure for electron microscopy of Golgi-impregnated nerve cells. *Neurosci. Lett.* **32,** 1.

Bradley, T. J. (1981). Improved visualization of apical vesicles in chloride cells of fish gills using an osmium quick-fix technique. *J. Exp. Zool.* **217,** 185.

Bray, D. F., and Wagenaar, E. B. (1978). A double staining technique for improved contrast of thin sections from Spurr-embedded tissue. *Can. J. Bot.* **56,** 129.

Bretschneider, A., Burns, W., and Morrison, A. (1981). Pop-off technic. The ultrastructure of paraffin-embedded sections. *Am. J. Clin. Pathol.* **76,** 450.

Brown, J. N. (1975). The avian erythrocyte: a study of fixation for electron microscopy. *J. Microsc. (Oxford)* **104,** 293.

Brown, J. N. (1983a). A fluid exchange apparatus for the processing of biological specimens for electron microscopy. *Microsc. Acta* **87,** 329.

Brown, J. N. (1983b). An improved apparatus for staining large numbers of electron microscope ultrathin sections simultaneously. *Micron* **14,** 69.

Brown, R. M., Jr., and Arnott, H. J. (1971). A photographic method for producing true three-dimensional electron micrographs. *Protoplasma* **72,** 105.

Bülow, F. A. von, and Høyer, P. E. (1983). Use of epoxy resin slides for handling unfixed cryostat sections intended for histochemistry at the ultrastructural level. *Histochem. J.* **15,** 825.

Burke, C. N., and Croxall, G. (1983). Variation in composition of media and reagents used in the preparation of cell cultures from human and other animal tissues: Dulbecco's, Earles', and Hank's balanced salt solutions. *In Vitro* **19,** 693.

Butler, J. K. (1974). A precision hand trimmer for electron microscope tissue blocks. *Stain Technol.* **49,** 129.

Butler, J. K. (1980). The use of Ralph–Bennett glass knives for specimen block trimming. *Stain Technol.* **55**, 323.

Byers, B., and Goetsch, L. (1975). Behavior of spindles and spindle plaques in the cell cycle and conjugation of *Saccharomyces cerevisiae. J. Bacteriol* **124**, 511.

Callahan, W. P., and Horner, J. A. (1964). The use of vanadium as a stain for electron microscopy. *J. Cell Biol.* **20**, 350.

Callas, G. (1982). Osmium vapor fixation of pulmonary surfactant. *Anat. Rec.* **203**, 301.

Campbell, R. D., and Hermans, C. O. (1972). A rapid method for resectioning 0.5–4.0 μ epoxy sections for election microscopy. *Stain Technol.* **47**, 115.

Campbell, R. D. (1981). A method for making ribbons of semithin plastic sections. *Stain Technol.* **56**, 247.

Cañete, M., and Stockert, J. C. (1981). Polychromatic Giemsa staining for Epon semi-thin sections. *J. Microsc. (Oxford)* **122**, 321.

Capco, D. G., Krochmalnic, G., and Penman, S. (1984). A new method of preparing embedment-free sections for transmission electron microscopy: applications to the cytoskeletal framework and other three-dimensional networks. *J. Cell Biol.* **98**, 1878.

Cardamone, J. J., Jr. (1982). A simple and inexpensive multiple grid staining device. *EMSA Bull.* **12**, 78.

Carlemalm, E., Garavito, R. M., and Villiger, W. (1982). Resin development for electron microscopy and an analysis of embedding at low temperature. *J. Microsc. (Oxford)* **126**, 123.

Carlstedt, T. (1977). Observations on the morphology at the transition between the peripheral and the central nervous system in the cat. I. A preparative procedure useful for electron microscopy of the lumbosacral dorsal rootlets. *Acta Physiol. Scand., Suppl.* No. 446, 5.

Carson, J., Lee, R. M. K. W., and Forrest, J. B. (1982). A simple method for the handling and orientation of small specimens for electron microscopy. *J. Microsc. (Oxford)* **126**, 201.

Castejón, O. J., and Castejón, H. V. (1972). Application of alcian blue and OS-DMEDA staining. *Rev. Microsc. Electron.* **1**, 227.

Causton, B. E. (1981). Resins: toxicity, hazards, and safe handling. *Proc. R. Microsc. Soc.* **16**, 265.

Chan, E. C. S., Gomersall, M., and Bernier, J. (1974). The negative staining of "difficult" bacteria like *Arthrobacter globiformis* for electron microscopy. *Can. J. Microbiol.* **20**, 901.

Chan-Curtis, V., Beer, M., and Koller, T. (1970). Cytochemical localization of nucleic acids by acriflavin–phosphotungstate complex for fluorescence microscopy and electron microscopy. *J. Histochem. Cytochem.* **18**, 609.

Chang, J. H. T. (1972). Fixation and embedding, *in situ*, of tissue culture cells for electron microscopy. *Tissue Cell* **4**, 561.

Chang, S. H., Merguer, W. J., Pendergrass, R. E., Buler, R. E., Berezesky, I. K., and Trump, B. F. (1980). A rapid method of cryofixation of tissues in situ for ultracryomicrotomy. *J. Histochem. Cytochem.* **28**, 47.

Chew, E. C., Riches, D. J., Lam, T. K., and Hou Chan, H. J. (1983). A fine structural study of microwave fixation for tissues. *Cell Biol. Int. Rep.* **7**, 135.

Chiba, T., and Murata, Y. (1982). Preservation of catecholamine granules by the fixative containing nitrogen mustard N-oxide. *J. Electron Microsc.* **31**, 419.

Chien, K. (1980). *In situ* embedding of cell monolayers on untreated glass surfaces for vertical and horizontal ultramicrotomy. *Proc.—Annu. Meet., Electron Microsc. Soc. Am.* **38**, 644.

Chien, K., Van de Velde, R. L., and Heusser, R. C. (1982). A one-step method for reembedding paraffin embedded specimens for electron microscopy. *Proc.—Annu. Meet., Electron Microsc. Soc. Am.* **40**, 356.

Childress, S. A., and McIver, S. B. (1983). Improvements in semithin sectioning techniques. *Bull. Microsc. Soc. Can.* **11**, 4.

Chiovetti, R. (1978). Encystment of *Naegleria gruberi*. I. Preparation of cysts for electron microscopy. *Trans. Am. Microsc. Soc.* **97**, 244.

Coalson, J. (1983). A simple method of lung perfusion fixation. *Anat. Rec.* **205**, 233.

Codling, B. W., and Mitchell, J. C. (1976). A method of embedding tissue culture preparations *in situ* for transmission electron microscopy. *J. Microsc. (Oxford)* **106**, 103.

Coffey, M. D., Palevitz, B. A., and Allen, P. J. (1972). The fine structure of two rust fungi, *Puccinia helianthi* and *Melampsora lini*. *Can. J. Bot.* **50**, 231.

Cogliati, R., and Gautier, A. (1973). Mise en évidence de l'ADN et des polysaccharides à l'aide d'un nouveau reactif "de type Schiff". *C. R. Hebd. Seances Acad. Sci.* **276**, 3041.

Coleman, J. R., and Moses, M. J. (1964). DNA and the fine structure of synaptic chromosomes in the domestic rooster (*Gallus domesticus*). *J. Cell Biol.* **23**, 63.

Connelly, P. S. (1977). Reembedding of tissue culture cells for comparative and quantitative electron microscopy. *Proc.—Annu. Meet., Electron Microsc. Soc. Am.* **35**, 552.

Constantin, L. L., Franzini-Armstrong, C., and Podolsky, R. J. (1964). Localization of calcium-accumulating structures in striated muscle fibers. *Science (Washington, D.C.)* **147**, 158.

Daddow, L. Y. M. (1983). A double lead stain method for enhancing contrast of ultrathin sections in electron microscopy: a modified multiple staining technique. *J. Microsc. (Oxford)* **129**, 147.

Dae, M. W., Heymann, M. A., and Jones, A. L. (1982). A new technique for perfusion fixation and contrast enhancement of fetal lamb myocardium for electron microscopy. *J. Microsc. (Oxford)* **127**, 301.

Dalton, A. J. (1955). A chrome-osmium fixative for electron microscopy. *Anat. Rec.* **121**, 281.

Dalton, A. J., and Zeigel, R. F. (1960). A simple method of staining thin sections of biological material with lead hydroxide for electron microscopy. *J. Biophys. Biochem. Cytol.* **1**, 409.

Darley, J. J., and Ezoe, H. (1976). Potential hazards of uranium and its compounds in electron microscopy: a brief review. *J. Microsc. (Oxford)* **106**, 85.

Davidson, J. P., Faber, P. J., Fischer, R. G., Mansy, S., Peresie, H. J., Rosenberg, B., and Van Camp, L. (1975). Platinum–pyrimidine blues and related complexes. A new class of antitumor agents. *Cancer Chemother. Rep.* **59**, 287.

Day, T. (1984). Formvar films. *Proc. R. Microsc. Soc.* **19**, 77.

De Boer, J., and Brakenhoff, G. J. (1974). A simple method for carbon film thickness determination. *J. Ultrastruct. Res.* **49**, 224.

Davison, E., and Colquhoun, W. (1985). Ultrathin Formvar support films for transmission electron microscopy. *J. Electron Microsc. Tech.* **2**, 35.

Dietrich, H. F., and Fontaine, A. R. (1975). A decalcification method for ultrastructure of echinoderm tissue. *Stain Technol.* **50**, 351.

Dijk, F., Oosterbaan, J. A., and Hulstaert, C. E. (1982). Purification of glutaraldehyde by vacuum fractional distillation. *Ultramicroscopy* **9**, 421.

Doane, F. W., and Anderson, N. (1977). Electron and immunoelectron microscopic procedures for diagnosis of viral infections. *In* "Comparative Diagnosis of Viral Diseases" (E. Kurstak and C. Kurstak, eds.), pp. 505–539. Academic Press, New York.

Dougherty, M. M., and King, J. S. (1984). A simple, rapid staining procedure for methacrylate embedded tissue sections using chromotrope 2R and methylene blue. *Stain Technol.* **59**, 149.

Drahos, V., and Delong, A. (1960). A simple method for obtaining perforated supporting membranes for electron microscopy. *Nature (London)* **186**, 104.

Dubochet, J., and Kellenberger, E. (1972). Selective adsorption of particles to the supporting film and its consequences on particle counts in electron microscopy. *Microsc. Acta* **72**, 119.

Dubochet, J., Groom, M., and Mueller-Neuteboom, S. (1982). The mounting of macromolecules for electron microscopy with particular reference to surface phenomena and the treatment of support films by glow discharge. *Adv. Opt. Electron Microsc.* **8**, 107.

Dulbecco, R., and Vogt, M. (1954). Plaque formation and isolation of pure lines with poliomyelitis viruses. *J. Exp. Med.* **99**, 167.

Dunbar, A. (1981). The preservation of soluble material on the surface and in cavities of the pollen wall of Campanulaceae and Penta-fragmataceae. *Micron* **12**, 47.

Duncan, G. H., and Roberts, I. M. (1981). Extraction of virus particles from amounts of material for electron microscope serology. *Micron* **12**, 171.

Dwarte, D. M., and Vesk, M. (1982). Cytochemical localization of biliproteins with silicotungstic acid. *J. Microsc. (Oxford)* **126**, 197.

Earle, W. R. (1934). A technique for adjustment of oxygen and carbon dioxide tensions, and hydrogen ion concentration, in tissue cultures planted in Carrel flasks. *Arch. Exp. Zellforsch.* **16**, 116.

Earle, W. R. (1943). Production of malignancy in vitro. IV. The mouse fibroblast culture and changes seen in the living cell. *J. Natl. Cancer Inst.* **4**, 165.

Eddy, E. M., and Ito, S. (1971). Fine structural and radioautographic observations on dense perinuclear cytoplasmic material in tadpole oocytes. *J. Cell Biol.* **49**, 90.

Edwards, C. A., Walker, G. K., and Avery, J. K. (1984). A technique for achieving consistent release of Formvar film from clean glass slides. *J. Electron Microsc. Tech.* **1**, 203.

Edwards, H. A., and Harrison, J. B. (1983). An osmoregulatory syncytium and associated cells in a fresh water mosquito. *Tissue Cell* **15**, 271.

Eisenman, E. A., and Alfert, M. (1982). A new fixation procedure for preserving the ultrastructure of marine invertebrate tissues. *J. Microsc. (Oxford)* **125**, 117.

Elgjo, R. F. (1976). Platelets, endothelial cells, and macrophages in the spleen. An ultrastructural study on perfusion-fixed organs. *Am. J. Anat.* **145**, 101.

Elsner, P. R. (1971). A simple, reliable method for preparing perforated Formvar films. *Proc.— Annu. Meet., Electron Microsc. Soc. Am.* **29**, 460.

Eppig, J. J., Leiter, E. H., and Waymouth, C. (1976). Culture of cells in BEEM capsules: a new technique for electron microscopic study of monolayer cultures. *In Vitro* **12**, 65.

Eskelinen, S., and Saukko, P. (1982). The use of slowly increasing glutaraldehyde concentrations preserves the shape of erythrocytes under the influence of an osmotic pressure gradient or detergents. *J. Ultrastruct. Res.* **81**, 403.

Estis, L. F., and Haschemeyer, R. H. (1980). Electron microscopy of negatively stained and unstained fibrinogen. *Proc. Natl. Acad. Sci. U.S.A.* **77**, 3139.

Fabergé, A. C., and Oliver, R. M. (1974). Methylamine tungstate, a new negative stain. *J. Microsc. (Paris)* **20**, 241.

Fahmy, A. (1967). An extemporaneous lead citrate stain for electron microscopy. *Proc.—Annu. Meet., Electron Microsc. Soc. Am.* **25**, 148.

Fairén, A., Peters, A., and Saldanha, J. (1977). A new procedure for examining Golgi impregnated neurons by light and electron microscopy. *J. Neurocytol.* **6**, 311.

Farragiana, T., and Marinozzi, V. (1979). Phosphotungstic acid staining of polysaccharides containing structures on epoxy embedded tissues. *J. Submicrosc. Cytol.* **11**, 263.

Feldherr, C. M. (1974). The binding characteristics of the nuclear annuli. *Exp. Cell Res.* **85**, 271.

Ferguson, D. J. P., and Anderson, T. J. (1981). A technique for identifying areas of interest in human breast tissue before embedding for electron microscopy. *J. Clin. Pathol.* **34**, 1187.

Ferrer, J. M., Tato, A., and Stockert, J. C. (1984). Blue molybdenum oxides: a stain for light and electron microscopy. *J. Microsc. (Oxford)* **134**, 221.

Flood, P. R. (1970). Preliminary experiments with iodine in electron opaque stains for ultrathin sections. *Proc. Int. Congr. Electron Microsc., 7th, Grenoble, Fr.* **1**, 431.

Forssmann, W. G., Siegrist, G., Orci, L., Girardier, L., Picket, R., and Rouiller, C. (1967). Fixation par perfusion pour le microscope électronique essai de generalisation. *J. Microsc. (Paris)* **6**, 279.

Forssmann, W. G., Ito, S., Weihe, E., Aoke, A., Dym. M., and Fawcett D. W. (1977). An improved perfusion fixation method for the testis. *Anat. Rec.* **188**, 307.

Franchi, L. L., and Symons, D. (1980). A simple aid in the evaluation of glass knives. *J. Microsc. (Oxford)* **120**, 15.

Franks, F. (1977). Biological freezing and cryofixation. *J. Microsc. (Oxford)* **111**, 3.

Frederik, P. M., and Busing, W. M. (1980). Crystals and artifacts in frozen-thin sections. *Electron Microsc.* **2**, 712.

Frederik, P. M., and Busing, W. M. (1981). Strong evidence against section thawing whilst cutting on the cryoultratome. *J. Microsc. (Oxford)* **122**, 217.

Frederik, P. M., Bomans, P. H. H., Busing, W. M., Odselius, R., and Hax, W. M. A. (1984). Vapor fixation for immunocyto-chemistry and x-ray microanalysis on cryoultramicrotome sections. *J. Histochem. Cytochem.* **32**, 636.

Friend, D. S. (1969). Cytochemical staining of multivesicular body and Golgi vesicles. *J. Cell Biol.* **41**, 269.

Fryer, P. R., Wells, C., and Ratcliffe, A. (1983). Technical difficulties overcome in the use of Lowicryl 4KM EM embedding resin. *Histochemistry* **77**, 141.

Fujita, H., Komatsu, M., and Nakajima, T. (1978). Use of nitrogen mustard N-oxide for the fixation of electron microscopic tissues. *Histochemistry* **58**, 49.

Fukami, A., and Adachi, K. (1964). On an adhering method of thin film specimens to specimen grids. *J. Electron Microsc.* **13**, 26.

Furness, J. B., Costa, M., and Blessing, W. W. (1977). Simultaneous fixation and production of catecholamine fluorescence in central nervous tissue by perfusion with aldehydes. *Histochem. J.* **9**, 745.

Furness, J. B., Heath, J. W., and Costa, M. (1978). Aqueous aldehyde (Faglu) methods for the fluorescence histochemical localization of catecholamines and for ultrastructural studies of central nervous tissue. *Histochemistry* **57**, 285.

Furtado, J. S. (1970). The fibrin clot: a medium for supporting loose cells and delicate structures during processing for microscopy. *Stain Technol.* **45**, 19.

Garland, C. D., Nash, G. V., and McMeekin, T. A. (1982). The preservation of mucus and surface-associated microorganisms using acrolein vapor fixation. *J. Microsc. (Oxford)* **128**, 307.

Gautier, A., Cogliati, R., Schreyer, M., and Fakau, J. (1973). Ultrastructural cytochemistry: a new specific stain for DNA and polysaccharides. *Experientia* **29**, 771.

Gbewonyo, A. J. K. (1982). Rapid and reliable method for diagnostic electron microscopy of feces. *J. Microsc. (Oxford)* **126**, 191.

Gerrits, P. O., and Smid, L. (1983). A new, less toxic polymerization system for the embedding of soft tissues in glycol methacrylate and subsequent preparing of serial sections. *J. Microsc. (Oxford)* **132**, 81.

Giammara, B. L. (1981). The Grid-All: a multiple grid handling, staining, and storage device for use in electron microscopy. *Proc.—Annu. Meet., Electron Microsc. Soc. Am.* **39**, 552.

Giammara, B. L., and Hanker, J. A. (1982). Epoxy slide embedment for LM. TEM. STEM and HVEM cytochemistry. *Proc.—Annu. Meet., Electron Microsc. Soc. Am.* **40**, 358.

Gicquaud, C., Turcotte, A., and St-Pierre, S. (1983). Peptides from *Amanita virosa:* viroidin and viroisin are more effective than phalloidin in protecting actin *in vitro* against osmic acid. *Eur. J. Cell Biol.* **32**, 171.

Gillett, R., Jones, G. E., and Partridge, T. (1975). Distilled glutaraldehyde: its use in an improved fixation regime for cell suspensions. *J. Microsc. (Oxford)* **105**, 325.

Gilloteaux, J., and Naud, J. (1979). The zinc-iodide-osmium tetroxide staining-fixative of Maillet: Nature of the precipitate studied by x-ray microanalysis and detection of Ca^{2+}-affinity subcellular sites in a tonic smooth muscle. *Histochemistry* **63**, 227.

Gorycki, M. A., and Oberc, M. A. (1978). Cleaning diamond knives before or during sectioning. *Stain Technol.* **53**, 51.

Gorycki, M. A., and Sohm, E. K. (1979). Modifying a Porter–Blum MT-2 ultramicrotome knife holder to accept Ralph knives. *Stain Technol.* **54**, 293.

Goto, K., and Sato, H. (1980). Formation of cell-sized single-layered liposomes in a simple system of phospholipid, ethanol and water. *Tohoku J. Exp. Med.* **131**, 399.

Gowans, E. J. (1973). An improved method of agar pelleted cells for electron microscopy. *Med. Lab. Technol.* **30**, 113.

Greany, P. D., and Rubin, R. E. (1971). Serial sectioning of tissues in glycol methacrylate by supplementary embedding in a paraffin-plastic matrix. *Stain Technol.* **46**, 216.

Gregory, D. W., and Pirie, J. S. (1973). Wetting agents for biological electron microscopy. I. General considerations and negative staining. *J. Microsc. (Oxford)* **99**, 261.

Grinnell, F., Tobleman, M. Q., and Hackenbrock, C. R. (1976). Initial attachment of baby hamster kidney cells to an epoxy substratum. *J. Cell Biol.* **70**, 707.

Grönblad, M. (1983). Improved demonstration of exocytotic profiles in glomus cells of rat carotid body after perfusion with glutaraldehyde fixative containing a concentration of potassium. *Cell Tissue Res.* **229**, 627.

Grote, M., Fromme, H. G., and Sinclair, N. J. (1983). The use of cetylpridinium chloride to preserve water-soluble surface material in pollen for electron microscopy. *Micron* **14**, 29.

Grund, S., Eichberg, J., and Asmussen, F. (1982). A specific embedding resin (PVK) for fine cytological investigations in the photoemission electron microscope. *J. Ultrastruct. Res.* **80**, 89.

Guatelli, J. C., Porter, K. R., Anderson, K. L., and Boggs, D. P. (1982). Ultrastructure of the cytoplasmic and nuclear matrix of human lymphocytes observed using high voltage electron microscopy and embedment-free sections. *Biol. Cell* **43**, 69.

Gulati, D. K., and Akers, S. W. (1977). Improved trough for ultramicrotomy. *Stain Technol.* **52**, 351.

Hallam, N. D. (1976). Anhydrous fixation of dry plant tissue using non-aqueous fixatives. *J. Microsc. (Oxford)* **106**, 337.

Hama, K., and Kosaka, T. (1981). Neurobiological applications of high voltage electron microscopy. *Trends Neurosci.* **4**, 193.

Hammond, G. W., Hazelton, P. R., Chuang, I., and Klisko, B. (1981). Improved detection of viruses by electron microscopy after direct ultracentrifuge preparation of specimens. *J. Clin. Microbiol.* **14**, 210.

Handley, D. A., Alexander, J. T., and Chien, S. (1981a). Rotary beveling for *in situ* thin-section microscopy of cell cultures. *Proc.—Annu. Meet., Electron Microsc. Soc. Am.* **39**, 550.

Handley, D. A., Alexander, J. T., and Chien, S. (1981b). The design and use of a simple device for rapid quench-freezing of biological samples. *J. Microsc. (Oxford)* **121**, 261.

Hanks, J. H. (1948). The longevity of chick tissue cultures without renewal of medium. *J. Cell. Comp. Physiol.* **31**, 235.

Hanks, J. H., and Wallace, R. E. (1949). Relation of oxygen and temperature in the preservation of tissues by refrigeration. *Proc. Soc. Exp. Biol. Med.* **71**, 196.

Harris, J. R. (1982). The production of paracrystalline two-dimensional monolayers of purified protein molecules. *Micron* **13**, 147.

Harris, W. J. (1962). Holey films for electron microscopy. *Nature (London)* **196**, 499.

Hartmann, R. (1984). A new embedding medium for cryo-sectioning eggs of high yolk and lipid content. *Eur. J. Cell Biol.* **34**, 206.

Hassell, J., and Hand, A. R. (1974). Tissue fixation with diimidoesters as an alternative to aldehydes. I. Comparison of crosslinking and ultrastructure obtained with dimethylsuberimidate and glutaraldehyde. *J. Histochem. Cytochem.* **22**, 223.

Hatta, T., and Francki, R. I. B. (1981). Identification of small polyhedral virus particles in thin sections of plant cells by an enzyme cytochemical technique. *J. Ultrastruct. Res.* **74,** 116.

Haudenschild, C., Baumgartner, H. R., and Studer, A. (1972). Significance of fixation procedure for preservation of arteries. *Experientia* **29,** 828.

Hausen, K., and Wolburg-Buchholz, K. (1980). An improved cobalt sulfide–silver intensification method for electron microscopy. *Brain Res.* **187,** 462.

Hawes, C. R., and Horne, J. C. (1983). Staining plant cells for thick sectioning: uranyl acetate, copper and lead citrate impregnation. *Biol. Cell.* **48,** 207.

Hawes, C. R., Juniper, B. E., and Horne, J. C. (1983). Electron microscopy of resin-free sections of plant cells. *Protoplasma* **115,** 88.

Hawkes, J. W. (1974). The structure of fish skin. I. General organization. *Cell Tissue Res.* **149,** 147.

Hayat, M. A., and Giaquinta, R. (1970). Rapid fixation and embedding for electron microscopy. *Tissue Cell* **2,** 191.

Hayat, M. A. (1975). "Positive Staining for Electron Microscopy." Van Nostrand-Reinhold, New York.

Hayat, M. A. (1978). "Introduction to Biological Scanning Electron Microscopy." University Park Press, Baltimore, Maryland.

Hayat, M. A. (1981a). "Principles and Techniques of Electron Microscopy: Biological Applications," Vol. 1. 2nd ed., International Pub. House, Berkeley Heights, New Jersey.

Hayat, M. A. (1981b). "Fixation for Electron Microscopy." Academic Press, Inc., New York.

Hayat, M. A. (1986). "Staining for Electron Microscopy." John Wiley and Sons, New York. In press.

Heinen, E. (1977). Cis-dichloro-diammine platinum (II) as stain for electron microscopic preparations. *Histochemistry* **51,** 257.

Helander, K. G. (1984). The Ralph Knife in practice. *J. Microsc. (Oxford)* **135,** 139.

Helin, H., Pasternack, A., and Rantala, I. (1979). Rapid processing of renal glomeruli for electron microscopy. *J. Clin. Pathol.* **32,** 516.

Hendricks, H. R., and Eestermans, I. L. (1982). Electron dense granules and the role of buffers: artifacts from fixation with glutaraldehyde and osmium tetroxide. *J. Microsc. (Oxford)* **126,** 161.

Henry, E. C. (1977). A method for obtaining ribbons of serial sections of plastic embedded specimens. *Stain Technol.* **52,** 59.

Herbert, J. J., Hensorling, T. P., Jack, T. J., and Berni, R. J. (1972). An indirect staining technique for cellulose fibrils. *Microscope* **20,** 161.

Hernandez, W., Rambourg, A., and Leblond, C. P. (1968). Periodic acid-chromic acid-methanamine silver technique for glycoprotein detection in the electron microscope. *J. Histochem. Cytochem.* **16,** 507.

Heuser, J. E., Reese, T. S., Dennis, M. J., Jan, Y., Jan, L., and Evans, L. (1979). Synaptic vesicle exocytosis captured by quick freezing and correlated with quantal transmitter release. *J. Cell Biol.* **61,** 275.

Hill, P. K., de la Torre, J. C., Thompson, S. M., Bullock, D. F., and Rosenfield-Wessells, S. (1982). A comparative study of EM fixation procedures for the adult rat spinal cord based on regional blood flow. *J. Neurosci. Methods* **5,** 23.

Hinton, D. E. (1975). Perfusion fixation of whole fish for electron microscopy. *J. Fish. Res. Board Can.* **32,** 416.

Hitchborn, J. H., and Hills, G. J. (1965). The use of negative staining in the electron microscopic examination of plant viruses in crude extracts. *Virology* **27,** 528.

Hoelke, C. W. (1975). Preparation and use of holey carbon microgrids in high resolution electron microscopy. *Micron* **5,** 307.

Höglund, S. (1968). Some electron microscopic studies on the satellite tobacco necrosis virus and its IgG-antibody. *J. Gen. Virol.* **2**, 427.

Hopwood, D., Coghill, G., Ramsay, J., Milne, G., and Kerr, M. (1984). Microwave fixation: its potential for routine techniques, histochemistry, immunocytochemistry, and electron microscopy. *Histochem. J.* **16**, 1171.

Horisberger, M., and Vonlanthen, M. (1977). Location of mannan and chitin on thin sections of budding yeasts with gold markers. *Arch. Microbiol.* **115**, 1.

Horisberger, M., and Vonlanthen, M. (1980). Ultrastructural localization of soybean agglutinin on thin sections of *Glycine max* (soybean) Var. Altona by the gold method. *Histochemistry* **65**, 181.

Horne, R. W. (1965). Negative staining methods. *In* "Techniques for Electron Microscopy" (D. Kay, ed.), p. 328. Blackwell, Oxford.

Horne, R. W., and Pasquali-Ronchetti, I. (1974). A negative staining-carbon film technique for studying viruses in the electron microscope. I. Preparation procedures for examining icosahedral and filamentous viruses. *J. Ultrastruct. Res.* **47**, 361.

Horowitz, J. (1981). A new preparative technique for studying dry seeds of *Pisum sativum* with the aid of transmission electron microscopy. *Micron* **12**, 139.

Huet, C., and Garrido, J. (1972). Ultrastructural visualization of cell coat components by means of wheat germ agglutinin. *Exp. Cell Res.* **75**, 523.

Humphreys, C. (1976). High voltage electron microscopy. *In* "Principles and Techniques of Electron Microscopy: Biological Applications". (M. A. Hayat, ed.), Vol. 6. Van Nostrand-Reinhold, New York.

Humphreys, W. J. (1977). Health and safety hazards. *SEM* **1**, 537.

Hunziker, E. B., Herrmann, W., and Schenk, R. K. (1982). Improved cartilage fixation by ruthenium hexamine trichloride (RHT). *J. Ultrastruct. Res.* **81**, 1.

Hwang, Y.-C. (1970). A modification for orientation by the use of silicone rubber molds for reembedding tissue in epoxy resins. *J. Electron Microsc.* **19**, 189.

Issidorides, M. R., and Katsorchis, T. (1981). Dispersed and compact chromatin demonstrated with a new EM method: phosphotungstic acid-hematoxylin block-staining. *Histochemistry* **73**, 21.

Ito, U., and Inaba, Y. (1977). Electron microscopic observation of cerebrospinal fluid cells: a new method for embedding of CSF cells. *J. Electron Microsc.* **19**, 265.

Janisch, R. (1974). Oriented embedding of single-cell organisms. *Stain Technol.* **49**, 157.

Jensen, O. A., Prause, J. U., and Laursen, H. (1981). Shrinkage in preparatory steps for SEM. *Albrecht von Graefes Arch. Klin. Exp. Ophthalmol.* **215**, 233.

Jewell, G. G. (1981). The microstructure of orange juice. *Scanning Electron Microsc.* **3**, 593.

Johansen, B. V. (1974). Bright field electron microscopy of biological specimens. II Preparation of ultrathin carbon support films. *Micron* **5**, 209.

Johansson, O. (1983). The Vibratome–Ralph knife combination: a useful tool for immunohistochemistry. *Histochem. J.* **15**, 265.

Johnson, N. F., and Ibe, K. (1981). Sectioning and imaging hard mineral fibers in biological tissues. *J. Microsc. (Oxford)* **122**, 87.

Johnson, P. C. (1976). A rapidly setting glue for resectioning and remounting epoxy embedded tissue. *Stain Technol.* **51**, 275.

Junquera, P., and Went, D. F. (1981). A simple method enabling standardized handling of small biological objects for light and electron microscopic preparation. *Experientia* **37**, 534.

Kaláb, M. (1981). Electron microscopy of milk products: a review of techniques. *Scanning Electron Microsc.* **3**, 453.

Kalimo, H., and Pelliniemi, L. J. (1977). Pitfalls in the preparation of buffers for electron microscopy. *Histochem. J.* **9**, 241.

Karnovsky, M. J. (1961). Simple methods for staining with lead at high pH in electron microscopy. *J. Biophys. Biochem. Cytol.* **11,** 729.

Karnovsky, M. J. (1965). A formaldehyde–glutaraldehyde fixative of high osmolarity for use in electron microscopy. *J. Cell Biol.* **27,** 137A.

Karp, R. D., Silcox, J. C., and Sombyo, A. V. (1982). Cryoultramicrotomy: evidence against melting and the use of a low temperature cement for specimen orientation. *J. Microsc (Oxford)* **125,** 157.

Kawamoto, K., Hirano, A., and Herz, F. (1980). Simplified in situ preparation of cultured cell monolayers for electron microscopy. *J. Histochem. Cytochem.* **28,** 178.

Keene, D. R. (1984). A method for producing dust-, streak- and hole-free Formvar films in laboratories having high atmospheric humidity. *Stain Technol.* **59,** 56.

Kellenberger, E., and Arber, W. (1957). Electron microscopical studies of phage multiplication. I. A method for quantitative analysis of particle suspensions. *Virology* **3,** 245.

Kellenberger, E., and Bitterli, D. (1976). Preparation and counts of particles in electron microscopy: application of negative stain in the agarfiltration method. *Microsc. Acta* **78,** 131.

Kellenberger, E., Carlemalm, E., Villiger, W., Roth, J., and Garavito, R. M. (1980). "Low Denaturation Embedding for Electron Microscopy of Thin Sections." Chem. Werke Lowi, D-8264 Waldkraiburg, West Germany.

Kim, C. K., Pfister, R. M., and Somerson, N. L. (1977). Electron microscopy of *Mycoplasma pneumoniae* microcolonies grown on solid surfaces. *Appl. Environ. Microbiol.* **34,** 591.

Kim, S., and Martin, G. M. (1981). Preparation of cell-size unilaminar liposomes with high captured volume and defined size distribution. *Biochim. Biophys. Acta* **646,** 1.

Kindblom, L.-G. (1983). Light and electron microscopic examination of embedded fine-needle aspiration biopsy specimens in the preoperative diagnosis of soft tissue and bone tumors. *Cancer (Philadelphia)* **51,** 2264.

King, D. G., Kammlade, N., and Murphy, J. (1982). A simple device to help reembed thick plastic sections. *Stain Technol.* **57,** 307.

Kobayasi, T. (1983). A simple staining device for grids. *Proc. R. Microsc. Soc.* **18,** 220.

Komnick, H. (1962). Elektronemikroskopische Lokalisation von Na^+ und Cl^- in Zellen und Geweben. *Protoplasma* **55,** 414.

Kraft, L. M., Joyce, K., and D'Amelio, E. D. (1983). Removal of histological sections from glass for electron microscopy: use of Quetol 651 resin and heat. *Stain Technol.* **58,** 41.

Kristt, D. A., and Trythall, D. (1982). A procedure for sampling specific portions of identified neurons for ultrastructural examination. *Neurosci. Lett.* **31,** 1.

Kuchler, R. J. (1977). Milieu for maintaining and growing animal cells in vitro. In "Biochemical Methods in Cell Culture and Virology," pp. 99–142. Dowden, Hutchinson & Ross, Stroudsburg, Pennsylvania.

Kuhn, J. (1981). A simple method for the preparation of cell cultures for ultrastructural investigation. *J. Histochem. Cytochem.* **29,** 84.

Kushida, H. (1966). Staining of the thin section with lead acetate. *J. Electron Microsc.* **15,** 93.

Lake, J. A. (1979). Practical aspects of immune electron microscopy. In "Enzyme Structure," Part H (C. H. W. Hirs and S. N. Timasheff, eds.), Methods in Enzymology, Vol. 61, pp. 250–257. Academic Press, New York.

Landis, W. J., Paine, M. C., and Glimcher, M. J. (1977). Electron microscopic observations of bone tissue prepared anhydrously in organic solvents. *J. Ultrastruct. Res.* **59,** 1.

Langenberg, W. G. (1982a). Fixation of plant virus inclusions under conditions designed for freeze-fracture. *J. Ultrastruct. Res.* **81,** 184.

Langenberg, W. G. (1982b). Silicone additive facilitates epoxy plastic sectioning. *Stain Technol.* **57,** 79.

Ledingham, J. M., and Simpson, F. O. (1972). The use of *p*-phenylenediamine in the block to enhance osmium staining for electron microscopy. *Stain Technol.* **47**, 239.

Lee, C. S., and Outteridge, P. M. (1981). Leucocytes of sheep colostrum, milk, and involution secretion, with particular reference to ultrastructure and lymphocyte sub-populations. *J. Dairy Res.* **48**, 225.

Lehmann, V. H., and Schultz, D. (1982). Die Verwendung der Semisimultan-Fixierung zur Darstellung der Feinstruktur spezialisierter Mooszellen. *Mikroskopie* **39**, 285.

Leknes, I. L. (1985). An improved method for transferring semithin epoxy sections from the microtome knife to microscope slides. *Stain Technol.* **60**, 58.

Lenard, J., and Singer, S. J. (1968). Alterations of the conformation of proteins in red blood cell membranes and in solution by fixatives used in electron microscopy. *J. Cell Biol.* **37**, 117.

Lethias, C., Garrone, R., and Mazzorana, M. (1983). Fine structure of sponge cell with freeze-fracture and conventional thin section methods. *Tissue Cell* **15**, 523.

Leunissen, J. L. M., Elbers, P. F., Leunnissen-Bijvelt, J. J. M., and Verkleij, A. J. (1984). An evaluation of cryosectioning of fixed and cryoprotected rat liver. *Ultramicroscopy* **12**, 345.

Lewis, P. R. (1983). Fixatives: hazards and safe handling. *Proc. R. Microsc. Soc.* **18**, 164.

Lickfield, K. G. (1976). Transmission electron microscopy of bacteria. *Methods Microbiol.* **9**, 127.

Lickfield, K. G., and Menge, B. (1980). Close-to-life chemical fixation of spatial site of intracellular phage particles. *J. Ultrastruct. Res.* **72**, 206.

Locke, M., and Krishnan, N. (1971). Hot alcoholic phosphotungstic acid and uranyl acetate as routine stains for thick and thin sections. *J. Cell Biol.* **50**, 550.

Locke, M., and Huie, P. (1977). Bismuth staining for light and electron microscopy. *Tissue Cell* **9**, 347.

Login, G. R. (1978). Microwave fixation versus formalin fixation of surgical and autopsy tissue. *Am. J. Med. Technol.* **44**, 435.

Loomis, T. A. (1979). Formaldehyde toxicity. *Arch. Pathol. Lab. Med.* **103**, 321.

Luft, J. (1971). Ruthenium red and violet. II. Fine structural localization in animal tissues. *Anat. Rec.* **171**, 369.

Lumb, W. V. (1963). "Small Animal Anesthesia." Lea & Febiger, Philadelphia, Pennsylvania.

Lutke-Schipholt, L. F., and Stadhouders, A. M. (1981). A simple method for the collection of cells from cerebrospinal fluid for electron microscopy. *Ultrastruct. Pathol.* **2**, 175.

McCombs, W. B., McCoy, C. E., and Holton, O. D. (1980). Electron microscopy for rapid viral diagnosis. *Tex. Soc. Electron Microsc. J.* **11**, 9.

McFarland, W. N., and Klontz, G. W. (1969). Anesthesia in fishes. *Fed. Proc., Fed. Am. Soc. Exp. Biol.* **28**, 1535.

McKinney, R. V. (1969). Facilitation of sealing metal troughs to glass knives by use of an alcohol hand torch and dental baseplate wax, *Stain Technol.* **44**, 44.

McLean, I. W., and Nakane, P. K. (1974). Periodate-lysine-paraformaldehyde fixative. A new fixative for immunoelectron microscopy. *J. Histochem. Cytochem.* **22**, 1077.

McLean, M. R., Limoni, A., Garancis, J. C., and Hause, L. L. (1982). A method for the preparation of platelets for transmission electron microscopy. *Stain Technol.* **57**, 113.

Malech, H. L., and Albert, J. P. (1979). Negative staining of protein macromolecules: a simple rapid method. *J. Ultrastruct. Res.* **69**, 191.

Manston, J., and Katchburian, E. (1984). Demonstration of mitochondrial mineral deposits in osteoblasts after anhydrous fixation and processing. *J. Microsc. (Oxford)* **134**, 177.

Marchese-Ragona, S. P., and Johnson, S. P. S. (1982). A simple method for the progressive infiltration of resin into a dehydrated biological sample. *Proc. R. Microsc. Soc.* **17**, 311.

Marchese-Ragona, S. P. (1984). Ethanol, an efficient coolant for rapid freezing of biological material. *J. Microsc. (Oxford)* **134**, 169.

393

Marek, L. F., and Kelley, R. O. (1983). A simple technique for the visualization of whole mount cytoskeletons with transmission electron microscopy. *Anat. Rec.* **207**, 365.

Mascorro, J. A., Ladd, M. W., and Yates, R. D. (1976). Rapid infiltration of biological tissues utilizing *n*-hexenyl succinic anhydride (HXSA)/vinyl cyclohexene dioxide (VCD), an ultra-low viscosity embedding medium. *Proc.—Annu. Meet., Electron Microsc. Soc. Am.* **34**, 346.

Mather, J., Stanbridge, C. M., and Butler, E. B. (1981). Method for the removal of selected cells from cytological smear preparations for transmission electron microscopy. *J. Clin. Pathol.* **34**, 1355.

Mattson, J. C., Borgerding, P. J., and Craft, D. L. (1977). Fixation of platelets for scanning and transmission electron microscopy. *Stain Technol.* **52**, 151.

Maunsbach, A. B. (1966). The influence of different fixatives and fixation methods on the ultrastructure of rat kidney proximal tubule cells. II. Effects of varying osmolality, ionic strength, buffer system, and fixative concentration of glutaraldehyde solutions. *J. Ultrastruct. Res.* **15**, 283.

Maupin, P., and Pollard, T. D. (1983). Improved preservation and staining of HeLa cell actin filaments, clathrin-coated membranes, and other cytoplasmic structures by tannic acid–glutaraldehyde–saponin fixation. *J. Cell Biol.* **96**, 51.

Mawhinney, W. H. B., and Ellis, H. A. (1983). A technique for plastic embedding of mineralized bone. *J. Clin. Pathol.* **36**, 1197.

Mazia, D., Sale, W. S., and Schatten, G. (1974). Polylysine as an adhesive for electron microscopy. *J. Cell Biol.* **63**, 212a.

Mereau, M., Dive, D., and Vivier, E. (1982). A new rapid method for collection and preparation of cell suspensions for electron microscopy. *Experientia* **38**, 282.

Merkus, E., Dam, B. A. D., and Boers-van der Sluijs, F. P. (1974). Use of a Spurr embedding for electron microscopy of the cell walls of Ascomycetes. LKB application Note 154.

Miller, M. F., Allen, P. T., and Dmochowski, L. (1973). Quantitative studies of oncornaviruses in thin sections. *J. Gen. Virol.* **21**, 57.

Miller, M. F. (1979). A universal sedimentation virus particle counting procedure. *Proc.—Annu. Meet., Electron Microsc. Soc. Am.* **37**, 56.

Miller, M. F., and Rdzok, E. J. (1981). Improved virus particle counting by direct sedimentation onto EM grids. *Proc.—Annu. Meet., Electron Microsc. Soc. Am.* **39**, 404.

Millonig, G. (1961). A modified procedure for lead staining of thin sections. *J. Biophys. Biochem. Cytol.* **11**, 736.

Millonig, G. (1964). Study on the factors which influence preservation of fine structure. *In* "From Molecule to Cell" (P. Buffa, ed.), p. 347. Consiglio Naz. Ric., Rome.

Møller, J. C., Skriver, E., Olsen, S., and Maunsbach, A. B. (1982). Perfusion fixation of human kidneys for ultrastructural analysis. *Ultrastruct. Pathol.* **3**, 375.

Monga, G., Canese, M. G., and Bussolati, G. (1972). Electron microscopical demonstration of sulfate mucopolysaccharides in mouse tracheal cartilage with diaminobenzidine-osmium tetroxide technique. *Histochem. J.* **4**, 205.

Morris, R. E., Ciraolo, G. M., Cohens, D. A., and Bubel, H. C. (1980). *In situ* fixation of cultured mouse peritoneal exudate cells: comparison of fixation methods. *In Vitro* **16**, 136.

Moyne, G. (1973). Feulgen-derived techniques for electron microscopical cytochemistry of DNA. *J. Ultrastruct. Res.* **45**, 102.

Mrena, E. (1980). A modification of negative staining for the study of isolated microsomes. *Philips Bull.* **EM114**, 6.

Müller, G., and Peters, D. (1963). Substrukturen des Vaccinevirus, dargestellt durch Negativkontrastierung. *Arch. Gesamte Virusforsch.* **13**, 435.

Müller, M., Meister, N., and Moor, H. (1980). Freezing in a propane jet and its application in freeze-fracturing. *Mikroskopie* **36**, 129.

Münch, G. (1964). Simplified preparation method for carbon replicas and carbon films for specimen support in electron microscopy. *Rev. Sci. Instrum.* **35**, 524.

Muñoz-Guerra, S., and Subirana, J. A. (1982). Crosslinked polyvinyl alcohol: a water soluble polymer as embedding medium for electron microscopy. *Mikroskopie* **39**, 346.

Nagington, J., Newton, A. A., and Horne, R. W. (1964). The structure of orf virus. *Virology* **23**, 461.

Nanci, A., Bai, P., and Warshawsky, H. (1983). The effect of osmium postfixation and uranyl and lead staining on the ultrastructure of young enamel in the rat incisor. *Anat. Rec.* **207**, 1.

Narang, H. K., and Codd, A. A. (1979). A low-speed centrifugation technique for the preparation of grids for direct virus examination by electron microscopy. *J. Clin. Pathol.* **32**, 304.

Narang, H. K. (1982). Embedding bacteria and tissue culture cells for electron microscopy. *Med. Lab. Sci.* **39**, 87.

Neiss, W. F. (1983). Extraction of osmium-containing lipids by section staining for TEM. *Histochemistry* **79**, 245.

Nelson, B. K., and Flaxman, B. A. (1973). Use of high intensity illumination to aid alignment of knife edge and block face for ultramicrotomy. *Stain Technol.* **48**, 13.

Nermut, M. V. (1972). Negative staining of viruses. *J. Microsc. (Oxford)* **96**, 351.

Nermut, M. V. (1977a). Negative staining in freeze-drying and freeze-fracturing. *Micron* **8**, 211.

Nermut, M. V. (1977b). Freeze-drying for electron microscopy. *In* "Principles and Techniques of Electron Microscopy: Biological Applications" (Hayat, M. A., ed.), Vol. 7, p. 79. Van Nostrand-Reinhold, New York.

Nermut, M. V. (1982a). Advanced methods in electron microscopy. *In* "New Developments in Practical Virology" (C. Howard, ed.), pp. 1–58. Liss, New York.

Nermut, M. V. (1982b). The 'cell monolayer technique' in membrane research. *Eur. J. Cell Biol.* **28**, 160.

Nickerson, P. A. (1983). Lipid droplets in the adrenal cortex of the rat. Preservation after tannic acid–paraformaldehyde–glutaraldehyde fixation and extraction during staining. *Tissue Cell* **15**, 975.

Nicklas, R. B., Brinkley, B. R., Pepper, D. A., Kubai, D. F., and Rickards, G. K. (1979). Electron microscopy of spermatocytes previously studied in life: methods and some observations on micromanipulated chromosomes. *J. Cell Sci.* **35**, 87.

Nicklas, R. B., Kubai, D. F., and Hays, T. S. (1982). Spindle microtubules and their mechanical associations after micromanipulations in anaphase. *J. Cell Biol.* **95**, 91.

Nunes, J. F. M., Soares, J. O., and Alves de Matos, A. P. (1979). Micro-buffy coats of whole blood: a method for the electron microscopic study of mononuclear cells. *Stain Technol.* **54**, 257.

Ohtsuki, Y., Dmochowski, L., Seman, G., and Bowen, J. M. (1978). *In situ* embedding method of cells grown in BEEM capsules for immunoelectron microscopic studies of oncornaviruses. *J. Histochem. Cytochem.* **26**, 149.

Oliveira, L., Burns, A., Bisalputra, T., and Yang, K.-C. (1983). The use of an ultralow viscosity medium (VCD/HXSA) in the rapid embedding of plant cells for electron microscopy. *J. Microsc. (Oxford)* **132**, 195.

Olson, L. W., and Eden, U. M. (1977). A glass bead treatment facilitating the fixation and infiltration of yeast and other refractory cells for electron microscopy. *Protoplasma* **91**, 417.

Olson, L. W. (1978). Preparation of difficult spores, cells, and sporangia for electron microscopy. *In* "Lower Fungi in the Laboratory" (M. S. Fuller, ed.). Dep. Bot., Univ. of Georgia, Athens.

Paavola, L. G. (1977). The corpus luteum of the guinea pig. Fine structure at the time of maximum progesterone secretion and during regression. *Am. J. Anat.* **150**, 565.

Palmer, E. L., Martin, M. L., and Gray, W. G., Jr. (1975). The ultrastructure of disrupted herpesvirus nucleocapsids. *Virology* **18**, 445.

Parker, R. C. (1961). Balanced salt solutions and pH control. *In* "Methods of Tissue Culture," pp. 53–61. Harper (Hoeber), New York.

Paul, J. R. (1975). Media for culturing cells and tissues. *In* "Cell and Tissue Culture," 5th ed., pp. 95–99. Churchill–Livingston, New York.

Paul, R. N., and Egley, G. H. (1983). Techniques for preparing seeds with water-impermeable coats for light and electron microscopy. *Stain Technol.* **58**, 73.

Payne, C. M., and Satterfield, V. G. (1980). A simple procedure for the preparation of rosetted cells for electron microscopy. *J. Clin. Pathol.* **33**, 505.

Payne, C. M., Nagle, R. B., and Borduin, V. (1984). An ultrastructural cytochemical stain specific for neuroendocrine neoplasms. *Lab. Invest.* **51**, 350.

Pease, D. C. (1975). Micronets for electron microscopy. *Micron* **6**, 85.

Pease, D. C. (1982). Unembedded, aldehyde-fixed tissue, sectioned for transmission electron microscopy. *J. Ultrastruct. Res.* **79**, 250.

Penso, G., and Balducci, D. (1963). Tissue culture media. *In* "Tissue Cultures in Biological Research," pp. 68–100. Elsevier, New York.

Pentz, S., Vergani, G., Amthor, S., Horler, H., and Rich, I. (1981). A method for electron microscopic preparation of cultured cells (monolayer) in a new test chamber (TCSC-1). *Microsc. Acta* **84**, 117.

Pentz, S., Amthor, S., and Vergani, G. (1983). Vertical sections of cultivated anchorage-dependent cells for electron microscopy. *J. Microsc. (Oxford)* **129**, 233.

Perera, F., and Petito, C. (1982). Formaldehyde: a question of cancer policy. *Science (Washington, D.C.)* **216**, 1285.

Peters, K.-R. (1980). Improved handling of structural fragile cell-biological specimens during electron microscopic preparation by the exchange method. *J. Microsc. (Oxford)* **118**, 429.

Pfeiffer, S. W. (1981). Use of ethanol to facilitate trimming of tissue blocks embedded in Spurr's medium. *Stain Technol.* **56**, 328.

Pfeiffer, S. W. (1982). Use of hydrogen peroxide to accelerate staining of ultrathin sections. *Stain Technol.* **57**, 137.

Pitman, R. M., Tweedle, C. D., and Cohen, M. J. (1972). Branching of central neurons: intracellular cobalt injection for light and electron microscopy. *Science (Washington, D.C.)* **176**, 412.

Popescu, L. M., Diculescu, I., Zelck, U., and Ionescu, N. (1974). Ultrastructural distribution of calcium in smooth muscle cells of guinea pig taenia coli. A correlated electron microscopic and quantitative study. *Cell Tissue Res.* **154**, 357.

Portmann, R., and Koller, Th. (1976). *Proc. Eur. Congr. Electron Microsc., 6th, Jerusalem* **2**, 546.

Pudney, J., and Singer, R. H. (1980). Intracellular filament bundles in whole mounts of chick and human myoblasts extracted with Triton X-100. *Tissue Cell* **12**, 595.

Rambourg, A. (1967). An improved silver methenamine technique for the detection of periodic acid-reactive complex carbohydrates with the electron microscope. *J. Histochem. Cytochem.* **15**, 409.

Redmond, B. L., and Bob, C. (1984). A fixation/dehydration/infiltration apparatus that minimizes human exposure to harmful chemicals. *J. Electron Microsc. Tech.* **1**, 97.

Reedy, M. K. (1976). Preservation of x-ray patterns from frog sartorius muscle prepared for electron microscopy. *Biophys. J.* **16**, 126a.

Reichelt, R., König, T., and Wangermann, G. (1977). Preparation of microgrids as specimen supports for high resolution electron microscopy. *Micron* **8**, 29.

Reissig, M., and Orrell, S. A. (1970). A technique for the electron microscopy of protein-free particle suspensions by the negative staining method. *J. Ultrastruct. Res.* **32**, 107.

Reynolds, E. S. (1963). The use of lead nitrate at high pH as an electron opaque stain in electron microscopy. *J. Cell Biol.* **17**, 208.

Ribi, W. A. (1976). A Golgi-electron microscope method for insect nervous tissue. *Stain Technol.* **51**, 13.

Rice, S. J., and Phillips, A. D. (1980). Rapid preparation of fecal specimens for detection of viral particles by electron microscopy. *Med. Lab. Sci.* **37**, 371.

Richards, J. G., and Da Prada, M. (1977). Uranaffin reaction: a new cytochemical technique for the localization of adenine nucleotides in organelles storing biogenic amines. *J. Histochem. Cytochem.* **25**, 1322.

Richards, P. R. (1980). A technique for obtaining ribbons of epoxy and other plastic sections for light microscope histology. *J. Microsc. (Oxford)* **122**, 213.

Richardson, R. L., Hinton, D. M., and Campion, D. R. (1983). An improved method for storing and using stains in electron microscopy. *J. Electron Microsc.* **32**, 216.

Ringo, D. L., Brennan, E. F., and Cota-Robles, E. H. (1982). Epoxy resins are mutagenic: implications for electron microscopists. *J. Ultrastruct. Res.* **80**, 280.

Rittenburg, J. H., Bayer, R. C., Gallagher, M. L., and Leavitt, D. F. (1979). A rapid technique for preparing microorganisms for transmission electron microscopy. *Stain Technol.* **54**, 275.

Riva, A. (1974). A simple and rapid staining method for enhancing the contrast of tissues previously treated with uranyl acetate. *J. Microsc. (Paris)* **19**, 105.

Roberts, I. M. (1975). Tungsten coating—a method of improving glass microtome knives for cutting ultrathin sections. *J. Microsc. (Oxford)* **103**, 113.

Rohde, C. J. (1965). Serial sections from plastic-embedded specimens: arthropods in methacrylate. *Stain Technol.* **40**, 43.

Roland, J. C., Lembi, C. A., and Morré, D. J. (1972). Phosphotungstic acid-chromic acid as a selective electron-dense stain for plasma membranes of plant cells. *Stain Technol.* **47**, 195.

Romanovicz, D. K., and Hanker, J. S. (1977). Wafer embedding; specimen selection in electron microscopic cytochemistry with osmiophilic polymers. *Histochem. J.* **9**, 317.

Rømert, P., and Matthiessen, M. E. (1975). Fixation of fetal pig liver for electron microscopy. I. The effect of various aldehydes and of delayed fixation. *Anat. Embryol.* **147**, 243.

Ross, A., Sumner, A. T., and Ross, A. R. (1981). Preparation and assessment of frozen hydrated sections of mammalian tissue for electron microscopy and X-ray microprobe analysis. *J. Microsc. (Oxford)* **121**, 261.

Ross-Canada, J., Becker, R. P., and Pappas, G. D. (1983). Synaptic vesicles and the nerve–muscle preparation in resinless sections. *J. Neurocytol.* **12**, 817.

Rostgaard, J., and Tranum-Jensen, J. (1980). A procedure for minimizing cellular shrinkage in electron microscope preparation: a quantitative study on frog gall bladder. *J. Microsc. (Oxford)* **119**, 213.

Roth, J., Bendayan, M., Carlemalm, E., Villiger, W., and Garavito, M. (1981). Enhancement of structural preservation and immunocytochemical staining in low temperature embedded pancreatic tissue. *J. Histochem. Cytochem.* **29**, 663.

Ruiter, D. J., Mauw, B. J., and Beyer-Boon, M. E. (1979). Ultrastructure of normal epithelial cells in Papanicolaou-stained cervical smears. *Acta Cytol.* **23**, 507.

Russo, M. A. (1977). A new tube for preparing the buffy coat for electron microscopy. *Stain Technol.* **52**, 178.

Ryder, T. A., and Mackenzie, M. L. (1981). Routine preparation of seminal fluid specimens for transmission electron microscopy. *J. Clin. Pathol.* **34**, 1006.

Ryter, A., and Kellenberger, E. (1958). Étude au microscope électronique de plasma contenant de l'acide desoxyribonucleique. I. Les nucléotides des bacteries en croissance active. *Z. Naturforsch., B* **13B**, 597.

Safa, A. R., and Tseng, M. T. (1983). Isolation and preparation of colonies grown on soft agar for ultrastructural investigation. *J. Microsc. (Oxford)* **130**, 119.

Saikawa, M., and Kobayashi, K. (1974). A simple method for electron microscopic preparation of the extracellular protoplast. *J. Electron Microsc.* **23**, 311.

Sandström, B. (1970). Liver fixation for electron microscopy by means of transparenchymal perfusion with glutaraldehyde. *Lab. Invest.* **23,** 71.

Sargent, G. F., Sims, T. A., and McNeish, A. S. (1981). The use of polystyrene microcarriers to prepare cell monolayers for transmission electron microscopy. *J. Microsc. (Oxford)* **122,** 209.

Sato, T. (1967). A modified method for lead staining of thin sections. *J. Electron Microsc.* **16,** 133.

Schmidt, N. J. (1964). Tissue culture methods and procedures for diagnostic virology. *In* "Diagnostic Procedures for Viral and Rickettsial Diseases" (E. H. Lennette and N. J. Schmidt, eds.), pp. 78–176. Am. Public Health Assoc., New York.

Schofield, B. H., Williams, B. R., and Doty, S. B. (1975). Alcian blue staining of cartilage for electron microscopy. Application of the critical electrolyte concentration principle. *Histochem. J.* **7,** 139.

Schroeder, H. E., Rossinsky, K., and Müller, W. (1980). An established routine method for differential staining of epoxy-embedded tissue sections. *Microsc. Acta* **83,** 111.

Schulz, A. (1977). A reliable method of preparing undecalcified human bone biopsies for electron microscopic investigation. *Microsc. Acta* **80,** 7.

Schwab, M. E., and Thoenen, H. (1978). Selective binding, uptake, and retrograde transport of tetanus toxin by nerve terminals in the rat iris. An electron microscope study using colloidal gold as a tracer. *J. Cell Biol.* **77,** 1.

Schwartz, I. R. (1982). A simple technique for osmicating and flat embedding large tissue sections for light and electron microscopy. *Stain Technol.* **57,** 52.

Séchaud, J., and Kellenberger, E. (1972). Electron microscopy of DNA-containing plasma. IV. Glutaraldehyde-uranyl acetate fixation of virus-infected bacteria for thin sectioning. *J. Ultrastruct. Res.* **39,** 598.

Seed, L. J. (1980). The preparation of fragile protoplasts for electron microscopy. *J. Microsc. (Oxford)* **120,** 109.

Seligman, A. M., Wasserkrug, H. L., Deb, Ch., and Hanker, J. S. (1968). Osmium containing compound with multiple basic or acidic groups as stains for ultrastructure. *J. Histochem. Cytochem.* **16,** 87.

Shands, J. W. (1968). Embedding free floating cells and microscopic particles: serum albumin coagulation-epoxy resin. *Stain Technol.* **43,** 15.

Shepard, N., and Mitchell, N. (1980). Vapor staining of electron microscope grids using BEEM capsules. *Stain Technol.* **55,** 259.

Shepard, N., Delvin, E. E., and Mitchell, N. (1981). Ruthenium red-*p*-phenylenediamine staining of monolayers to facilitate handling and selection of specific cells for transmission electron microscopy. *Stain Technol.* **56,** 143.

Shigenaka, Y., Watanabe, K., and Kaneda, M. (1973). Effects of glutaraldehyde and osmium tetroxide on hypotrichous ciliates, and determination of the most satisfactory fixation methods for electron microscopy. *J. Protozool.* **20,** 414.

Shimizu, S., Mizorogi, F., and Suzuki, T. (1978). An improved method of preparing free cells for electron microscopy. *J. Electron Microsc.* **27,** 181.

Shinagawa, Y., and Shinagawa, Y. (1978). Melamine resin as water-containing embedding medium for electron microscopy. *J. Electron Microsc.* **27,** 13.

Shires, T. K., Johnson, M., and Richter, K. M. (1969). Hematoxylin staining of tissues embedded in epoxy resins. *Stain Technol.* **44,** 21.

Sigee, D. C. (1976). A resin slide technique to select fixed embedded cells for transmission electron microscopy. *J. Microsc. (Oxford)* **108,** 325.

Silvester, N. R., Marchese-Ragona, S. P., and Johnston, D. N. (1982). The relative efficiency of various fluids in the rapid freezing of protozoa. *J. Microsc. (Oxford)* **128,** 175.

Simpson, T. L., and Vaccaro, C. A. (1974). An ultrastructural study of silica deposition in the freshwater sponge *Spongilla lacustris. J. Ultrastruct. Res.* **47,** 296.

Skaer, H. B., Franks, F., Asquith, M. H., and Echlin, P. (1977). Polymeric cryoprotectants in the preservation of biological structure. III. Morphological aspects. *J. Microsc. (Oxford)* **110**, 257.

Slot, J. W., and Geuze, H. J. (1982). Ultracryotomy of polyacrylamide embedded tissue for immunoelectron microscopy. *Biol. Cell.* **44**, 325.

Smith, A. R., and Wren, M. J. (1983). A Ralph knife holder assembly for use with the Sorvall Porter–Blum MT-1 ultramicrotome. *Stain Technol.* **58**, 235.

Smith, P. R. (1981). A trough designed to facilitate the coating of electron microscope grids. *Phillips Electron Opt. Bull.* **115**, 13.

Smits, H. T. J., Linders, P. W. J., Stols, A. L. H., and Stadhouders, A. M. (1983). Simple procedure for the production of carbon-polymer support grids for use in x-ray microanalysis. *J. Microsc. (Oxford)* **132**, 129.

Soma, L. R., ed. (1971). "Textbook of Veterinary Anesthesia." Williams & Wilkins. Baltimore, Maryland.

Somerson, N. L., Senterfit, L. B., and Hamparian, V. V. (1973). Development of a *Mycoplasma pneumoniae* vaccine. *Ann. N.Y. Acad. Sci.* **225**, 425.

Sommer, J. R. (1977). To cationize glass. *J. Cell Biol.* **75**, 245a.

Spindler, M. (1978). An improved method for fixing attached cells for transmission electron microscopy. *Trans. Am. Microsc. Soc.* **97**, 127.

Spoerri, P. E., Dresp, W., and Heyder, E. (1980). A simple embedding technique for monolayer neuronal cultures grown in plastic flasks. *Acta Anat.* **107**, 221.

Stephenson, P. M. (1982). Collection of epoxy resin sections for light microscopy using a large water-filled trough. *Med. Lab. Sci.* **39**, 183.

Stockert, J. C. (1977). Sodium tungstate as a stain in electron microscopy. *Biol. Cell.* **29**, 211.

Stolinski, C., and Gross, M. (1969). A method for making thin large surface area carbon supporting films for use in electron microscopy. *Micron* **1**, 340.

Street, C. H., and Mize, R. R. (1983). A simple microcomputer-based three dimensional serial section reconstruction system (MICROS). *J. Neurosci. Methods* **7**, 359.

Sturrock, R. R. (1984). Identification of mitotic cells in the central nervous system by electron microscopy of reembedded semithin sections. *J. Anat.* **138**, 657.

Swift, J. A. (1968). The electron histochemistry of cystine-containing proteins in thin transverse sections of human hair. *J. R. Microsc. Soc.* **88**, 449.

Swinehart, P. A., Bentley, D. L., and Kardong, K. V. (1976). Scanning electron microscopic study of the effects of pressure on the luminal surfaces of the rabbit aorta. *Am. J. Anat.* **145**, 137.

Szczesny, T. M. (1978). Holder assembly for "Ralph" type glass knives. *Stain Technol.* **53**, 48.

Takagi, M., Parmley, R. T., Denys, F. R., and Kageyama, M. (1983). Ultrastructural visualization of complex carbohydrates in epiphyseal cartilage with the tannic acid–metal salt methods. *J. Histochem. Cytochem.* **31**, 783.

Tandler, B., Sherman, J. M., and Boat, T. F. (1983). Surface architecture of the mucosal epithelium of the cat trachea. I. Cartilagenous portion. *Am. J. Anat.* **168**, 119.

Taylor, T. N., and Rothwell, G. W. (1982). Studies of seed fern pollen: development of the exine in monoletes (Medullosales). *Am. J. Bot.* **69**, 570.

Thiéry, G., and Rambourg, A. (1976). A new staining technique for studying thick sections in the electron microscope. *J. Microsc. Biol. Cell.* **26**, 103.

Thiéry, J. P. (1967). Mise en évidence des polysaccharides sur coupes fines en microscopie électronique. *J. Microsc. (Paris)* **6**, 987.

Thurston, E. L. (1978). Health and safety hazards in the SEM laboratory: update 1978. *SEM* **2**, 849.

Timár, J., Gyapay, G., and Lapis, K. (1979). Acridine orange staining of the mammalian fibroblast cell coat. *Histochemistry* **64**, 189.

Todd, W. J., and Burgodorfer, W. (1982). Rapid processing of biopsy specimens for examination by electron microscopy. *Am. J. Clin. Pathol.* **77**, 95.

Trett, M. W. (1980). A technique for the handling and orientation of small specimens for electron microscopy. *Proc. R. Microsc. Soc.* **15**, 169.

Trett, M. W., and Crouch, J. A. (1984). The collection of plastic films on specimen support grids for transmission electron microscopy. *Proc. R. Microsc. Soc.* **19**, 250.

Troyer, D., and Wollert, B. (1982). Trimming selected areas of embedments for electron microscopy—dead simple. *Stain Technol.* **57**, 289.

Tsuji, S., Alameddine, H. S., Nakanishi, S., and Ohoka, T. (1983). Molybdic and tungstic heteropolyanions for "ionic fixation" of acetylcholine in cholinergic motor nerve terminals. *Histochemistry* **77**, 51.

Umrath, W. (1974). Cooling bath for rapid freezing in electron microscopy. *J. Microsc.* **101**, 103.

Valentine, R. C., Shapiro, B. M., and Stadtman, E. R. (1968). Regulation of glutamine synthetase. XII. Electron microscopy of the enzyme from *Escherichia coli*. *Biochemistry* **7**, 2143.

Venable, J. M., and Coggeshall, R. (1965). A simplified lead citrate stain for use in electron microscopy. *J. Cell Biol.* **25**, 407.

Verna, A. (1983). A simple quick-freezing device for ultrastructural preservation: evaluation by freeze-substitution. *Biol. Cell.* **49**, 95.

Vincent, R. A., Gilbert, L. M., Doty, S. B., and Merz, T. (1975). A blaze-dry spreading procedure for the electron microscopy of chromosomes from acid alcohol-fixed human lymphocytes. *Stain Technol.* **50**, 233.

Vonnahme, F. J. (1980). An improved method for transparenchymal fixation of human liver biopsies for scanning electron microscopy. *Proc. SEM* p. 177.

Wake, K. (1974). Cytochemistry of the lipid droplets containing vitamin A in the liver. *In* "Electron Microscopy and Cytochemistry" (E. Wisse, W. Th. Daems, I. Molenaar, and P. van Duijn, eds.). North-Holland Publ., New York.

Walker, M. H., and Roberts, E. M. (1982). A thin film albumin method for encapsulating single cells for ultramicrotomy. *Histochem. J.* **14**, 999.

Wallstrom, A. C., and Iseri, O. A. (1972). Ultrasonic cleaning of diamond knives. *J. Ultrastruct. Res.* **41**, 561.

Walton, J. (1979). Lead aspartate, an en bloc contrast stain particularly useful for ultrastructural enzymology. *J. Histochem. Cytochem.* **27**, 1337.

Walton, J., Yoshiyama, J. M., and Vanderlaan, M. (1982). Ultrastructure of the rat urothelium in en face section. *J. Submicrosc. Cytol.* **14**, 1.

Warchol, J. B., Herbert, D. C., and Rennels, E. G. (1974). An improved fixation procedure for microtubules and microfilaments in cells of the anterior pituitary gland. *Am. J. Anat.* **141**, 427.

Ward, R. T. (1977). Some observations on glass-knife making. *Stain Technol.* **52**, 305.

Warmbrodt, R. D., and Fritz, E. (1981). Embedding plant tissue with plastic using high pressure: a new method for light and electron microscopy. *Stain Technol.* **56**, 299.

Watson, M. L. (1958). Staining of tissue sections for electron microscopy with heavy metals. II. Application of solutions containing lead and barium. *J. Biophys. Biochem. Cytol.* **4**, 727.

Watson, M. L., and Aldridge, W. G. (1964). Selective electron staining of nucleic acids. *J. Histochem. Cytochem.* **12**, 96.

Weakley, B. S. (1979). A variant of the pyroantimonate technique suitable for localization of calcium in ovarian tissue. *J. Histochem. Cytochem.* **27**, 1017.

Weakley, B. S. (1981). "A Beginner's Handbook in Biological Transmission Electron Microscopy." Churchill–Livingston, London.

Weber, G. (1977). Electron microscopy of free-floating cells: thin-layering technique, and selection of individual cells for ultramicrotomy. *Stain Technol.* **52**, 25.

Weidmer-Bridel, J., Vogel, A., and Hedinger, Ch. (1978). Elastic fibers in the tunica propria of the seminiferous tubules. *Virchows. Arch. B* **27**, 267.

Wells, B. (1974). A convenient technique for the collection of ultrathin serial sections. *Micron* **5**, 79.

Wells, B., Horne, R. W., and Shaw, P. J. (1981). The formation of two-dimensional arrays of isometric plant viruses in the presence of polyethylene glycol. *Micron* **12**, 37.

Westfall, J. A. (1961). Obtaining flat serial sections for electron microscopy. *Stain Technol.* **36**, 36.

Westphal, C., and Frösch, D. (1984). Electron-phase-contrast imaging of unstained biological materials embedded in a water-soluble melamine resin. *J. Ultrastruct. Res.* **88**, 282.

Whitaker, A. M. (1973). Mammalian cell culture media. *In* "Animal Tissue Culture. Advances in Techniques" (G. D. Wasley, ed.), pp. 1–19. Williams & Wilkins, Baltimore, Maryland.

White, J. G., and Krivit, W. (1967). An ultrastructural basis for the shape changes induced in platelets by chilling. *Blood* **30**, 625.

Wigglesworth, V. E. (1981). The distribution of lipid on the cell structure: an improved method for the electron microscope. *Tissue Cell* **13**, 19.

Williams, M. G., and Adrian, E. K. (1977). The use of elemental iodine to enhance staining of thin sections to be viewed in the electron microscope, *Stain Technol.* **52**, 269.

Williamson, F. A. (1984). A rapid-access system for the storage of small samples under liquid nitrogen. *J. Microsc. (Oxford)* **134**, 125.

Willingham, M. C., and Yamada, S. S. (1979). Development of a new primary fixative for electron microscopic immunocytochemical localization of intracellular antigens in cultured cells. *J. Histochem. Cytochem.* **27**, 947.

Wu, M., and Waddell, J. (1982). Transmission electron microscopic study of Polytene chromosome 2R from *Drosophila melangaster*. *Chromosoma* **86**, 299.

Wyatt, J. H. (1970). Coating of electron microscope grids. *J. Electron Microsc.* **19**, 283.

Yamamoto, N., and Yasuda, K. (1977). Use of a water soluble carbodiimide as a fixing agent. *Acta Histochem.* **10**, 14.

Yamamoto, N., Yamashita, S., and Yasuda, K. (1980). New embedding method for immunohistochemical studies using acrylamide gel. *Acta Histochem. Cytochem.* **13**, 601.

Yuan, L. C., and Gulyas, B. J. (1981). An improved method for processing single cells for electron microscopy utilizing agarose. *Anat. Rec.* **201**, 273.

Yunghans, W. N., Clark, J. E., Morré, D. J., and Clegg, E. D. (1978). Nature of the phosphotungstic acid-chromic acid (PACP) stain for plasma membranes of plants and mammalian sperm. *Cytobiologie* **17**, 165.

Zeikus, J. A., and Aldrich, H. C. (1975). Use of hot formaldehyde fixative in processing plant-parasite nematodes for electron microscopy. *Stain Technol.* **50**, 219.

Zetterqvist, H. (1956). The ultrastructural organization of the columnar absorbing cells of the mouse jejunum. An electron microscopic study including some experiments regarding the problem of fixation and an investigation of vitamin A deficiency. Page 83. Dissertation Karolinska Inst., Stockholm.

Zierold, K. (1984). The morphology of ultrathin cryosections. *Ultramicroscopy* **14**, 201.

Zingsheim, H. P., and Plattner, H. (1976). Electron microscopic methods in membrane biology. *Methods Membr. Biol.* **7**, 1.

Index

401